KU-247-461

Reclaimed Land
Erosion Control, Soils and Ecology

Land Reconstruction and Management
Vol 1, 2000

Essays towards improving the geoecological self-sustainability of lands reconstructed after surface coal mining

Reclaimed Land
Erosion Control, Soils and Ecology

Editor
Martin J. Haigh

A.A. BALKEMA/ROTTERDAM/BROOKFIELD/2000

UNIVERSITY OF HERTFORDSHIRE
HATFIELD CAMPUS LRC
HATFIELD AL10 9AD 306075

BIB
9054107936

CLASS
333.765153 REC

LOCATION
OWL

BARCODE
440534285l

Authorization to photocopy items for internal or personal use, or the internal or personal use of specific clients, is granted by A.A. Balkema, Rotterdam, provided that the base fee of US$.1.50 per copy, plus US$0.10 per page is paid directly to Copyright Clearance Center, 222 Rosewood Drive, Danvers, MA 01923, USA. For those organizations that have been granted a photocopy license by CCC, a separate system of payment has been arranged. The fee code for users of the Transactional Reporting Service is: 90 5410 793 6 US$1.50 + US$0.10.

A.A. Balkema, P.O. Box 1675, 3000 BR Rotterdam, Netherlands
Fax: +31.10.4135947; E-mail: balkema@balkema.nl;
Internet site: http://www.balkema.nl

Distributed in USA and Canada by
A.A. Balkema Publishers, Old Post Road, Brookfield, VT 05036-9704, USA
Fax: 802.276.3837; E-mail: Info@ashgate.com

ISSN 1389-2541
ISBN 90 5410 793 6

© 2000 Copyright reserved

This book is for Tamsin and Anna

Acknowledgements

Thanks go to the British Council and the Earthwatch Institute (Oxford), whose support and encouragement built the community of researchers represented here. Special thanks also go to Allison Flege, Department of Geography, University of Cincinnati, Ohio, USA, for major assistance in preparation and proof reading

The Land Reconstruction and Management Series
and
Volume 1: Reclaimed Land, Erosion Control, Soils and Ecology
are official publications of the
World Association of Soil and Water Conservation (Europe)

'Land Reconstruction and Management' Series Preface

Reclaimed Land: Erosion Control, Ecology and Soils is presented as the first volume in a series of monographs and thematic collections of review and research papers. These are to be produced by the international community of researchers working under the broad heading of *'Land Reconstruction and Management'*. It is hoped to pitch this series somewhere between series like *Advances in Agronomy* and the *FAO Conservation Guides*. The primary objective is to produce work that offers insights, deeper than those of the advanced textbooks, which are of immediate value to the researcher and lasting value practical to the innovative practitioner. Although sponsored by the *World Association of Soil and Water Conservation (Europe)*, the series will address a wider spectrum of land management issues than are traditionally included within the remit of soil and water conservation or erosion and sediment control.

However, the series will do more than simply publish works that provide an overview of the state of play in a key area of applicable research. It also aims to address the evolving conceptual underpinning of the applied sciences and technologies involved in practical land reconstruction and management. It is created to provide a platform for new philosophies and the ideals of particular international communities, 'Schools', of applied researchers and practitioners. Hence, this volume was conceived from the outset as a combination of 'handbook' and 'manifesto' from a community of 'Greens' working for land reclamation. It is hoped that future volumes might do the same service for similar groups. These groups might well include, for example, the *'Land Husbandry Movement'*, which is striving to change the structure, orientation and practice of soil and water conservation, and the *'Headwater Control Movement,'* which hopes to restructure current thinking on watershed management and landscape reconstruction in highlands and headwaters. These groups share the belief that it is necessary to work within nature and with the aim of creating self-sustainable environmental management systems that benefit local communities of land users.

This first volume covers an array of key issues within current thinking on the conservation of land that has been reclaimed after surface mining for coal. This is a new field of concern. In the past, land reclamation after coal mining was conceived as an act, a once and for all action. Today, the huge tracts of degrading and low quality reclaimed land testify to the need to consider land reclamation as a continuing process. This book's authors argue that this process continues until the 'reclaimed land' attains a condition of self-sustaining self-control. In this, the book's thesis reflects the evolving impact of current government-level thinking on the responsibilities that attach to mining contractors following mine closure. It also reflects the mining industry's growing concern about the long term liabilities associated with reclaimed land, the ultimate product of surface mining, which are emerging in many national regulatory frameworks, not least in North America.

The book contains work connected to the search for self-sustainability in reconstructed landscapes. Its authors seek strategies for the creation of the landforms, soils, drainage-channels and vegetation systems that will protect the land from degradation and survive through the long term. It also contains work connected to the stabilisation of newly reclaimed lands, so that long-term self-sustainable systems may become established. Especially, it tries to take a cool view of the problems and options for the management of reclaimed lands.

This book, and those that will follow, all describe work in progress. They offer consciously a perspective that is evolving and incomplete. In the case of the current volume, many more topics would have to be added to create a general overview of the reconstruction and management of surface-coal-mine disturbed lands. There is room for at least one more volume. This might address, in greater detail, topics such as: agricultural reclamation/management; reclamation with grass; biogeochemistry; subsidence/settlement/geotechnical stability; ruderal and microbial ecology; as well as strategies for improved mechanised land reclamation practice, and the important socio-economic aspects of self-sustainable land management. The special problems and literature of the brown coal and lignite fields of Eurasia need special consideration, so too the special problems of land reclamation in arid, saline and tropical environments, Australia and Brazil for example. There is also a need to extend the perspective beyond coal mining. So, as Series Editor, may I issue a challenge to you the reader? If you think that you and your colleagues can help roll this heavy load at little further up the hill, please do contact me, Martin Haigh. Equally, if you are the keeper of a body of knowledge that fits within the general remit of this Series and if you espouse an approach, which is not alien to its general philosophy, then please do contact. The focus of this series is the practical reconstruction and management of landscapes damaged by development. Its philosophy is a delicate shade of green.

Contents

List of Contributors

Martin J. Haigh, Editor
Oxford Brookes University
Oxford OX3 OBP, England

Anderson, P.G.
Golder Associates Ltd.
Calgary, AB
Canada T2P OW1

Asensio, Enrique
Instituto Pirenaico de Ecologia,
CISC
Jaca, Spain

Bender, Michael J.
Golder Associates Ltd.
Calgary, AB
Canada T2P OW1

Flege, Allison
Department of Geography
McMicken College,
University of Cincinnati
Cincinnati, Ohio, USA

Gentcheva-Kostadinova, Svetla
University of Forestry
bul. 'Kl. Ohridski'
1150 Sofia, Bulgaria

Ghosh, Rekha
Indian School of Mines
Dhanbad, Bihar, India

Keys, Marie José
Syncrude Canada Ltd.
Calgary, AB
Canada T2P OW1

Kilmartin, Marianne P.
Open University in the South
Oxford, U.K.

Kostakinov, Stanimir
Faculty of Forestry
Belgrade University
Kneza Viseslava 1
11030 Belgrade, Yugoslavia

Long, Dejiang
Golder Associates Ltd.
Calgary, AB
Canada T2P OW1

McKenna, Gord
Syncrude Canada Ltd.
Calgary, AB
Canada T2P OW1

Nicolau, José-Manuel
Area de Ecologia
Universidad de Alcala de
Henares
Madrid, Spain

Rankovič, Renand
Faculty of Forestry
Belgrade University
Kneza Viseslava 1
11030 Belgrade, Yugoslavia

Sansom, Ben
Knight Piésold Ltd,
Kanthack House,
Station Road,
Ashford, Kent
England

Sawatsky, Les
Golder Associates
Calgary, AB
Canada T2P OW1

Stanojević, Gradimir
Faculty of Forestry
Belgrade University
Kneza Viseslava 1
11030 Belgrade, Yugoslavia

Vućković, Milvoj
Faculty of Forestry
Belgrade University
Kneza Viseslava 1
11030 Belgrade, Yugoslavia

Zheleva, Elena
University of Forestry
bul. 'Kl. Ohridski'
1150 Sofia, Bulgaria

Zlatić, Miodrag
Faculty of Forestry
Belgrade University
Kneza Viseslava 1
11030 Belgrade, Yugoslavia

1

The Aims of Land Reclamation

Martin J. Haigh

Abstract

The greatest environmental impact of surface mining is the production of reclaimed land. While most impacts of active mining are transitory, reclaimed lands remain forever. Recent years have seen serious discussion of the ethics and realities of contemporary land restoration. Some contend that, since restored land is an artifact, it is no better than any other technological product. It is not 'nature' and of little intrinsic value. Hence, land reclamation resolves as another inherently futile attempt to show that human technology can control nature. Certainly, while sustainable reclamation aims to create land which can be managed and sustained indefinitely, unsustained reclaimed lands, degraded, actively degrading and contaminated 'reclaimed' lands, can be found across large areas of Europe. In reality, those engaged in ecological reconstruction do not want, and cannot cope with, the responsibility of perpetual repair and maintenance. Today, the true measure of land reclamation success is conceived to be the degree to which the reclaimed land can look after itself. Self-sustaining geoecological systems are not created by human artifice. Consequently, practical land reclamation reduces to a variety of active or passive supplications for the return of Nature. Its aim is to return the land to the biocybernetic, self-creating control of Nature. Self-sustainable reclaimed land develops from its own internal autopoietic, self-referenced processes. Since such a system is by definition natural and cannot be called artificial, it is an ethically apt objective.

Land Reconstruction and Management Vol. 1, 2000, pp 1–20
ISSN 1389-2541
ISBN 90 5410 793 6
A.A. Balkema, Rotterdam, The Netherlands

LAND RECLAMATION

As the tide of heavy industry ebbs from Europe and North America, large tracts of land are left mantled with the debris of the industrial strand-line. The problems of decommissioning former coal-mine sites and of recycling the land they have used and left waste are being viewed with greater critical attention. Since the environmental awakenings of the 1960s, communities, governments and finally the industries themselves have begun to consider the long-term problems posed by mined-out lands and to become involved in attempts to recommission these 'brown field' sites for new uses.

Between, 1940–1971, mining affected 0.16% of USA's land. Just 40% of the land was reclaimed. Coal mining accounted for 40% of the area mined and 70% of the area reclaimed during 1930–1971 (Doyle, 1976: 3). Such figures help reinforce an ecological message. In the long-term, society's welfare depends on its ability to reuse the land without degrading it. Duane Smith (1987) describes the rise of ecological concern as 'An Environmental Whirlwind', which struck American mining from the 1960s. Manifestations included the Clean Air Act of 1963, Water Quality Act of 1965, and the capstone National Environmental Policy Act of 1969, which created the US Environmental Protection Agency (Smith, 1987: 136). By the time of the first Federal land reclamation legislation, Title 30 in 1977, 2.3 million hectares had been disturbed by surface mining (SCS, 1981: 18).

Inappropriately, land reclamation research has been slow to develop. In America, the Bibliography on Mined Land Reclamation lists 4 articles from 1951 against 98 in 1976 (Smith, 1987: 188). In 1981, the British Government regretted that its 'opencast executive had only recently begun to carry out and sponsor research into the ecological impact of opencast extraction and the effect of operations on, for example soil structures and biological processes' (HMSO, 1981: 93). At length, environmental agencies started to worry about 'land consumption', often defined as the loss of agricultural land to other uses. The Czech and Slovak Republics believe they have lost 5% of their productive land since 1970 (Department of Environment, Prague, 1990). Current figures for Federal Germany run at 90 hectares each day (Rothkirch and Klinger, 1994). At the same time, the Rhine Lignite Field covers 2500 km^2 and only 11% has so far been worked (Hartung, 1995). In the 1980s, Britain was losing 2000 hectares/year to surface coaling (HMSO, 1981: 81).

However, as in the case of the recycling of other materials used by industry, recycling industrial land is more discussion than practice. The enterprise is in its infancy. Land reclamation has yet to mature beyond its initial 'propaganda' phase. The successes of the enterprise are trumpeted. Its oversights, failings and outright failures are ignored.

One problem is human preconception. Until recently, the belief existed that technology can solve all ills. In the former Socialist Countries, and across much of the Developing World, technological pessimism and the environmentalist alarums from the West were disregarded. They were condemned as the bourgeois indulgence of rich people and, sometimes, as a conspiracy to prevent others from attaining high standards of living (cf. Oizerman, 1983: 356; Sharma, 1992).

In fact, it is precisely because these prosperous societies are determined to protect the quality of their own living quarters that reclamation and remediation have become a vogue. More communities are in revolt against the pollution of their environment and against the loss of its natural and cultural capital. In some cases, these concerns have grown to the point where they threaten the existence of mining industries. In 1992, the Confederation of UK Coal Producers published a leaflet portraying its surface mine workers as an endangered species (Haigh, 1993). A tide of legislation and litigation is flowing from the United States (Carr and Thomas, 1994). Its flotsam includes a new ingredient in land reclamation research—money. A remediation and reclamation industry is growing upon these resources.

Despite this new research, reclamation remains a far from certain business. Success is unpredictable and land reclamation failure is common. Scherer (1995: 379) describes the current enterprise as a monumental expression of faith. Independently, I use the phrase 'act of faith' to describe current practice in European land reclamation (Haigh, 1993).

This is misplaced faith, argues the philosopher Katz (1996: 222). It represents faith in the human power to control the world. Many agree that faith in the power of technology has lulled society into treating natural processes with less respect than they require. Certainly, faith in technology is encouraged by industry propagandists and Governments. Many agencies publish statistics on the number of hectares of land treated by reclamation. All conjure the illusion that once land is reclaimed, reclaimed it remains. None publish areas for 'reclaimed lands' where the reclamation has failed. However, the percentage of land in this category could be very high. In reality, the title 'reclaimed land' identifies an administrative category rather than a physical fact. Just as official 'forest' is land classified as forest, not necessarily land with lots of trees, so 'reclaimed land' is not necessarily anything more than wasteland.

Wales provides a case in point. The Welsh Development Agency claims to have reclaimed 10,000 hectares at an average cost of more than £34,000/hectare (Bridges, 1992). In Wales, 30 years of intensive land reclamation activity has transformed the landscapes of the northern outcrop of the South Wales Coalfield. Formerly, it was dominated by declining industry and coal spoil tips. Now, it is a mixture of 'reclaimed' spoil tips and low-quality modern development. However, it is still regarded as a '100 km belt of ecological scar tissue' (Walley, 1994). It is a measure of progress in Wales that, today, the most serious land reclamation problems are found on land that is 'reclaimed', in some cases several times over (Haigh, 1992). The coalfields of Wales are distinguished by low-quality and actively degrading 'reclaimed' lands. Similar problems are widespread in England and, apparently, in many other coalfields across Europe.

This book attempts to move land reclamation towards a new era of critical introspection and ecological awareness. It argues that the enterprise has matured enough to be able to learn from its failures as much as its successes. These lessons have importance for those who hope for better things. However, at present, the objectives of the whole reclamation enterprise may need rethinking. Currently, a gulf divides land reclamation practitioners from those who conceive ecological restoration in theoretical terms (Scherer, 1995 : 359). There is a need to clarify the aims and objectives of land reclamation.

History of Land Reclamation

The area of land damaged by industry has been expanding rapidly since the Nineteenth Century. In Wales and England, 78,000 ha are said to be affected by mineral working or related disposal of mineral waste, although the total including opencast coaling may be twice as great (Hood and Moffat, 1995). In Poland, industrial devastation affects over 93,000 ha (1990 data), of which two thirds are due to mining (Carter and Turnock, 1996: 118). In the Czech and Slovak Republics, 35,000 hectares are 'devastated by mining' (Department of Environment, Prague, 1990). In Russia's middle Urals, 35,000 ha have been disturbed by surface mining, including just 2227 ha by coal (Khokhryakov et al., 1995). In Bulgaria, more than 16,000 ha have been destroyed by surface coaling, of which 12,700 require treatment (Malakov, 1993). An additional 13,000 ha may be affected by the dumping of industrial wastes (Carter and Turnock, 1996: 49). A further 20,000 ha are threatened by future coal mining (Carter and Turnock, 1996: 49). In total, 3.5% of the national area is involved. Of course, most statistics are produced by official agencies who tend to minimise the scale of the problems. In Britain and elsewhere, environmentalists suggest estimates of land degradation which are much greater; for example, 'Ecoglasnost' suggested that 44% of Bulgaria is damaged by industrial wastes and pollution (Carter and Turnock, 1996: 50). However, whatever numbers are accepted, the inescapable conclusion is that a large area of land is being damaged and, potentially, wasted because of mining.

In the very early years of land reclamation, the emphasis was simply on 'tidying up'. There are many examples of this kind of work: the city park of Pernik, Bulgaria, the late Bargoed Tip in Wales, Buffalo Rock State Park in Illinois (USA). Spoil mounds were crudely reshaped or flattened or simply replanted with trees and/or grass. Land reclamation from these periods tended to stress the greening of undervegetated wastes.

Later, the emphasis shifted to returning the land to a productive use. Large proportions of the lands reclaimed were to be reconstructed and recultivated as farmland or forest. In Poland, 85% of the land reclaimed between 1985 and 1990 was restored for agriculture (Carter and Turnock, 1996: 118). As in Bulgaria, the bulk of land not suitable for agricultural recultivation was turned to forestation (Government of Poland, 1988).

Eventually, 'making safe' became part of the equation. This involved identifying and remediating serious chemical contamination. It involved the prevention and control of water pollution problems such as ARD/AMD (Acid Rock Drainage/Acid Mine Drainage). Half of the 500 million tonnes of waste rock generated each year by Canadian surface mining includes acid-producing sulphide ores (NR Can, 1995). The problem is widespread across the World but chemical pollution is not the only problem. In Wales, the Aberfan Disaster of 1966, which killed 112 school children, was caused by the failure of a coal-spoil tip and made geotechnical stability an issue. In 1989, another failure brought tons of rubble to within 5 metres of Tredegar School. In this case, the failure involved 'reclaimed land'.

Reflecting these concerns, it is possible to recognise three different approaches to land reclamation. These differ in the degree to which they try to ignore, work against, or work within natural processes.

- Cosmetic land reclamation strategies simply disguise the problem. Their impact is temporary, but they aim to work long enough for another solution to emerge. The dream is that society, industry, or nature will come to the rescue of the project before its weakness is exposed. Cosmetic measures include the temporary drains or thin (but expensive) topsoil layers which are employed in the 'once and for all' reclamation of many surface-mine disturbed lands.
- Sustainable land reclamation strategies employ engineering measures to contain and control problems caused by the disruption of natural systems. These structures work, so long as they are sustained by repair and management, until they are destroyed by an 'extreme event' that exceeds their design specification, such as a 'one in a hundred-and-one years' storm event. Such defensive measures include the buttress walls that reinforce landslide-prone slopes and the bunds, dykes and armoured channelways that control stream flow and flooding where runoff has been overconcentrated, the fabrics that waterproof the base of flood attenuation ponds and other water features, and the topsoils that are restored on lands that are to be managed for arable agriculture thereafter.
- Self-sustainable land reclamation strategies use ecological measures that try to work with and within natural processes. They aim to lend

a helping hand to the re-establishment of Nature. Their ambition is to help self-create a dynamic, evolving, organic control-system that transcends the crude artifice of the reclamation process. Such measures include forest recultivation, some of the strategies described as 'bioengineering', as well as the modern engineering of landscape analogues for slopes and watercourses (Sawatsky, 1995 and this book).

Cosmetic Reclamation

In many nations, reclamation thinking remains embedded in civil engineering. Lands are reclaimed as civil engineering projects. Slopes are treated like road embankments or public parks. Rectilinear batters are constructed and their surfaces veneered with thin layers of topsoil and treated with commercial seed mixtures. Drainage is controlled with pipes, french drains, riprap and/or concrete. Accelerated runoff is mitigated by concrete drop-structures and in geotextile-lined stilling ponds.

Aftercare is presumed, although, in reality, some engineered artifacts are created with built-in irreparable obsolescence. Thin topsoil covers compact and wear away and must be replaced if they are not replenished by pedogenesis. Drains clog with sediment and become overgrown. Concrete structures crack and become undermined by erosion. They have a lifespan measurable in decades, perhaps more with careful maintenance. After that, they must be reworked and reconstructed.

Frequently, the architects of these cosmetic projects feel that their work is temporary. It is undertaken as a stop-gap, to fill a space before new industrial or urban land uses return to rework the abandoned land. This notion is written large in community arguments for land reclamation (cf. Blaenau Gwent Borough Council, 1991; Pipkov, 1993). Reclamation effects an environmental improvement that will encourage new enterprises to move onto the site. Of course, frequently, this does not happen. In Wales, to date, just 10% of the land reclaimed by the Welsh Development Agency has been redeveloped for industry, commerce or housing. The Agency believes that a further 10% 'may be' developed in similar ways (Griffiths, 1992). However, this leaves over 3/4 of the land unoccupied. These figures ignore the work in Wales of the former British Coal Opencast Executive, the nationalised surface-coaling agency. The proportion of their lands restored to anything other than pasture, forestry and parkland is far smaller.

Sustainable Reclamation

Sustain v.t. (L. sustinere sus for sub, under, and teneo, I hold) to support, to hold suspended, to keep from sinking, to keep alive, to furnish sustenance for, to nourish to aid effectually, to keep from ruin....

Land reclamation is forever. Naturally, this realisation dawned early on those involved with the relatively short-term land use that is surface coal mining. Most surface coal mines operate for 5–20 years and, compared to deep mines, all have a relatively large land-take. The industry engages in an unceasing search for new land to work. Inevitably, much of this new land lies close to the remains of earlier workings. The industry's successes and failures in land reclamation remain in the eye of the local community and provide a context for their political representatives (cf. Haigh, 1995). They affect the degree to which such communities resist the industry's consumption of new lands.

Surface mining is conceptualised as a once-and-for-all operation. In Wales, this is not entirely the case. Advances in surface mining technology and changes in the price of minerals, allow access to deeper reserves and permit larger overburden/mineral ratios than in earlier years. Nevertheless, there remains the general conception that surface mining is a unique, short-duration event. Surface mining is not restricted to high-value lands or associated with the development of urban communities. In many instances, the land must be returned to permanent agricultural, forest, rough grazing or pastureland uses. In sum, mining may come and go but reclaimed land remains. It is a permanent and ever-increasing landscape feature.

This realisation has sparked a search for better land reclamation quality. Smith (1987) describes the rise of environmental concern in US mining as a collision between three groups—the hardliners within the industry, the NGO environmentalists outside and an emergent group of environmentally conscious individuals within the industry (Smith, 1987: 136). The lever for the environmentally aware within the industry was the appreciation that it would be easier to market any proposal for a new mine if the community were happy about the outcome from earlier mines. In sum, these communities wanted their land back when the miners had finished. Even in the vastness of Canada, the Surface Mining Act requires that reclamation should return affected sites to a condition as near the original as possible—and for which the original or adjacent land's condition provides the benchmark (Railton et al., 1995).

Thus, a new style of land reclamation negotiation began to evolve and affect environmental decision-making. Developers began to offer 'sweeteners' for permission to open a new mine in a community. These included large payments to local property owners, the offer of new recreational or economic facilities to the community, improved farmlands to agriculturalists, and the restoration of the legacy of previous industrial damage to local planning authorities. In Wales, the proposal to opencast the site of the former British Iron Works, near Pontypool is being represented to the community as a project for 'environmental improvement' not coal mining. Similar arguments were employed for another mine at Blaenavon, the Kays and Kears (1994–1996), now known as Garn Lakes, which is being developed as a park.

In Wales, such problems arise because the national Government does not pay the full cost for the reclamation of 'orphan', abandaned, or otherwise derelicted industrial sites, and because most local authorities in industrial areas have large tracts of historical dereliction within their jurisdiction. In 1971, planners in South-east Wales assessed their problems. The Welsh Valleys had been 'desecrated' by deforestation, mining, industrial working and spoil tips, which had 'completely destroyed the visual amenities' (Rogers, 1971: 3). One example was the 'extremely derelict' 88 ha of the first opencast coal mine in Wales at Pwll Du (1942–1947) (Rogers, 1971: 31). Rogers suggests that if the surrounding area could be worked for coal, then negotiations could be entered into with a view to the whole area 'being restored concurrently with the opencast restoration' (Rogers, 1971: 31). When, much later, mining became imminent, the Welsh Development Agency (WDA) welcomed 'the reclamation of the Pwll Du site by opencast coal extraction because it would eliminate the very extensive tract of dereliction without cost to the public purse' (Griffiths, 1990).

Newly privatised coal companies, however, remain very unhappy about this approach. At the 1994 meeting of the UNECE's Committee on Coal, both Spanish and Czech delegates recommended that the responsibility for liabilities from past mining should rest with the State. Mining companies should not be expected to pay for infrastructure, training or land improvement, which could add anything from 3–20% to costs. They should have to do no more than return the land reclaimed to a standard equivalent to that of the original (UNECE, 1994).

Quality of Land after Reclamation

Nevertheless, it remains both the boast and the promise of many mining agencies that the land they create after coal mining is of a quality as high, and in some cases higher, than that which it replaces. This promise is reinforced by the creation of showcase recreational sites, lakes and woodland such as Bryn Bach Park, Wales, the similar development opposite the Czech Army Mine near Most, and nature preserves such as Durridge Bay, Northumberland, England. More controversially, it is manifested in a great deal of sponsored research that tries to highlight instances where reclaimed land is as productive as undisturbed agricultural or forest land in the same area. In fact, the soils on reclaimed land are less robust and far lower quality than those they replace. They are easily damaged by tillage and are prone to compaction and erosion. They have a relatively low nutrient status and a reduced ability to recycle and incorporate organic residues.

Land reclamation aims to replace an initially sterile (sometimes toxic or acidic spoil) by a productive soil. Current practice involves achieving this by burying the spoil beneath a carpet of topsoil—perhaps preserved from the pre-mine environment—perhaps manufactured from 'soil-forming' materials, and certainly boosted by chemical additives and fertilisers. This masking of mine spoils with topsoil is a common practice and, in theory, is sustainable.

However, natural soils are complex living systems that evolve over many centuries as a partnership between substrate, local climate, drainage and topographic location, and in response to land usage. Natural soils are capable of a sophisticated self-regulation involving nutrient supply, chemical buffering, the preservation of soil density, porosity, aeration and water holding capacity, detoxification, and self-regeneration (Van Breeman, 1993). They are tightly linked integrations of physical, chemical and, most importantly, biological processes (Lal, Hall and Miller, 1989: 52; Van Breemen, 1993).

In reclamation, as in agriculture, the key to soil quality is the soil organic system. This system is subject to rough treatment during the processes of land reconstruction. New artificial topsoils, perhaps the remnants of soils scraped up and stored from the pre-mine environment, are laid down like a carpet over a subsoil or spoil from which that soil is entirely divorced—indeed which it is often intended to mask—and which may have unfavorable physical, chemical and hydrological properties (Zhengqi Hu et al., 1993; Kilmartin, 1989 and this book). Little attention is paid to topographic and microclimatic context or to the subtleties and welfare of the living soil system.

The resultant soil differs dramatically from that it replaces. The unusual qualities of mine spoils make them difficult to classify in the standard FAO or US Comprehensive System of Soil Classification taxonomies (Gentcheva, 1994; Smith and Sobek, 1978). Some workers group them with the Entisols as Spolic Entisols or Spolents. However, this is not widely supported.

There is a huge variation in the soil characteristics of reclaimed surface mine sites, even coal-mines. Lignite spoils have radically different properties to those derived from hard coals. However, these hard coal spoils may relate part of the problem. They often contain a large percentage of cobbles that hard coal are mine stones. These are often very weakly consolidated shale and siltstone fragments which, being unstable, act partly like soil and partly like rock. They disintegrate on wetting and so it is unclear how they can be assigned to a particle size class (Schafer et al., 1980). These soils also often contain low chroma mottles and sulphuric horizons, often without evidence of a high water table—a necessary requirement for such features in natural soils (Schafer et al., 1979; Fanning et al., 1977). Statistical discriminant analyses of soil profile descriptions prepared in Wales for the proposed Pwll Du Opencast Coal Mine (cf. Haigh, 1995) found that the soils on local reclaimed lands had fewer horizons, more clay and more cobbles (Kilmartin and Haigh, 1997).

This new situation stresses the soil ecological system. The populations of the organisms, critical to the capacity of soils to recover from disturbance and translocation, are all depleted. There are fewer aerobic bacteria, vesicular arbuscular mycorrhizal endophytes, and macroscopic soil organisms especially earthworms (Harris and Birch, 1989; Fresquez and Aldon, 1984; Filtcheva et al., 1999; Scullion, 1992). Enzymatic activity is also eliminated (Uzbek, 1992). The new soils are a fragile system at the beginning of a long process of adapting to a new situation (Filtcheva et al., 1999).

In some environments, the quality of the substrate, vitality of the soil organic system and skill of land management are such that, despite the disruptions of collection, storage, amendment and deposition, the living system survives, adapts or regenerates naturally over time. Equally, in others, unfavourable influences from the subsurface, surface soil erosion, inadequate surface vegetation cover and/or inappropriate land usage, combine to overwhelm it (Haigh, 1995ab, 1992).

When this happens the soil system collapses and, since the structural integrity of the soil is preserved by organic processes, the soil itself also collapses. The result is soil compaction, low fertility, poor water holding properties and a decline in the soil as a habitat for life (cf. OSM-RE, 1988; Felton, 1992; Haigh, 1995). Unless the soil is managed very carefully, soil degradation develops progressively (Haigh, 1992).

The fertility of such reclaimed land depends entirely upon the layer of applied topsoil (Halvorsen et al., 1986). If this is deep, well managed and increases, the land thrives (cf. Zhengqi Hu et al., 1993, Banov et al., 1994). If it is shallow, abused and subjected to trafficking, the land degrades (Becher, 1985). One reclamation agency, Rheinbraun in Koln, Germany, offers training to farmers who move on newly reclaimed lands. However, for the most part, farmers are left to learn by trial, and frequently, by destructive error. It is not unusual for the topsoils on reclaimed sites to perform very poorly under tillage or for one ill-advised pass of a tractor to cause gully erosion (BCC-OE, 1988; Haigh, 1992).

Soil erosion has been labelled the 'Quiet Crisis' (Brown and Wolf, 1984). All soils may undergo erosion. Normally, this is balanced by pedogenesis—soil creation. However, the rate of topsoil creation on reclaimed land can be very small. The net loss of a thin layer of topsoil, perhaps a millimetre or few millimetres thick, from either reclaimed or agricultural land, can consume such a topsoil in a few decades. Below lie subsoils and/or spoils, which may be incapable of supporting vegetation.

Ethical Foundations of Sustainable Land Reclamation

In Europe, the ethics of the school of sustainable land reclamation thinking within the surface mining enterprise seem perched on four foundations.

- First, land restoration is a duty, though not necessarily its own. The industry agrees that the lands damaged through surface mining should be reconstructed as a part of its contract with its host community.
- Second, reclaimed land is a 'tabla rasa' upon which anything may be written at the request of the 'end-user'. Most reclamation agencies

claim that it is possible to reconstruct much damaged land to whatever specification society demands.

- Third, the duty of preservation towards sustainably reclaimed land ends when that land is passed back to its owners or on to new owners. The industry does not see its responsibility as perpetual, which is one reason why America's 'superfund' legislation and its broad definition of liability has caused shockwaves in the business (Carr and Thomas, 1994).

- Fourth, it is society that has the right to determine how land should be restored.

Philosophers have suggested that one of society's aims in land reclamation should be to 'make restitution to nature through restoring, or compensating for, natural systems which have been damaged' (Katz, 1992). This is not the case. As a rule, land reclamation is negotiated as a social contract. The motivation is anthropocentric, never biocentric. The industry may agree that land reclamation is a duty. However, it sees society, rather than the mining agency, as having moral responsibility for the environments that have been damaged (UNECE, 1994).

In Europe, the old state-controlled industries could say with justice that it was their duty to produce low-cost energy and that their work was in the national interest (BCC, 1988). Their success was measured in terms of tons produced and economic gain, not in land reclaimed.

Newer private coaling companies may transfer their allegiance to shareholders but they also have no doubt that their function remains to produce coal and to produce profits. Land reclamation is their concern only to the degree it is made their concern through their contract with the local landholders and their government. They argue that where restoration is governed by regulations, the contractor will satisfy the letter of those rules. Where the end-product is governed by environmental standards, the contractor will meet those standards (Haigh, 1995). Where the local community requires a period of aftercare and land restored for a particular end use, then that will become part of the cost calculations for the project. If the project goes ahead, then it is because the (reputable) contractor believes that these costs and these duties can be met.

The problem is that all reclamation agencies presume that they are capable of meeting their reclamation obligations. In the 'front office', where production and sales rule, there is absolute faith in the power of the reclamation technologist's ability, not merely to subdue and control nature, but to beat it into the shape required and keep it there sustainably. Sustainable land reclamation thinking is constructed upon belief in the 'technological fix' (Katz, 1992). This is the myth, so Heidegger (1971: 14–15) argues, that allows our society to see everything as a resource which 'challenges' us 'to set upon' it and exploit it. Even the concept of 'progress' becomes defined as technological control for the rule of nature.

Most of today's reclaimed landscapes are engineered artifacts. In common with all artifacts, they are not supposed to repair, reproduce or otherwise seek to preserve themselves. In these important respects, they differ from nature's living systems. They may be sustained by regular human intervention but they are not, in general, designed to become self-sustaining.

SUSTAINABILITY: THE NIGHTMARE

Since UNCED, Rio, there has been much talk about sustainability. Sustainability is promoted as a good thing, a target for land-users everywhere. However, sustaining is an active process. It requires an input: sustenance, effort, assistance. This is not a goal which attracts those in land reclamation.

Land reclamation is normally employed as a full stop. It is about the closure of an enterprise, a mine or an industry. Some land-uses—not least mineral extraction and industrial—are inherently ephemeral and will be replaced by reclaimed land. The last thing anyone involved in mine closure requires is a legacy of perpetual responsibility, or even residual liability, for preserving or sustaining the outcome of that closure. In land reclamation, sustainability is not a goal. It is a nightmare.

This nightmare is receiving reinforcement from legislative and legal frameworks emerging around the world and starting in North America. The American system ranks with the most demanding. In the United States, CERCLA, the Comprehensive Environment Response Compensation and Liability Act of 1980, produced to deal with contaminated lands, is based on the principle that the polluter pays for clean-up. It imposes three drastic standards of liability. Liability is strict—there is no need to prove negligence or criminality. Liability is joint and several—everyone responsible can be held individually liable for clean up costs. Liability is retroactive—it applies to historical situations (Carr and Thomas, 1994). In sum, it is no longer possible to rip, restore and run—residual responsibility for the site remains. Similar ideas creep into the thinking of EU nations. In Canada, the Environmental Protection Act of 1988 also dramatically increased the responsibility of mine operators for the environmental impacts of their work. Today, mindful of the persistence of AMD (Acid Mine Drainage) problems, they are talking about a 30-year period of post-mine closure monitoring (Railton et al., 1995). Even in the UK, the nationalised British Coal Opencast Executive began to offer aftercare periods exceeding a decade as part of their negotiations towards a project (BBC2 Wales, 1992).

Self-sustainable Land Reclamation

If reclamation agencies do not wish the failures of 'sustainable' engineering works to return and haunt them, if they do not wish to remain managers for all the lands they have ever worked, then they must seek solutions that do not rely on perpetual sustenance. At present, only one environmental system is capable of self-creation, self-regeneration and self-improvement. It is the same system that is responsible for many of the longer term successes of early reclamation projects. It is nature.

Cynically, much current land reclamation reduces to little other than different approaches to the problem of waiting for nature. 'Cosmetic' land reclamation aims to disguise the problem temporarily. The practitioner's dream is that nature will come to the rescue before the weaknesses are exposed. 'Sustainable' land reclamation strategies employ engineering to contain the problems caused by interference with the natural system. The ambition is to stave off the consequences of environmental disturbance for as long as necessary—until Nature reaches an accommodation with the interference or the site can be formally returned to natural control.

By contrast, instead of passively 'waiting for nature,' self-sustainable land reclamation strives to lend a helping hand to the re-establishment of natural processes. The ambition is to promote the self-creation of a natural control system and to plan for the emergence of a new natural system.

This is not the same as recreating the original landscape. Once upon a time, some American land restoration regulators required that land should be restored to its original contour. In Britain, the new land was supposed to 'blend in to surrounding contours' (HMSO, 1981). It proved very difficult to imitate nature in this fashion. Natural landscape systems are too complicated, too capricious, too little understood, and may remain so for the foreseeable future (Lovelock, 1993). In addition, surface coal-mine spoils are new materials with new dynamic properties. They require new landscapes.

So today, a new form of forgery is emerging. Reclamation specialists try to copy 'appropriate' landscape analogues from the natural environment and try to use these as building blocks in the wastes they would restore (Sawatsky, 1995). Sawatsky and Beckstead (1996) note that, in contrast to many engineered drainage systems, natural systems are often self-healing. The key to designing channels with this capacity involves replacing the rigid bed and banks of most engineered channels with mobile beds of natural armour—large boulders—which can be moved, removed and replaced by extreme events—and channels which can move laterally and vertically to adjust their dynamic energy balance. Their Canadian Oilsands Mine reclamation included analogue channels, designed with floodways and meander trains on a 30-metre wide and 5-metre deep mattress of gravel with 25–50% cobbles. The cobbles provide enough coarse material to rearmour the channel after an occasional washout, while not preventing channel adjustment. Sawatsky et al. (1996: 132) stress the necessity of 'designing for dynamic end-state conditions'. 'Sustainable' reclamation engineering stresses construction and tolerances.

By contrast, self-sustainable reclamation stresses not landscape making, but rather the 'wild-becoming' (Devdorijani 1954). The aim is to create an environment that can become naturalised without major disruption.

However, many of the activities included by self-sustainable reclamation go further than the imitation of natural systems and their self-healing properties. They are designed to the nurseries for those systems. As suggested by the current European Union's Directive on Environmental Assessment, they aim to 'protect...the reproductive capacity of the ecosystem' (85/337/EEC; cf. Malcom, 1993).

The only way to guarantee the reproductive capacity of reclaimed land is to develop its living system. Self-creation and self-preservation are characteristic properties of all living systems. The problem of land reclamation reduces to fostering the regeneration of a living system in disturbed lands.

The reclamation structures created for this task have the same function as a child's cradle. They are created in order to be superseded and discarded as the occupant matures. In this case, the occupant is the self-creating natural system of Nature itself.

These techniques recapture the spirit of the oldest land reclamation strategies known to society: long rotation forest fallowing. Its strategy is that, when land is damaged to the point its soils are no longer productive, it is rested under forest.

Trees are deep rooting species, with large associated populations of soil organisms and a large production of organic material. Recultivating with appropriate trees helps reduce the tendency for soil compaction, and leads to the generation of deeper, more free-draining soils, with better water-holding properties. If these soils provide a better environment for life, the soil ecosystem continues to thrive, creating a self-catalysing hypercycle of positive environmental change, and also better drainage and water-holding properties that reduce the potential for soil erosion.

Eventually, when the soils have regenerated, the land can be re-used. In the traditional agricultural systems of South-east Asia's tropical steeplands, the practice is to allow the land to regenerate as forest for periods of 50 years or more (cf. Haigh, 1990). Of course, these systems are applied to soils that have suffered relatively minor degradation and where the ecological building blocks, soil organisms, remain in the local environment. This is not the case in many reclaimed lands, where soil life may be almost totally absent. Nevertheless, the strategies of South-east Asia's shifting cultivators provide the basis and the name for one, recently patented, Western strategy for recultivation: Temperate Taungya (Watson, 1994). The idea also has a respectable history in reclamation research. The pioneering writings of Arlen Grandt argue that there are few uses of the land more intensive than surface mining and that after such intensive use, certainly, the land should be permitted to rest and recover.

Inadvertently, Bulgaria's policy, of encouraging the direct forest recultivation of reclaimed lands that cannot be put to agricultural use, has proved the point. At Pernik, foresters and soil scientists recultivate steep, raw, mine-spoil slopes directly through dense forestation. Here, on the site of the former Al. Milenov Mine, forests cultivated for 15 years on terraced surface coal-mine spoils generated healthy open-textured low density soils (Gentcheva-Kostadinova and Haigh, 1988). More recent research at Maritsa-Iztok has shown that forestation can quickly convert soil loss to soil growth and contributes more to organic accumulation in the soil profile than any land use apart from agriculture (Haigh et al., 1995; Banov et al., 1994).

The problem is that the regeneration of biological soil-forming systems can be very slow. In Wales, as elsewhere, reclamation agencies still seek a quick fix, most recently promoting the spraying of sewage sludge to improve soil fertility. This seems like a model 'Green' solution and an excellent example of recycling. Unfortunately, it is also very dangerous. Sewage sludge is routinely laced with heavy metals and other contaminants. Spraying it onto industrial and coal-mine spoils, which are also, inherently, loaded with metals and other toxins, risks the long-term contamination of these lands (cf. Weavers, 1992; Central Office of Information Films, 1993; Hangyel and Benesoczky, 1992).

Restoration to Nature?

Of course, some philosophers question the basic assumption of self-sustainable reclamation. 'Who could possibly believe that a land developer or strip mining company would actually restore nature to its original state...' (Katz, 1992: 233). They argue that restored ecosystems, and by extension reclaimed lands, are imitations of nature and of much less worth than the originals they replace (Elliot, 1982). They are not the real thing and even the best are no better than art forgeries. Elliot dismisses the claim that nature can be restored as mere propaganda, designed to undermine the arguments of the opponents of a development (Elliot, 1982: 81). Scherer (1995) suggests that restored ecosystems have 'functional equivalence' with 'natural systems', which is the aspiration for Sawatsky's landscape analogies (Sawatsky and Beckstead, 1996). Katz counters that even if the creation is 'dynamic' it is still art not nature (Katz, 1996: 223).

Actually, this is a flawed argument. The philosophers assess restored Nature in allopoietic, anthropocentric terms. They commodify it by discussing the degree to which it is appreciated and can be valued by society. Certainly, in the case of reclaimed land, the ultimate goal is usually allopoietic. Welsh Development Agency billboards, often scaled to dwarf the work they advertise, declare that the WDA is 'Reclaiming land for New Uses'. The professed aim is to create new lands for further abuse, to dominate and enslave nature for human purposes. Nevertheless, in the new, 'Green', ecological age of land reclamation, new uses are not the immediate concern. Rather, it is the replacement of technological management by natural control. The essence of nature is that it does not require to be looked after. It is self-creating, self-preserving and self-referenced or autopoietic (Jantsch, 1980). So, the practical goal of land reclamation is to foster a system that is robust enough to look after itself. Self-sustaining behaviour is the particular property of living systems, the signature of Nature and the goal of successful land reclamation.

Baird Callicott (1993) calls anthropocentrism 'the original sin'. A natural system has value only in human perception. Even Holmes Rolston's notion of 'autonomous intrinsic value' requires a valuer and the author warns against any species which 'takes itself as absolute and values everything else in nature to its potential to produce value for itself' (Rolston, 1994). However, nature is not a commodity, it has no value, only Self. A totally autopoietic system cannot be a forgery. It can only Be. It cannot be 'art'. It can only be Nature'. It matters not why a new baby was conceived or for what intent, the result is still a human being in its own right. It is still more so with a Natural system; it cannot care about the intention behind its creation. It exists. The worst that can be said here is that, on reclaimed land, the new natural system is in its infancy. However, the term 'infant' is not necessarily a pejorative.

THE GOAL OF LAND RECLAMATION: PROJECT CLOSURE

So, contrary to official proclamation and the argument of academic philosophers, the real goal of land reclamation is neither the domination nor the manufacture of Nature. Those engaged in ecological reconstruction do not wish to create sustainable artifacts because they do not want the responsibility of sustaining them through the long-term. They do not own the God-like qualities needed to create Nature. The best that they can do, and their real aim, is to help a self-sustaining system regenerate by itself.

Their dream remains to discover a land reclamation strategy that allows true closure—one that on completion, can be walked away from and forgotten forever. Today, land reclamation research teams seek the grail of self-sustainability. Unfortunately, the only agency routinely capable of creating self-sustainable lands is Nature. Since such a system cannot be created by human artifice, practical land reclamation strategies resolve as a variety of supplications for the return of Nature. Successful restoration is deemed to have been achieved when the land returns to the autopoietic, biocybernetic, self-creating control of Nature.

References

Baird Callicott, J. 1993. Toward a global environmental ethic. *Bucknell Review* 37(2): 30–40.

Banov, M., Hristov, B., Filtcheva, E. and Georgiev, B. 1994. Humus accumulation and its quality composition in reclaimed lands, pp. 279–285. *Sbornik Nauchni Dokladi, Jubileina Nauchna Conferentsiya 125 Godini BAN i 65 Godini Institute za Gorata*: Institut za Gorata i Bulgarska Acadamiiya na Naukite Sofia: 376 pp.

BBC (Wales) 1992. *Brave New World—Heads of the Valleys. How Green?* BBC2 (Wales). March 12th, 1992, 20.00–20.35.

BCC.1988. *Opencast Coal Mining in Great Britain.* British Coal Opencast Executive, Mansfield, 23 pp.

BCC/OE. 1988. *Ten Years of Research—What Next? A Seminar on Land Restoration Investigation and Techniques.* British Coal Opencast Executive, Mansfield, 90pp.

Becher, H.H. 1985. Compaction of arable soils due to reclamation or off-road military traffic. *Reclamation and Revegetation Research* 4: 155–164.

Blaenau Gwent Borough Council. 1991. *Pwll Du Public Inquiry: Proofs of Evidence.* Cumbran, Gwent County Hall.

Bridges, E.M. 1992. Quality of land restoration: an introduction. *Land Degradation and Rehabilitation* 3(3): 153–156.

Brown, L.R. and Wolf, E.C. 1984. *Soil Erosion: Quiet Crisis in the World Economy.* Worldwatch Institute, Washington D.C., Paper 60: 50 pp.

Carr, J.M. and Thomas, J.R. 1994. Civil liability for environmental damage: a comparison of the United States and European Union approaches, pp. 9–22. In: *Environmental Restoration Opportunities Conference (Munich, Germany). Proceedings* (Deutsche Aerospace/A.D.P.A., Arlington, Va.), 476 pp.

Carter, F.W. and Turnock, D. 1996. *Environmental Problems in Eastern Europe* (2e). Routledge, London, 291 pp.

Central Office of Information Films, 1993. *Perspective Repairing the Damage.* Central Office of Information, London: Video 28 mins.

Department of the Environment, Prague. 1990. *The Environment in Czechoslovakia.* Prague: State Commission for Scientific and Technological Development and Investments, 104 pp.

Devdarijani, A.S. 1954. Anthropogenniya formy reliefa. Voprosy Geografi-geomorfologi (Moskva), pp. 117–120.

Doyle, W.S. 1976. Strip mining of coal: environmental solutions. Noyes Data Corporation, Park Ridge, NJ, *Pollution Technology Review* 27, 352 pp.

Elliot, R. 1982. Faking nature. *Inquiry* 25: 81–93.

Erikson, D.L. 1995. Policies for the planning and reclamation of coal-mined landscapes—an international comparison. *J. Environmental Planning and Management* 38(4): 453–467.

Fanning, D.S., Wagner, D.P., Darmody, R.G. and Foss, J.E. 1977. Sulfuric materials and sulfuric horizons in minesoils and other upland situations. *Abstracts Soil Sc. Soc. Amer. Ann. Mtg.*, Los Angeles, p. 125.

Felton, G.K. 1992. Soil hydraulic properties of reclaimed prime farmland. *Amer. Soc. Agric. Eng., Trans.* 35(3): 871–877.

Filtcheva, E., Noustorova, M., Gentcheva-Kostadinova, Sv. and Haigh, M.J. 1999. Impact of forestation on organic accumulation and microbial action in surface coal-mine spoils, Pernik, Bulgaria, *Ecological Engineering* 13 (in litt.).

Fresquez, P.R. and Aldon, E.F. 1984. Distribution of fungal genera in stockpiled topsoil and coal-mine spoil overburden. *United States Department of Agriculture, Forest Service Research Note* RM-477, 4 pp.

Gentcheva, Sv. 1994. Classification of anthropogenic soils. *Forestry Ideas/Lesobudska Misul* 1/1994: 87–96.

Gentcheva-Kostadinova, Sv. and Haigh, M.J. 1988. Land reclamation and afforestation research on the coal-mine disturbed lands of Bulgaria. *Land-Use Policy* 5(1): 94–102.

Government of Poland. 1988. Overcoming environmental problems in opencast mining. United Nations Economic Commission for Europe, Working Party on Coal, Meeting of Experts (Karlovy Vary, Czechoslovakia), 7, Item 12 (UNECE/WP1/SEM 2./R.39): 15 pp.

Griffiths, D.G. 1990. Possible Opencasting Operation at Pwll Du (Letters: 11.1.1990/30.1.1990). Cardiff: Welsh Development Agency, 2 pp.

Griffiths, D.G. 1992. Land reclamation in Wales: aims and strategy of the Welsh Development Agency. *Land Degradation and Rehabilitation* 3(3): 157–160.

Haigh, M.J. 1990. Shifting agriculture (jhum) and environmental devastation: the search for a solution, pp. 371–376. In: N.K. Sah, S.D. Bhatt and R.K. Pande (eds.). *Himalaya: Environment, Resources and Development*. Shree Almora Book Depot; Almora, U.P., India, 492 pp.

Haigh, M.J. 1992. Degradation of "reclaimed" lands previously disturbed by coal mining in Wales: causes and remedies. *Land Degradation and Rehabilitation* 3(3): 169–180 (and 154).

Haigh, M.J. 1993. Surface mining and the environment in Europe. *Int. J. Surface Mining, Reclamation and Environment* 7(3): 91–104.

Haigh, M.J. 1995. Surface mining in the South Wales environment, pp. 675–682. In: R.K. Singhal, A. Mehotra, J. Hadjigeorgiou, and R. Poulin (eds.). *Mine Planning and Equipment Section '95*. Balkema, Rotterdam, 1117 pp.

Haigh, M.J. 1995b. Soil quality standards for reclaimed coal-mine disturbed lands: a discussion paper. *Int. J. Surface Mining, Reclamation and Environment* 9(4): 187–202.

Haigh, M.J., Gentcheva-Kostadinova, Sv. and Zheleva, E. 1995. Forest-biological erosion control on coal-mine spoil banks in Bulgaria. *Int. Erosion Control Assoc., Proc.*, 26: 383–394.

Halvorsen, G., Melsted, S.W., Schroeder, S.A., Smith, C.M. and Pole, M.W. 1986. Topsoil and subsoil requirements for reclamation of non-sodic mined land. *Soil Sci. Soc. Amer. J.* 50: 419–422.

Hangyel, L. and Benesoczky, J. 1992. Research on the effects of municipal effluents on the reclamation of spoil heaps at opencast coal mines. *UNECE, Symp. Opencast Coal Mining and Environment, ENERGY/WP.1/SEM.2/R.17*: 8 pp.

Harris, J.A. and Birch, P. 1989. Soil micobial activity in opencast coal-mine restorations. *Soil Use and Land Management* 5(4): 155–160.

Hartung, M. 1995. Opencast mine planning creates the prerequisite for environmental compatibility, pp. 683–689. In: R.K. Singhal, A. Mehotra, J. Hadjigeorgiou and R. Poulin (eds.). *Mine Planning and Equipment Selection '95*. Balkema, Rotterdam, 1117 pp.

Heidegger, M. 1971. The question concerning technology, pp. 14–17. In: W. Lovitt (trans): *The Question Concerning Technology and Other Essays*. Harper and Row, New York.

HMSO. 1981. *Coal and the Environment*. Her Majesty's Stationary Office, London, 112 pp.

Hood, R. and Moffat, A. 1995. Reclamation of opencast spoil using alder. *Nat. Env. Res. Coun.* (NERC, UK), *News*, October 1995: 12–14.

Jantsch, E. 1980. *The Self-organising Universe*. Pergamon, Oxford.

Katz, E. 1992. The big lie: human restoration of nature. *Research in Philosophy and Technology* 12: 231–241.

Katz, E. 1996. The problem of ecological restoration. *Environmental Ethics* 18: 222–224.

Khokhryakov, A.V., Dementyev, I.V. and Albrecht, V.G. 1995. Environmental approach and the future of mining in the Urals-region of Russia, pp. 715–716. In: R.K. Singhal, A. Mehotra, J. Hadjigeorgiou and R. Poulin (eds.). *Mine Planning and Equipment Selection '95*. Balkema, Rotterdam, 1117 pp.

Kilmartin M.P. 1989. Hydrology of reclaimed opencast coal-lands: a review, *Int. J. Surface Mining* (3): 71–82.

Kilmartin, M.P. 1995. Rainfall/runoff on reclaimed opencast coal-mined land, pp. 4.25–4.30. In: A. Black and D. Johnson (eds.). *Fifth National Hydrology Symposium* (Wallingford: Institute of Hydrology), 405 pp.

Kilmartin, M.P. and Haigh, M. 1997. Initial statistical comparisons between the properties of natural and reclaimed soils, Blaenavon, Wales. *Forestry Ideas/Lesobudska Misul* 1/1997 (in press).

Lal, R., Hall, G. and Miller F. 1989. Soil degradation. *Land Degradation and Rehabilitation* 1: 51–69.

Lovelock, J.E. 1993. *Gaia: The Practical Science of Planetary Medicine*. Gaia Books, Stroud, England 186 pp.

Malcom, R. 1993. Assessing environmental impacts under the European directive on environmental assessment pp. 156–161. In: Y. Guerrier (ed.). *Values and the Environment* (Conf. Proc.) University of Surrey, Guildford.

Malakov. P. 1993. Country Report: Bulgarian Delegation. United Nations Economic Commission for Europe, Committee on Energy, Working Party on Coal, Workshop on Environmental Regulations in Opencast Mining Under Market Conditions, spoken report, in: UNECE Secretariat 1994: *Outcome of the Workshop on the Development of Environmental Regulations in Opencast Coal Mining under Market Conditions* (Most, Czech Republic 9–11, November 1993) *ENERGY/WP.1/R.31:* 14 pp.

NR Canada. 1995. *Sustainable Development and Minerals and Metals: An Issues Paper*. Ottawa: Natural Resources Canada: Mineral Strategies Branch.

Pipkov, N., 1993. Ecological Programme for Environmental Protection, Radnevo District Municipality: Proposal" Sofia: University of Forestry, Project Documents for the Civic Meeting at Dom Obchina Radnevo, Maritsa-Iztok, Bulgaria: April 5th, 1993.

Oizerman, T.I. 1983. Historical materialism and the ideology of 'technical pessimism', pp. 343–362. In: A.D. Ursul (ed.). *Philosophy and the Ecological Problems of Civilisation*. Progress, Moscow, 411 pp.

OSM-RE. 1988: *Solicitation EC680-PFP8-13577:* Office of Surface Mining and Reclamation and Enforcement. United States Department of the Interior, Wash., 34 pp.

Railton, J., Ferguson, H. and Dawson, R. 1995. North American regulatory system In: *Mine Site Reclamation Planning: Issues and Solutions*. AGRA Earth and Environmental, Calgary, S1: 1–35.

Rogers, A. 1971: *Comprehensive Report on the Extent of Derelict Land*. Pontypool: Monmouthshire Derelict Land Reclamation Joint Committee, 66 pp.

Rolston III, Holmes. 1994. Value in nature and the nature of value. *Royal Inst. Phil.* (London), *Suppl.* 36: 13–30.

Rothkirsch, U.G. and Klinger, V. (eds.). 1994. *Environmental Policy in Germany*. Federal Ministry for the Environment (BMU), Bonn, 111 pp.

Sawatsky, L.F. 1995. Sustainable landscape design. In: *Minesite Reclamation Planning: Issues and Solutions*. AGRA Earth and Environmental, Calgary, 400 pp.

Sawatsky, L.F. and Beckstead, G.R.E. 1996. Geomorphic approach for design of sustainable drainage systems for mineland reclamation. *Int. J. Surface Mining, Reclamation and Environment* 10(3): 127–130.

Sawatsky, L.F., Cooper, D.L., McRoberts, E. and Ferguson, H. 1996. Strategies for reclamation of tailings empoundments. *Int. J. Surface Mining, Reclamation and Environment* 10(3): 131–134.

Schafer, W.M., Nielsen, G.A., Dollkopf, D.J. and Temple, K. 1979. Soil genesis, hydrological properties, root characteristics and microbial activity of 1–50-year-old stripmine spoils. *USDA, Interagency Energy/Environment R&D Report*, EPA-600/7–79-100, 212 pp.

Schafer, W.M., Nielsen, G.A. and Nettleton, W.D. 1980. Minesoil genesis and morphology in a spoil chronosequence in Montana. *Soil Sci. Soc. Amer. J.* 44: 802–807.

Scherer, D. 1995. Evolution, human living and the practice of ecological restoration. *Environmental Ethics* 17: 359–379.

SCS. 1981. *America's Soil and Water: Condition and Trends*. Washington: United States Department of Agriculture: Soil Conservation Service, 33 pp.

Scullion, J. 1992. Re-establishing life in restored topsoils. *Land Degradation and Rehabilitation* 3(3): 161–168.

Sharma, B.D. 1992. Earth essay: on sustainability—the voice of the disinherited. *Sanctuary Asia* 12(3): 14–25.

Smith, D. 1987. *Mining America. The Industry and the Environment 1800–1980*. University of Colorado Press (1993), Niwot, Col., 210 pp.

Smith, R.M. and Sobek, A.A. 1978. Physical and chemical properties of overburdens, spoils, wastes and new soils, pp. 149–169. In F.W. Schaller and P. Sutton (eds.). *Reclamation of Drastically Disturbed Lands*. Amer. Soc. Agron., Madison Wisc.

UNECE. 1994. Outcome of the Workshop on Development of Environmental Regulations in Opencast Coal Mining under Market Conditions (Most, Czech Republic, November 1993). United Nations Economic Commission for Europe, Working Party on Coal, Meeting of Experts (Karlovy Vary, Czechoslovakia), 7, Item 12 (UNECE/WP1/R.31), 14 pp.

Uzbek, I.K.H. 1992. Enzymatic activity of recultivated soils. *Soviet Soil Science* 27(3): 41–46 (trans: *Pochvovdeniye* 3/1991: 91–96).

Van Breemen, N. 1993. Soils as biotic constructs favouring net primary productivity. *Geoderma* 57: 183–211.

Walley, C. 1994. Carving out a future? *Rural Wales*, Summer 1994: 22–24.

Watson, J.W. 1994. Temperate taungya: woodland establishment by direct seeding of trees under an arable crop. *Quart. J. Forestry* 88: 199–204.

Weavers, P. 1992. Sewage sludge as an agent in reclamation to forestry. UNECE, Symposium on Opencast Coal Mining and the Environment, ENERGY/WP.1/SEM.2/R.41:1.

Zhengqi Hu, Caudle, R. and Chong, S. 1993. Evaluation of farmland reclamation effectiveness based on reclaimed mine soil properties. *Int. J. Surface Mining and Reclamation* 6: 129–135.

2

Towards Minimising the Long-term Liability of Reclaimed Mine Sites

Les Sawatsky, Gord McKenna, Marie-José Keys, Dejiang Long

Abstract

Reducing long-term liability is becoming increasingly important as mine owners and financiers become aware of their obligations after mine closure and reclamation. Certification of reclaimed mine land does not necessarily relieve owners of their responsibility if failure or severe environmental impacts after closure are attributed to deficient reclamation planning, design and implementation. Therefore, it is necessary for owners to factor long-term liability into mine plans in an attempt to minimise their exposure. At many mines, end-of-mine landforms are inherently stable and subject to minimal long-term change and environmental degradation. Other mine land is subject to gradual or rapid evolution, which can eventually result in negative environmental impact and reduced land productivity. Neglecting these long-term impacts may result in significant unexpected financial obligations. If the owner attempts to control landscape evolution by conventional structural or operational measures, a substantial bond may be required to cover perpetual maintenance. An alternative is to build robust, self-healing reclamation landforms that replicate the dynamic evolution and character of natural systems, thereby minimising long-term liability.

INTRODUCTION

Long-term liability has become an important issue for mine operators as their legal obligations are being enforced through the regulatory process as well as by third-party interventions. Shareholders and bankers are

Land Reconstruction and Management Vol. 1, 2000, pp 21–36
ISSN 1389-2541
ISBN 90 5410 793 6
A.A. Balkema, Rotterdam, The Netherlands

beginning to insist on full disclosure of mine closure obligations and post-closure environmental impacts. This is a relatively recent phenomenon. It reflects a sharp contrast with mine ventures of previous decades when reduced land productivity and risk to the environment seemed to be treated as inevitable consequences of development. In Canada, tougher provincial and federal regulations have imposed new standards of reclamation practice and restoration of land productivity. Mine owners and directors have been made accountable for negligent planning and operating procedures to the extent that company directors can be held criminally responsible for decisions by their staff that lead to environmental degradation. Mine owners cannot escape the obligations of developing a sustainable mine closure system. Accordingly, it has become essential to include the long-term liabilities associated with mine closure in any inventory of current asset value.

For some sites, particularly existing mines built without sustainable reclamation strategies, it may not be possible to reduce long-term liabilities to zero, and some long-term maintenance of these sites may be unavoidable. However, by careful closure planning and by dividing the landscape into units or blocks, almost all of the landscape can be returned to an essentially 'maintenance free' condition by designing a landscape that will evolve at rates similar to those of surrounding natural areas. Despite such action, it is of course likely that these will remain some units that will require minor maintenance on an as-required basis, for which a financial bond may have to be posted by the mine.

This paper describes some important causes of long-term liability and compares several alternative approaches, which may be taken during mine operation or upon mine closure. These recommended approaches aim to minimise liability and reduce overall costs.

TYPES OF LONG-TERM LIABILITIES

Potential long-term liabilities associated with mine development are illustrated by the conditions on previously abandoned and reclaimed mines. There are many examples of mines abandoned earlier in the 1900s, which would result in large liabilities were they decommissioned in the 1990s. The poor environmental conditions on many of these abandoned mines are an impediment to the modern mining industry. Their existence makes it difficult for the industry to gain credibility with regulators and the public. Despite recent successes, stringent mine permit conditions and bonding for premature mine closure, the mining industry continues to be plagued by the legacy of abandoned mines, where liability has been transferred to the public as a result of bankruptcies.

Based on a tour in Canada of fifty-seven abandoned and partially reclaimed operating mines, McKenna and Dawson (1997) created an inventory of mine closure practices, physical performance of the reclaimed mine land and environmental impacts of reclaimed and abandoned mines. The inventory establishes several types of residual liability associated with mine closure. However, the greatest physical risks to the landscapes was associated with surface erosion (gullying) and the poor performance of drainage on the re-established site. The key problems, therefore, pertain to the various aspects of surface water hydrology that are affected by mine closure and reclamation.

Reduced Land Productivity

In Canada, current mine permit conditions commonly require that mine-disturbed land must be restored to previous levels of land productivity, often allowing some changes in land use. Such conditions aim to counter the problems caused by the mine closures of previous generations, which have produced less productive and sometimes effectively sterile landscapes. The intention is that a mine operator's inability to comply with permit conditions governing land productivity will result in a significant long-term liability.

Water Quality Impacts

Major water quality impacts of mine developments include acid rock drainage, leaching of heavy metals from waste-rock piles, and pore-water seepage from the tailings disposal areas. These impacts represent a significant cost for remediation either by engineered structures such as cutoff barriers, impervious covers and containment, or by water treatment. The last might include flushing waste rock upon mine closure to reduce long-term impacts, or continued water treatment, possibly in perpetuity. Such residual liabilities loom large, particularly if they are not addressed until the end of mine life.

Catastrophic Failure of Reclaimed Landforms

The catastrophic failure of reclaimed landforms is normally associated with geotechnical instability of waste dumps, mine cut slopes, deep excavations and dam failure. Unfortunately, the successful performance of mine-disturbed land during the first few years after mine closure may not be indicative of future stability because such failures can be affected by uncommon events such as earthquakes, increased piezometric levels caused by wet periods or extreme hydrologic events, and or by progressive changes in the physical configuration of landforms as a result of erosion and gullying. This type of occurrence affects land-use, land productivity and water quality. It is generally unexpected and costly.

Erosion and Gullying

Although often overlooked, soil erosion is a most frequent cause of landform deterioration, sediment accumulation and reduced aquatic habitat. Erosion is a progressive phenomenon with a cumulative impact that is governed by recurrent extreme hydrologic events. Erosion is episodic. Normal, year-to-year, landscape evolution may not be indicative of long-term trends. The degree of long-term liability associated with erosion varies depending on the rate of erosion, impact of the resultant landscape evolution and sediment yields. The cost of remediation can be very high if the end-of-mine landforms are not designed to accommodate the progressive forces of erosion.

Despite its seemingly small rate, its progressive nature makes the erosion of landforms after mine closure represent a high risk of environmental impact. Adverse changes in landform configuration that result from soil erosion include: slope failures due to toe erosion of vulnerable slopes, gullies that penetrate through protective covers, drainage channels that adjust to a new characteristic regime by incision, channel widening or aggradation, and hillslope degradation. Perhaps the most catastrophic impact of erosion is the breaching of dam embankments that results in the runout of tailings or impounded effluent.

Liability Related to Evolving Regulation

One of the greatest liabilities for a mine is changing retroactive environmental legislation. Working with the regulators, mines need to develop clear and measurable site-specific performance goals and an agreement that meeting these goals will absolve the mine from future changes in regulation.

WHEN TO ADDRESS LONG-TERM LIABILITY

The time to address long-term liability is 'now'. Ideally it is done in the prefeasibility stages. This is when the greatest savings and environmental protection can be achieved. For mines that are currently operating without a closure plan, closure planning should begin as soon as practicable. Experience has proven that closure planning uncovers suboptimal mine plans.

In the past, it may have been possible to ignore the issue of long-term liability. This is no longer an option, given the emergent regulatory climate and accountability to shareholders and lending agencies. Most mine operators now find it necessary to incorporate long-term liability into their internal financial reports as well as their mine environmental assessments for obtaining regulatory approval.

Nevertheless, in some jurisdictions it is still possible to delay preparation of detailed mine closure plans and to postpone reclamation activities. Even where conceptual mine closure plans are required for regulatory approval, mine operators may not prepare adequately for mine closure during operations. There is a strong temptation to delay detailed closure activities until mine closure to defer the cost of such work and to allow for those changes in mine development plans that commonly occur during mining. Radical mine plan changes can make a closure plan obsolete and as a result the effort invested in mine closure may appear to be wasted. Nevertheless, deficient mine closure planning during mine operation and disregard of mine closure costs and liabilities will almost certainly lead to deficient environmental management and increased costs.

A superior response to long-term liability is to be pro-active in minimising liabilities through mine planning that takes into full account mine closure costs and liabilities. The best time to minimise long-term liability is during pre-mine planning and during mine operation, when it is still possible to develop an economical landscape configuration by modest changes to mine plans and waste disposal operations. Measures to minimise long-term liability can be implemented at a much lower cost if planned in advance and if integrated into mine operations. Rehandling of material can be avoided by appropriate planning. The costs of land-forming can be minimised if mine closure work is conducted during periods of reduced equipment utilisation. If detailed closure planning is delayed until mine closure, such opportunities for major reductions in mine reclamation costs will be lost.

An example of the pro-active approach is to reduce the quantity of mine water releases requiring treatment. This can be achieved by controlling the size of mine waste-rock dumps, segregating the mine waste rock into areas subject to acid rock drainage (ARD) and not subject to ARD, landscape contouring to maximise water shedding (thereby minimising infiltration) and placing overburden dumps over ARD waste dump areas. However, the most appropriate time to implement such activity is during mine operation. Once again, the opportunity is lost if mine closure planning is delayed until the end of mining.

In sum, remedying long-term liabilities by physical rehabilitation of mine closure facilities after the end of mine operation can be more expensive than conducting such work during mine operation. Reconfiguring physical facilities after their construction involves double handling of materials and precludes the opportunity for more economical solutions. Post-closure remedial measures may require placement of cover materials from a new borrow area, which may be far more costly than the selective use of mine overburden material during mine operation. The high cost of rehabilitating mine closure facilities after cessation of mine operation is a positive incentive for closure planning during mine operation.

There are several further economic advantages to reclaiming mine-disturbed land progressively during mine operation, which is now required by mine operating permits in many jurisdictions. Firstly, progressive reclamation offers the opportunity to optimise reclamation methods based on experience and to monitor the effectiveness of specific strategies. The outcome may be substantial savings in the costs of reclamation. Secondly, progressive reclamation offers an extended performance record when monitoring may provide a guide to the costs of maintenance and sustainability. This may permit a more accurate costing of maintenance costs and financial bonding. Thirdly, it provides an opportunity to remedy any deficiencies of the reclamation plan in advance of post-project assessment, which will benefit the reputation of the mining enterprise.

STRUCTURAL SOLUTIONS REQUIRING CONTINUED OPERATION OR MAINTENANCE

A conventional method of minimising residual liability is to develop structural systems designed to avoid the negative impacts of mine closure. Structural solutions include impervious covers over waste rock, seepage cutoffs, riprap spillway channels, rock armour covers, dam embankments and various types of landscape configurations designed to control surface water, hydrology, salinity and water quality. Most structural solutions are semi-permanent, requiring occasional repair or replacement after a period of years. Any structural solutions built of concrete, steel or other man-made materials cannot be considered permanent and must be included in the category of continued operations and maintenance solutions. Similarly, many rigid systems such as rock armouring of stream channels may not be permanent, even if the erosion protection is designed for extreme events. The reason is that rigid systems are incompatible with any type of landscape subjected to damage as a result of the natural forces of weathering and erosion. Unexpected occurrences such as frost heave, slope failure, beaver dams, debris jams, ice plucking and root growth can dislodge individual stones and lead to deterioration.

Various types of structural covers over waste-rock dumps subject to ARD and leaching of heavy metals, can be developed to minimise infiltration. These include impervious covers and evapotranspiration (ET) covers that minimise deep percolation by water shedding and evapotranspiration (Fig. 1). These can be built at moderate cost if they are constructed during mining and utilise mine waste materials.

AN APPROACH FOR CONSTRUCTING SELF-SUSTAINING LANDSCAPES

The complexities and dynamics of landscapes (physical, biological and chemical) are multiple and impossible to understand fully. Nevertheless,

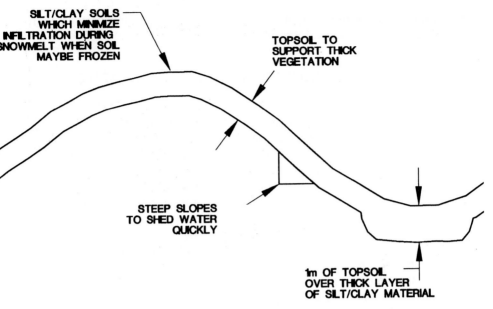

Fig. 1: Hydrologic cover configuration

the reclamation planner is charged with the responsibility of constructing sustainable landscapes that will continue to evolve and degrade slowly, while at the same time meeting goals such as physical stability, containment of wastes and various land-use objectives. With an imperfect understanding of nature and unable to predict future performance accurately, the reclamation planner nevertheless needs to adopt an approach that accommodates the inherent uncertainties of landscape development. Such an approach might include the following measures:

- patterning landscapes after natural analogues;
- designing the landscape with multiple lines of defence against certain failure models;
- designing robustness into the landscape, whereby the system becomes more stable with time;
- designing components and systems that are reliable and predictable;
- constructing 'fire-breaks' in the landscape that limit the transmission of certain failure modes (examples include use of large lakes as sediment traps and flow attenuators, offsetting receiving streams from major structures to lessen the risk of siltation and using chemically reactive barriers to arrest the flow of contaminants);
- designing conservatively, using proven technology where available;
- limiting the use of artificial materials such as pipes, gabions, fences etc.;

• releasing contaminants at acceptable levels over time rather than attempting to contain them forever.

SUSTAINABLE LANDSCAPE SOLUTIONS PATTERNED AFTER NATURAL ANALOGUES

The historical approach to configuring landscape for reclamation is to develop uniform slopes conforming to neat lines and grades. This lends itself to uniformity of design and construction but does not necessarily achieve the mine closure objectives of minimum erosion and long-term sustainability. Uniform landforms represent immature topography and are poised to evolve by accelerated erosion. In contrast, the development of a sustainable landscape for mine closure, involves the development of landforms that replicate natural landscapes. The replication of mature and relatively stable natural systems reduces the rate and risk of accelerated erosion. It also encourages replication of the self-healing erosion control systems that help preserve the stability of the natural analogue.

Examples of natural analogues for reclamation of mine-disturbed land are given below.

Mature (non-uniform) Topography

Uniform topography at tailings storage areas and mine waste-rock dumps is often incompatible with the goal of long-term sustainability. It is preferable to reconfigure the mine closure landscape to replicate mature topography (Fig. 2). This strategy acknowledges the evolutionary process of landscape development. Mature topography has already been subjected to the rapid erosion of its previous immature state and has developed to a state of relatively slow change. Mature topography is characterised by relatively short slope lengths and slopes that become more gentle as flow concentrates in the downslope direction. Flow paths are well defined in swales, which are deep enough to handle any extreme flood and to avoid spillage into adjacent swales or subbasins. Instead of uniform slopes, mature topography has variable slopes with hills and valleys. These serve to improve the aesthetic appearance, provide a wider range of habitats for wildlife and avoid the large surface flow rates typical of long, straight slopes.

Reduced Slope Length at Steep Slopes

Steep slopes may be acceptable as long as the contributing drainage areas are small. The allowable slope length and steepness are a function of the density of vegetation and root mass, soil erodibility and infiltration capacity of the soil. Some areas, such as the sand-dunes near Lake Athabasca (Canada), have steep slopes and minimal plant cover but are not subject to surface erosion by water. The reason is that the sand-dunes are composed of coarse sand, which allows high infiltration. The infiltration rates are greater than maximum rainfall intensities. As a result, surface runoff and surface erosion by water is minimal. Slopes covered by dense grasses, which develop a thick root mass, are also highly resistant to erosion. Allowable slopes at such areas may be steeper and longer than areas with a sparse vegetation cover. Cohesive soils are less erodible than sandy soils and therefore can support steeper slopes with larger catchment areas (cf. Nicolau and Asensio, this book).

Avoid Ponding on Terraces

One common misunderstanding regarding erosion control is that terraces prevent erosion. Whereas terraces intercept surface runoff during low intensity storms, erosion can only be controlled if the accumulation of surface water does not exceed the storage capacity on the terrace or if the resultant spillage is properly controlled by a spillway structure. Whereas terraces can prevent erosion in the short-term during normal hydrologic events, they can cause accelerated erosion during extreme events when their storage capacity is exceeded. The erosion damage caused by such uncontrolled spills can be very severe, as illustrated by many such failures in tropical rice production areas. Terraces can cause accelerated erosion even during normal hydrologic events if they are improperly maintained. Terracing represents an immature topography and is not presented in nature as an erosion control mechanism.

Self-sustaining tailings and submerged waste-rock containment facilities can be developed if planned in advance of mine closure and integrated into the overall mine plan. Design criteria include shallow ponds located far from any dam embankment, a large spill channel cut into bedrock or set at a non-erodible slope and a wide rock barrier between the pond and the containment dam.

PASSIVE CHANNEL EROSION PROTECTION MEASURES

Passive erosion protection measures minimise the risk of accelerated erosion by suitable configuration of the landforms at mine-disturbed land. This is accomplished without structural systems such as riprap, drop structures, or rigid linings. Passive erosion protection systems avoid high-velocity flows and large discharges over steep slopes. Like natural drainage systems, passive erosion protection systems are compatible with the gradual evolution of natural landscape, which is characterised by low rates of erosion.

Passive erosion protection includes measures whereby the size of drainage basins is controlled to avoid large discharges in a single channel

1. Mature Profile of Waterway

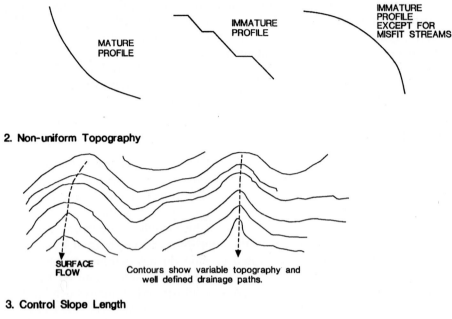

2. Non-uniform Topography

3. Control Slope Length

Fig. 2: Sustainable landscape patterned after natural analogues

on a steep slope. The use of cohesive soils mixed with gravel and cobbles beneath drainage courses provides for stream channel armouring and re-armouring in the event of large floods.

The characteristics of natural rivers provide ample guidance for the design of passive erosion protection systems at mine reclamation sites. However, the self-healing character of natural systems must be built into man-made channels by designers who appreciate the inherent stability of natural systems and who understand the geomorphological processes.

The resultant drainage systems, based on passive erosion control measures, will offer superior performance in the long run. The costs of such systems are not necessarily greater than the cost of conventional structural systems. Through appropriate planning, the costs of sustainable channels, which incorporate passive erosion control measures, may be less

4. Appropriate Drainage Density

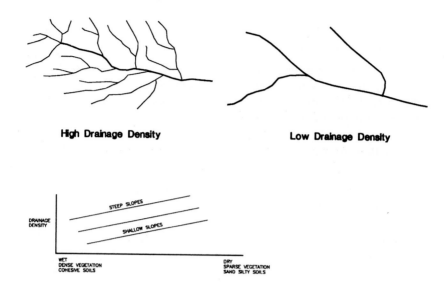

High Drainage Density

Low Drainage Density

5. Deep Valleys to Prevent Spillage

**Depth of valley exceeds
any extreme flow**

**Absence of valley
makes this configuration
vulnerable to spillage**

Fig. 2: *Contd.*

than those of conventional systems which are built with rigid structural erosion protection systems. Examples of passive erosion protection measures are given below.

Bouldery Ground Beneath Drainage Channels

Placement of waste rock or overburden material with high rock content beneath a drainage channel (see Fig. 4) introduces a self-healing capability. If the initial armour of a channel is removed by an extreme flood, channel degradation will expose other rocks and create a new armour layer. Like natural river systems, channels may not have to be repaired after extreme events such as the Probably Maximum Flood (PMF) or even the 100-year flood. The armour layer of natural streams is often non-erodible for events of 2- to 10-year recurrence and is subject to relatively small changes during more extreme events.

Suitable Drainage Density

The allowable slope length of overland flow can be described in terms of drainage density? which is the total length of active drainage channels per unit area (Schumm 1992, 1977). Drainage density can be qualitatively related to dependent parameters (Fig. 3).

Use of Regime Channels

Instead of providing channel armouring, the reclamation planner should design regime channels to replicate the dynamic character of natural channels. Some examples of natural channel features are given in Fig. 4. Channel geometry and pattern were selected from extensive data available from research by fluvial geomorphologists (Schumm 1977). Thus the planner has more than adequate data for selecting appropriate channel parameters to suit the required overall valley gradient and bed/bank materials. The parameters include channel depth, slope, width, sinuosity, meander wave length and width to depth ratio. The resultant regime

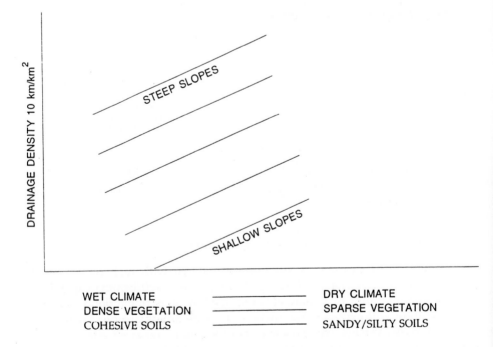

Fig. 3: Factors governing drainage density

channels, which are patterned after natural channel characteristics, will exhibit equilibrium conditions, thereby avoiding progressive channel degradation or aggradation. Regime channels are capable of handling extreme events. Erosion control is not necessary because the channels are designed to be dynamic and accommodate erosion. Flow capacity can be achieved by building drainage channels in well-defined swales or small valleys, just like natural drainage systems.

Floodplains to Attenuate Flows

To reduce flow velocities, the planner may follow the example of natural systems wherein streams are flanked by floodplains, as illustrated on Fig. 4. The floodplains provide extra flow capacity and storage to attenuate the peak flood flow.

Littoral Zone Vegetation and Shoreline Armouring

If the end-pit lakes are small or shallow (i.e., less than 2 metres), it may be possible to protect the shoreline by littoral zone vegetation, following the pattern of similar conditions in the natural environment, as illustrated in Fig. 4. If the end-pit lakes are large and deep, then it will be necessary to provide a large source of coarse materials at the shoreline and inland from the shoreline. This approach allows a degree of shoreline recession, which exposes coarse material to armour the beach. The alternative, to provide a relatively thin layer of riprap shore protection, may not be sustainable because such rigid shore protection measures are subject to failure caused by ice, design event experiences, subsidence, lake level fluctuations and undermining.

A principle reclamation concept is that reclaimed landforms such as drainage channels will change over time and that no attempt should be made to resist such change. Instead, every attempt should be made to anticipate change so that systems can be designed to accommodate change. Anticipation of changes enables the reclamation planner to build robust systems with second and third lines of defence.

CONCLUSION

The development of geomorphically mature reclamation landforms and drainage systems can lead to improved sustainability and reduced long-term liability. Such systems are designed by replicating natural analogues and offer robust, self-healing capabilities similar to systems in the natural environment. This geomorphic approach should be used to develop permanent walk-away schemes and can also be used to reduce the liability of perpetual maintenance reclamation systems.

1. Use Regime Channels

WIDTH = *function of mean flow rate and bank materail.*
DEPTH = *function of mean flow rate and bank material.*
GRADIENT = *function of flood discharge and sediment size.*
MEANDER WAVELENGTH = *function of mean annual flood and bed/bank material.*
SINUOSITY = *function of bed and bank material.*

2. Flood Plains to Reduce Velocity and Attenuate Flow

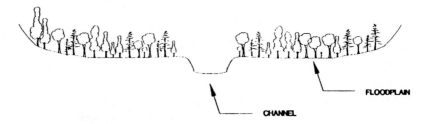

FLOODPLAIN

CHANNEL

3. Meanders to Reduce Gradient

4. Lakes and Wetlands to Attenuate Floods

Fig. 4: Passive channel erosion protection measures (Geomorphic Approach)

5. Bouldery Ground Beneath Channels for Self-healing

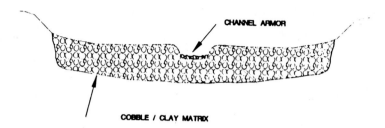

6. Sacrificial Zones of Armoring Material

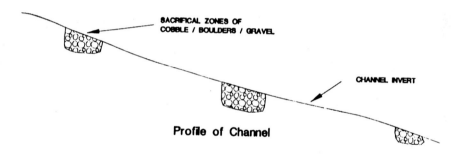

Profile of Channel

7. Lake Shoreline Protection by Littoral Vegetation

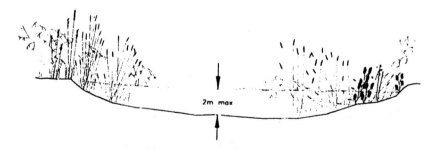

Fig. 4: *Contd.*

References

Gray, D.H. and Leiser, A.T. 1982. *Biotechnical Slope Protection and Erosion Control.* Krieger Publishing Company, FLA.

McKenna, G. and Dawson, R. 1997. Closure Planning Practice and Landscape Performance at 57 Canadian and US Mines (in litt.).

Schor, H.J. and Gray, D.H. 1995. Landform Grading. *ASCE J. Geotech Eng.* 121 (10): 729–734.

Schumm, S.A. 1977. *The Fluvial System.* John Wiley & Sons, Inc., NY.

Schumm, S.A. 1992. *Drainage Density: Problems of Prediction and Application. Unpublished Report* Calgary, Canada.

Schumm, S.A., M.D. Harvey, and C.C. Watson. 1984. *Incised Channels Morphology, Dynamics and Control.* Water Resources Publications, Colorado.

3

A Strategy for Determining Acceptable Sediment Yield for Reclaimed Mine Lands

*Michael J. Bender, Les Sawatsky, Dejiang Long and
P.G. Anderson*

Abstract

*This paper presents a procedure for determining acceptable sediment yield
and includes an example for mine reclamation planning. In the interest
of preserving or enhancing ecological integrity, landscape designers and
regulators tend to establish highly restrictive criteria governing erosion
at reclaimed mine lands. Some regulations call for non-erosive landscapes
even though such criteria may be unachievable. This neglects the fact
that all natural landscape is subject to erosion and that most natural river
basins yield substantial quantities of sediment. Non-erodible landscapes
are also undesirable in nature since the sediment process is a necessary
element of long-term landscape sustainability (stability or evolution).
Furthermore, an attempt to minimise sediment yield to negligible rates
may not address the real impacts on aquatic habitats which should be
mitigated, and may overlook the real ecological needs or natural basin
characteristics. An approach is offered, based on a holistic evaluation of
impacts in the determination of rational sediment yield criteria, relevant
to site and regional conditions, for mine land reclamation. Examples are
given, which show how sediment yield criteria are related to aquatic
systems and natural background rates of sediment delivery.*

INTRODUCTION

Sediment yield from disturbed land and suspended sediment
concentrations need to be regulated to protect downstream receiving

Land Reconstruction and Management Vol. 1, 2000, pp 37–49
ISSN 1389-2541
ISBN 90 5410 793 6
A.A. Balkema, Rotterdam, The Netherlands

streams and aquatic habitat and to avoid rapid degradation of upstream landscape. However, the consensual response to such regulation in many jurisdictions is to apply suspended sediment limitations uniformly, irrespective of local conditions and aquatic community and/or habitat sensitivities. Recent studies of aquatic habitat sensitivities have identified the important factors governing adaptability to increased sediment loads and in assimilative capacity to accommodate sediment. It is now possible to formulate rational sediment regulations based on the physical and biological systems that require protection.

This paper presents an outline of sediment yield conditions in natural and disturbed areas and a review of current sediment yield regulations. A rational method of deriving acceptable sediment yield from a disturbed basin is presented. The method involves the consideration of four factors: basin sediment yield, sustainability of upstream landscape, sustainability of downstream channel regime and preservation of downstream aquatic habitat.

SEDIMENT YIELD FROM NATURAL AND DISTURBED AREAS

The sediment yield of a river basin is jointly affected by its climatic, hydrologic and geomorphic characteristics, including precipitation, vegetation cover, runoff, land use, topography, drainage density, sediment storage, sediment transport capacity and soil erodibility.

Average sediment yield measurements reflect sediment contributions from channel and hillslope erosion. Reported measurements of sediment yield are normally given in units of tons per hectare or millimetres per year, which describe average erosion rates for an entire basin area. Some basins are controlled by channel degradation with little hillslope erosion. Other basins may be controlled by hillslope erosion with little contribution from channel erosion. The geomorphologic characteristics and vegetation cover of a basin affect the dominance of channel and hillslope erosion.

Natural Rates of Erosion

Global denudation is estimated in the range of 0.09 to 0.3 $mm \cdot yr^{-1}$ (Lal, 1994). The variation is due in part to the difficulty in compiling global statistics and in part to the inherent variability and difficulty in accurately measuring erosion rates. Fig. 1 illustrates the variability of sediment yield of various river basins. A selected list of rivers known to produce the greatest sediment, based on investigations by Walling (1994) and Peterson (1986), are also shown in this figure. Sediment yields from river basins are as high 20 $mm \cdot yr^{-1}$ at certain tributaries of the Huangho River in China.

Sediment Yield from Disturbed Lands

Disturbed lands generally exhibit higher sediment yield than natural landscapes, depending on land use. Typical erosion rates for various types of natural and disturbed lands are presented in Table 1 and Fig. 2. The impact of disturbances also varies over time. Generally, construction activities elevate the rate of erosion in the short-term, but may result in a long-term decrease in sediment yield. For example, urban areas after construction often display lower sediment yield than natural conditions.

Active mines and abandoned mines are common sources of increased sediment yield. Peterson et al. (1990) reports sediment yields ranging from 0.01 to 0.18 mm·yr^{-1} for McKinley mine in a semi-arid region. AGRA Earth & Environmental (1994) report sediment yields of 2.2 to 10.7 mm·yr^{-1} at Kidd Creek mine in Timmons, Ontario. Strip mining in the Beaver Creek Basin of southern Kentucky was reported to produce 23.9 mm·yr^{-1} erosion compared to 0.022 mm·yr^{-1} for unmined sites in the area (Collier et al., 1971).

Conventional Regulation of Sediment Yield

The conventional response, to elevated sediment discharge from disturbed areas such as mine lands in many jurisdictions of Canada and the US, is to specify maximum limits for erosion in the form of river sediment concentration. Suspended sediment concentration limits are often expressed in terms of the natural background sediment concentrations, arbitrary concentrations such as 10 to 50 mg·l^{-1}, or a similar limit expressed in terms of water turbidity. Such regulatory response may be motivated by several reasons, such as the decreased lifespans for downstream

Table 1: Sediment yield estimates for various land uses (EPA, 1973; Hittman Associates, 1976; Gray & Leiser, 1982)

Land Use	Sediment yield (mm· yr^{-1})
Forest	0.003
Grassland	0.03
Cropland	0.6
Harvested forest	1.6
Construction	6.4
Highway construction	2 to 10
Unmined watershed	0.02
Abandoned surface mines	0.3
Mined watershed	1.7
Active surface mines	6.4
Spoil bank	24
Haul road	51

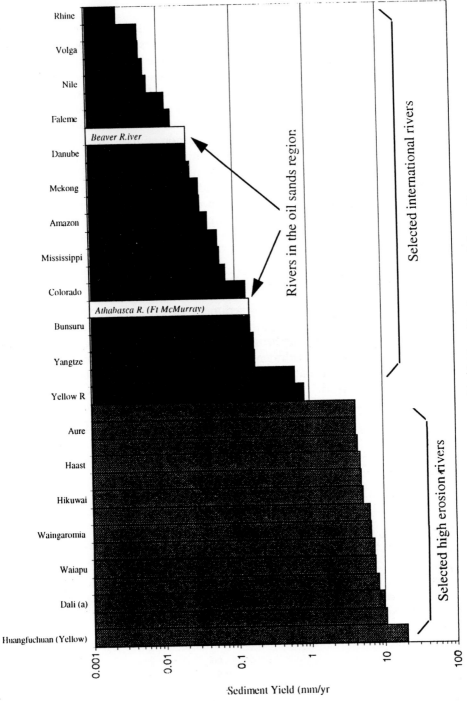

Fig. 1: Sediment yields of selected basins arround the world

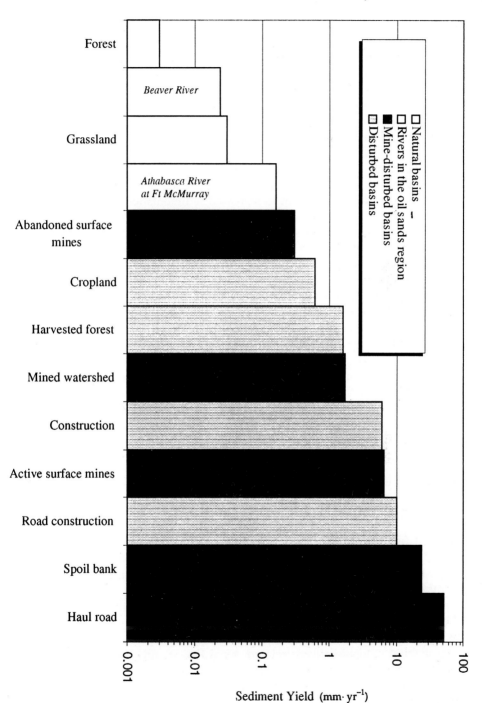

Fig. 2: Comparison of sediment yields of various types of Basins

reservoirs, or undesirable aggradation in navigable rivers. However, the governing factor for suspended sediment concentration regulations is normally the protection of fish stocks and habitat.

Whereas some regulation of allowable sediment yield and/or management practice for disturbed areas is essential for preserving downstream ecosystems, arbitrary limits do not address the complex life cycle requirements for aquatic resources or the sensitivities of certain habitats to elevated sedimentation. Impacts related to elevated sediment yield vary according to sediment particle sizes, toxicity and the sensitivity of different fish species. These may be sensitive to both the concentration and duration of elevated suspended sediment loadings (Newcombe and Jensen, 1996) and loss of habitat to increased sedimentation (Anderson et al., 1996). In addition, most existing regulations pertaining to sediment yield from disturbed lands do not consider the composition of sediment transported downstream, and the use of regulatory limits on turbidity may be irrelevant to the potential for habitat impairment related to sedimentation.

A rational, physically based, sediment yield criterion should be developed as an alternative to conventional regulations, which are based on seemingly arbitrary sediment concentration limits. A number of studies on the impacts of sedimentation on fish habitats have identified the key components that govern the degree of impact on the channel regime of receiving streams and aquatic habitat. The results provide insight for improved regulation of sediment yield and the development of regulatory criteria that offer greater flexibility depending on the grain size of sediment, fish species and fish habitat conditions. Use of a holistic approach may prove beneficial to developers and provide adequate protection for fish and aquatic habitats as well as developers.

BASIS FOR ESTIMATING ACCEPTABLE SEDIMENT YIELD

Aquatic Resource Management Goals

Resident fish populations of a river basin have gradually adapted to the natural concentration of suspended sediment, degradation processes, downstream aggradation and the occasional system-flushing due to flood events. Productive natural habitats also provide for the life cycle requirements of fish such as spawning, foraging or overwintering habitat. The management of potential changes in this environment must consider the specific needs of the various aquatic life cycles. This includes macrohabitat considerations related to water quality aspects such as dissolved oxygen, temperature, nutrients and toxins.

The goal for aquatic resource management is the maintenance or enhancement of a healthy aquatic environment for the preservation of valued fish populations. The long-term preservation of existing fish populations is dependent on natural aquatic environments that include erosion processes.

Dynamic Landscape Management

Landscapes are continually eroding according to topography, surficial geology, vegetation cover and climatic conditions. This erosion eventually contributes to sediment discharge in receiving streams. Measurements of sediment in streams are reported as sediment yields, averaged values that vary from year to year and are dominated by high flow events. Variations in the climatic conditions that affect erosion also result in variations in erosion rates.

The management of erosion on dynamic landscapes can employ one of two philosophical approaches. The first approach is to design structures to inhibit or intercept sediment. This includes settling basins, geotextiles, riprap and a variety of other artificial solutions. This approach may be necessary for urban environments, but requires unnecessarily intensive maintenance for large mine-reclaimed areas, especially after mine closure. Also, extreme hydrologic events may result in catastrophic failure of such structural systems if improperly designed or if left unattended following reclamation.

A second approach is to develop landscapes that are morphologically sustainable. Mature natural landscapes erode and evolve and are less likely to experience catastrophic failure. Mature natural landscapes respond to extreme hydrologic events without the need for extensive maintenance. Morphologically sustainable landscape designs create streams that are self-armouring and that evolve slowly after the fashion of a mature landscape (Sawatsky and Beckstead, 1995).

Determination of Natural Erosion Rates

The creation of a sustainable landscape is dependent on the interpretation of natural analogues that characterise natural conditions. This includes interpretation of natural erosion rates. Sediment yields have been studied in great detail in the literature. Schumm and Rea (1995) observe that higher precipitation/runoff produces lower sediment yield. Although water has erosive potential, lower yields are due to dense vegetation supported by high precipitation. High sediment yields in areas of low precipitation are due to a lack of vegetative cover and, perhaps, the non-cohesive nature of soils in these areas. For example, in a climate of high precipitation, high percentage forest cover produces low erosion rates while in an arid region, with little vegetation, produces very high erosion rates.

The many factors affecting erosion and the current state of available data contribute to the complexity and inaccuracy of predicting basin sediment yield. A reliable and practical approach to estimating basin sediment yield is to correlate the basin of interest with other basins, for which sediment yields have been calculated based on measurements and which have similar climatic, hydrologic and geomorphic conditions and similar histories (Schumm and Rea, 1995). If such correlation is constrained due to a lack of relevant data, measured sediment yields from other basins that are subject to more severe climatic, hydrologic, or geomorphic conditions, can be used to provide a conservative estimate of sediment yield from the basin of interest.

Pro-active Response

A detailed knowledge of natural erosion rates and aquatic habitat requirements are critical components in a pro-active approach to erosion control at disturbed areas. By encouraging natural erosion rates, sustainable solutions become feasible.

PROPOSED METHODOLOGY

A method for estimating acceptable sediment yield is outlined below in the form of four criteria. A study, which exemplifies the geomorphic approach, is presented for the oil-sands area of northern Alberta, Canada.

Basin Sediment Yield Criteria

The rate of net sediment yield from a drainage basin is an important criterion for establishing allowable sediment concentrations as it contributes to overall regional sediment yield. Basin sediment yield is affected by the design of drainage systems, contoured topography and vegetation throughout mine-reclaimed areas. Estimates of basin sediment yield are available from a number of basins throughout the region surrounding the oil-sands area, throughout Alberta and across North America. Published values consist of both direct measurements of suspended sediment load (assuming low bed loads—contributing only to stream bed erosion) and those from reservoir surveys.

Sediment yield data throughout the US and Canada range from 0.006 to 1 $mm \cdot yr^{-1}$, with an average sediment yield of 0.13 $m \cdot yr^{-1}$ (Golder Associates Ltd., 1996). Sediment yields in northern Alberta basins range from 0.002 to 1.8 $mm \cdot yr^{-1}$ while southern Alberta basins generally range from 0.002 to 0.6 $mm \cdot yr^{-1}$. One exception is the badland area of southern Alberta, which ranges from 0.5 to 3.4 $mm \cdot yr^{-1}$.

In the oil-sands region, sediment yield varies from 0.002 to 0.1 $mm \cdot yr^{-1}$. This yield for the Athabasca River, which flows along the mine lands, is relatively high (upstream of the mines) at 0.16 $mm \cdot yr^{-1}$ for the entire basin

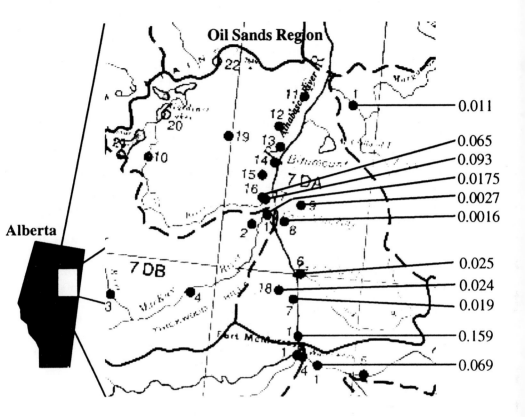

Fig. 3: Sediment yield in the soil sands region Note: All units in mm·yr^{-1}

area of 144,000 km^2. Other smaller basins in the oil-sands area produce sediment at an average rate of 0.038 mm·yr^{-1}.

Given these natural sediment yields and the small area of the mine-closure landscape compared to the drainage basin of the Athabasca River, it is possible to argue for a basin sediment yield criterion on the upper end of the range of sediment yields from local basins.

Table 2: Comparison of natural sediment yields near Syncrude (Carson, 1990)

Area	Sediment Yield (mm·yr^{-1})
Beaver River above Syncrude	0.024
Athabasca River	0.159
Fort McMurray area	0.038
North America	0.13
Southern Alberta	0.05

Landscape Sustainability Criteria

Erosion is an important criterion for establishing the sustainability of mine-closure landscapes. For the case of reclaimed open-pit mines, like those within the oil-sands area of northern Alberta, the final landscape, which can be defined in terms of contouring, revegetation and provision of drainage systems, may also be specified in terms of natural analogues. A sustainable landscape must:

* provide terrain suitable for appropriate terrestrial ecosystems;
* generate runoff with similar characteristics to the natural hydrologic regime of downstream receiving streams; and
* provide landscape features that are not susceptible to high rates of erosion such as through gullying.

Sustainable landscape features can be designed by examining the natural analogues for vegetation, drainage networks, valley slopes and other landscape textures. For example, natural drainage densities for specific soil properties have been identified in the oil-sands region of northern Alberta (Golder Associates Ltd., 1996). For a land slope of 0.005 m/m, recommended drainage densities for silty clay soils are approximately 700 m·km^2. Indeed the typical drainage density for silt-clay soils is around 700 m·km^2. Using a similar density in the construction of new lands, created with similar topography and soil type, might be expected to discourage the formation of new channels and allow erosion rates similar to those of the natural landscape.

Channel Regime Criteria

Streams receiving runoff from the oil-sands mining areas are the subject of concern related to sediment discharge and aquatic habitat. These receiving streams are adjusted naturally to a specific hydrologic regime. If the amount of sediment discharge of runoff from upland areas is changed significantly, the stream regime may alter. An increase in upland runoff may result in stream degradation and an increase in sediment discharge from upstream reclaimed areas may exceed the sediment-carrying capacity of the receiving stream. The result is a change in hydrologic regime, which may involve channel incisions, aggradation, or stream bank erosion. To maintain a morphologically sustainable stream, which evolves only slowly (naturally), allowable rates of sediment yield should be derived to achieve a suitable flow regime and sediment regime by designing reclamation drainage channels in accordance with applicable regime equations from the literature (Schumm, 1977).

Aquatic Habitat Criteria

The primary concern for an aquatic habitat is the extent of expected fish habitat utilisation in streams receiving sediment from reclaimed areas. This includes species-specific provision of spawning areas, forage habitat and overwintering habitat.

Sediment yield criteria should address the particle-size distribution of suspended sediment and the risk of catastrophic morphologic failure in the landscape. High flow events typically dominate the amount of sediment discharged to a stream and carried by a stream. The resultant impacts to aquatic habitats depend on the duration of elevated sediment concentrations, the concentration of suspended sand particles and depositional characteristics. Morphologic failure of a landscape often produces large amounts of sediment and major changes in the morphology and habitats of receiving channels.

To maintain an aquatic habitat downstream of reclaimed mine lands, the following criteria could be applied:

- Avoid significant changes to natural sediment concentrations in receiving streams. Significant changes can be defined in terms of background changes, say 2 to 10 times the average background concentration, depending on the fish species and the background level of suspended sediment concentration. Higher increases are allowable for streams with very low sediment concentrations (i.e., less than 10 mg/l). The allowable variation should be based on the observed variation of sediment concentration rates in a specific topography and climate. Sediment concentration increases should occur primarily during high flow events, when concentrations are normally high and typically of short duration. In this way, the proportion of sediment concentration change is restricted to short periods when sediment concentrations are naturally very high.
- Limit the risk of high suspended sand concentrations (particle size larger than 0.06 mm) to short durations so that the impacts to aquatic life are limited to temporary behavioural changes with no long-term effects. The allowable duration is dependent on the sediment concentration, grain size, species of fish and sensitivity of habitats to sediment deposition (Newcombe and Jensen, 1995; Anderson et al., 1966).
- Provide an equal or better habitat for life cycle needs (spawning, feeding, overwintering) according to the species.

CONCLUSIONS

Available sediment yield provides information that can be used to help define acceptable erosion rates. Sediment yield information can be combined with relevant habitat data to determine acceptable sediment discharge behaviour for disturbed river basins. Acceptable erosion rates may be achieved by incorporating sustainable landscape features in reclamation plans.

Rational criteria can be established for determining acceptable sediment yields from reclaimed land. The criteria should be based on natural rates of erosion and may be used as guidelines for determining landscape characteristics. The criteria should take into account downstream impacts on aquatic habitat, sediment carrying capacity of receiving streams, sustainability of reclaimed landscapes and channel regime.

The suspended sediment criteria so derived should result in morphologically sustainable landscape, negligible impacts on the channel regime of receiving streams and suitable aquatic habitat. Such criteria are believed to be superior to arbitrary sediment concentration limits, which are not necessarily related directly to quantifiable goals of sustainable landscape and productive aquatic habitat.

Industry is likely to benefit from the implementation of such an approach. It is likely that better management practices could be established and would be accepted by regulators as due diligence for impacts related to erosion. The ecological implications are not necessarily worse than conventional practice. In fact, this holistic landscape approach is expected to generate efficient engineering solutions which generally decrease ecological impacts.

References

AGRA Earth & Environmental Limited. 1994. Tailings Area Surface Hydrological Study, Phase II, Part II—Sediment Component. Report to Falconbridge Limited, Kidd Creek Division.

Anderson, P.G., Taylor, B.R. and Balch, G.C. 1996. Quantifying the Effects of Sediment Release on Fish and their Habitats. Canadian Department of Fisheries and Aquatic Sciences, Manuscript Report No. 2346.

Carson, M.A. 1990. Evaluation of Sediment Data for the Lower Athabasca River Basin Alberta. Environment Canada Water Resources Branch, Inland Waters Directorate, Reports.

Collier, C.C. et al. 1971. Influence of Strip Mining on the Hydrologic Environment of Parts of Beaver Creek Basin, Kentucky: 1955–1966. *United States Geological Survey. Professional Paper* 427-C;

EPA. 1973. Methods for Identifying and Evaluating the Nature and Extent of Nonpoint Sources of Pollutants. *US Environmental Protection Agency Report* EPA-4030/9-73-014.

Golder Associates Ltd. 1996. Inventory of Sediment Yield from Natural and Disturbed Surfaces. Technical Report No. 1, Supporting Studies for Mine Closure, Syncrude Canada Ltd.

Gray, D.H. and Leiser, A.T. 1982 *Biotechnical Slope Protection and Erosion Control.* Krieger Publishing Co. Inc., Malabar Florida.

Hittman Associates Inc. 1976. Erosion and Sediment Control. *Surface Mining in the Eastern US.* Volume 1. Planning. Volume II. Design. US Dept. Commerce, Environmental Protection Agency Report PB-261, 343 pp.

Lal, R. 1994. Soil erosion by wind and water: problems and prospects. In: R. Lal (ed.). *Soil Erosion Research Methods*. Soil and Water Conservation Society, Ankeny, IA.

MacDonald, D.D. and Newcombe, C.P. 1993a. Effects of suspended sediments on aquatic ecosystems. *N. Amer. J. Fisheries Management* 11: 72–82.

MacDonald, D.D. and Newcombe, C.P. 1993b. Utility of the stress index for predicting suspended sediment effects: response to comment. *North American J. Fisheries Management* 13 (4): 873–876.

Peterson R. 1986. Course notes. University of Alberta, Department of Civil Engineering (unpubl.)

Peterson, M.R., Watson, C.C. and Zevenbergen, L.W. 1990. Evaluation of alternative sediment control techniques at surface mines in the semiarid West. Planning, Rehabilitation and Treatment of Disturbed Lands, Billings Symp. (unpubl. paper).

Sawatsky, L. and Beckstead, G. 1995. Geomorphic approach for design of sustainable drainage systems for mineland reclamation. Canadian Soc. Civil Eng. Annual Conf. Ottawa (unpubl. paper).

Schumm, S.A. 1977. *The Fluvial System*. John Wiley and Sons, New York.

Schumm, S.A. and Rea, D.K. 1995. Sediment yield from disturbed earth systems. *Geology*, 23: 391–394.

Walling, D.E. 1994. Measuring Sediment yield from river basins. In R. Lal (ed.). *Soil Erosion Research Methods*. Soil and Water Conservation Society, Ankeny, IA.

4

Rainfall Erosion on Opencast Coal-mine Lands: Ecological Perspective

Jose-Manuel Nicolau and Enrique Asensio

Abstract

Land reclamation demands a holistic perspective that includes the development of a self-sustaining vegetation cover in addition to stable landforms and functioning soils. Erosion may be conceptualised as an abiotic process that delays ecological succession. It is controlled more by climate than by sediment transport, and on reclaimed sites by a combination of artificial and natural factors, notably overland flow. Several techniques are employed for modelling erosion on reclaimed lands. Most commonly used are WEPP and the (R)USLE; however, the new ecological models are more valuable for conceptualising the problem in reclamation.

INTRODUCTION

The ecological perspective on rainfall erosion in reclaimed landscapes derived from coal mining is not unrelated to the economic perspective. However, ecology introduces a wider perspective into the economy—an economy of Nature—which is to say an economy in which human society forms part of the system and not a higher/superior metasystem. Nowadays, it is recognised that ecological laws function whether they are taken into account or not and that it is impossible for mankind to operate economically while ignoring Nature. To do so only postpones what is usually called the 'ecological cost', which must be paid later, possibly putting other people at risk. This alone is a demand for a better framework of international laws and the movement of funds so that, where ecological

Land Reconstruction and Management Vol. 1, 2000, pp 51–73.
ISSN 1389-2541
ISBN 90 5410 793 6
A.A. Balkema, Rotterdam, The Netherlands

limitations are known, poorer countries are not forced to ignore them. This progressive development of the collective consciousness increasingly leads to recognition of the need for an economy with more respect for nature. This means that the ecological interrelationships, which underlie productive activity, must begin to change the concepts of profit and property.

For many years, the laws of several countries have incorporated regulations to control opencast mining. Many of these aim to prevent damage in and around the site during its operation and through reclamation. Reclamation involves processes of a longer timescale. In truth, reclamation 'reclaims' nothing—it only creates the conditions for nature to do its job. This is why it is important to know if the disturbance produced is within nature's capacity to return the altered landscape to a state of equilibrium, similar to that which existed before. All attempts to improve understanding of the ecological relationships involved in the process of coalmining and the restoration of the landscape are aimed at minimising, what we have called, the ecological cost and therefore optimising the production of the mines in the widest economic sense.

On site and off site, one of the ecological costs of mining and land reclamation is soil erosion. From the reclamation point of view, erosion on mined lands has two aspects. Erosion is a major factor limiting ecological succession and also one of the main processes contributing to the environmental impact of opencast mining on surrounding ecosystems, due to water pollution and sedimentation.

This article focuses on the first, i.e., on site, aspect. It begins with integration of landforms, soil and plants in a functional ecosystem. The role of erosion in delaying the evolution of reclaimed ecosystems is also analysed. Later, it examines the erosion processes that take place within reclaimed areas, emphasising their influence on ecosystem development. Finally, it considers the modelling of erosion. Modelling can be a useful tool for erosion control and land reclamation design but it is also interesting from the conceptual point of view.

ECOLOGICAL MEANING OF EROSION IN RECLAIMED LANDS

Reclamation: Integration of Landforms, Soil and Plants

Toy and Hadley (1987) established criteria to evaluate the intensity of the disturbance caused by human activities, based on depth of disturbance. According to such criteria, the disturbance of opencast mining is one of the most intense because plant communities, underlying soils and even the rock foundation of the landscape are affected. Mining operations remove vegetation and soil, so no lifeforms (propagules) remain *in situ*. In addition, geologic materials are also modified. This means that reclamation after mining involves building a whole new landscape. The process includes reconstruction of landforms (geology-geomorphology), formation of a new soil and development of a new functional biocoenosis (ecology). The three subsystems must be reclaimed in an integrated manner. Wade and Chambers (1992) point out that the new ecosystem's whole internal character must be determined initially by the reclamationist: topography, soils, hydrology, environmental and biological diversity, microclimate, soil quality and productivity, potential post-mining land uses and on-site aesthetic relationships.

Ecosystems reclaimed after mining should be self-sustaining, so a self-sustaining plant community must be established. Whitford and Elkins (1986) propose tackling the problem using a holistic approach, considering plants, animals, microbiota, soil and climate as interacting components in the transformation of the energy and in the nutrients cycle. According to Visser (1985), three ecological processes are needed for a community to maintain itself: primary production, decomposition and nutrient cycling. Most authors affirm that a functional soil is the key to achieving sustainability (Palmer, 1992: Beeby, 1993). Beeby argues that an ecosystem will only sustain itself by retaining or replacing the nutrients it receives. Decomposition processes are required to degrade dead tissues and make their nutrients available to the soil community and beyond. The storage and release of nutrients are crucial ecological functions that have to be developed in any soil before it can sustain a plant community. The requirement is not for a fully prepared soil in the pedological sense but one that is fully developed in the biological sense. This means that it possesses appropriate levels of organic matter and nitrogen, with associated mineralisation processes occurring actively (Bradshaw, 1988).

There is another component for self-sustainability. This is the geomorphic stability of reclaimed landscapes. Haigh (1996) observes that, in reclamation, little 'Little attention is paid to the topographic and microclimatic context of the living soil system. The geomorphic stability of reclaimed lands is especially important in semi-arid environments—where water availability for plants and soil erosion are the major factors limiting ecosystems development (Nicolau, 1996). Aldon and Oaks (1982) suggest manipulating the topographic contours to concentrate precipitation in arid areas. Toy and Hadley (1987) point out that from the geomorphic perspective, the goal of reclamation is the re-establishment of the balance or dynamic equilibrium between process and form, force and resistance. The development of this balance requires time during which an effective vegetation cover and root network can be generated, soil structure and profile properties develop, and hillslope and channel characteristics come into adjustment. Thus the self-sustainability of reclaimed landscapes derived from coal opencast mining must include stable landforms, a functional soil and productive biological communities with self-sustaining nutrient cycles and self-replacement of species with time.

Processes such as overland flow and sediment fluxes occur at the landform level. Infiltration/percolation and organic matter decomposition and mineralisation occur at the soil level. Photosynthesis/primary production take place at the plant community level. These three processes take place also at different timescales according to the nature of the forces involved. Thus the mechanical forces of the erosion act immediately, while the processes of soil reconstruction act more slowly because re-establishment of the organic matter cycle is required. Ecological succession is the frame wherein the organisation of such ecosystems takes place. The basic processes of succession include the arrival of propagules at the disturbed site, species establishment, alteration of the abiotic environment by the initial colonisers, competition and further changes in the environment and eventual stabilisation (Chambers and Wade, 1992).

Instability or lack of equilibrium, in any of the three basic components of the reclaimed landscape, prevents the development of the processes at whole/landscape level. Frequently there are no plant species adapted to carrying out photosynthesis and primary production. Bradshaw (1983) gives some examples in England of natural plant colonisation being very slow because the nearest source of appropriate plants was far away and those species had no means of long-distance dispersal. This problem can be solved readily by means of artificial seeding or planting, selecting those species adapted to the particular conditions of the site.

Usually, the most severe limitations on ecosystem development come from the soil subsystem. Haigh (1995) points out soil compaction and soil contamination as two common causes of degradation in reclaimed ecosystems. Compaction can make it difficult, if not impossible, for some reclamation species to germinate and grow due to the resistance against root penetration (Sabey et al., 1987). Soil compaction also causes a decline in water infiltration and water-holding capacity, with the result that more rainfall is converted into potentially erosive runoff (Gentcheva-Kostadinova et al., 1994). So rainfall erosion can be a consequence of high soil density. Adverse chemical properties (heavy metals, low pH, low fertility etc.) also limit plant germination and growth by different mechanisms.

Instability of the third component of the reclaimed ecosystems—geomorphology—is also quite common. Toy and Hadley (1987) explain that disequilibrium is produced by altering the balance between force and resistance in disturbed landscapes. Some forces increase as a result of disturbance: hillslope gradient causes increase in the downslope component of gravitational force. Changes in the characteristics of surface materials often result in an increase in the rate and volume of runoff generation. This, in turn, causes an increase in the tractive forces generated by overland flow during precipitation events; increase in hillslope gradient and length exacerbate this proces. Surface resistance is generally reduced as a result of land disturbance. The protective vegetation cover is altered.

Soil properties associated with erodibility are frequently modified in such a way as to increase the susceptibility of materials to impinging forces. Resistance provided by geologic structure and lithology may be reduced through fragmentation.

One of the consequences of such disequilibrium is rainfall erosion, which limits soil formation and plant community development. Rainfall erosion is produced by both geomorphic instability and/or adverse soil physical properties, so becoming one with the prime constraints on the development of reclaimed ecosystems. Once again, the main consequence is that integrated planning is necessary to co-ordinate the harmonious redevelopment of these three subsystems.

Ecological Role of Erosion

From an ecological perspective, erosion may be considered an abiotic mechanism of exploitation (Margalef, 1980). Exploitation is defined in ecology as the extraction of biomass from an ecosystem by means of grazing, agriculture, gravitational mechanisms in aquatic environments etc. In a broader sense, exploitation includes the loss not only of living biomass, but also of other elements involved in primary production: soil nutrients, litter, water, seeds etc. (Diaz Pineda, 1989). The latter is the type of exploitation accomplished by erosion. The loss of mineral particles, litter and micro-organisms prevents pedogenesis. Reduced water availability, often one of the most severe biological constraints in reclaimed landscapes can be caused by a decrease in soil depth that diminishes water storage capacity and infiltration under plant-covered soils. The washing away of lifeforms, such as seeds, rhizomes and other biological propagules is another mechanism of exploitation that limits revegetation (Young, 1992). Under severe erosion, even the physical substrate in which roots are anchored can be lost. So, erosion exploits the soil by extracting nutrients, water, biological propagules and substratum of the ecosystems. As a consequence, ecological succession is delayed, so ecosystems remain in the earlier stages of evolution (Margalef, 1968). Thornes (1984) affirms that soil erosion reduces the rate of plant growth throughout ecological succession in Mediterranean ecosystems, especially in the early phase.

Erosion has been described by many authors as a factor limiting the on-site development of reclaimed ecosystems. Haigh (1992) alludes to accelerated erosion as one factor causing the progressive deterioration of reclaimed grasslands in Welsh coalfields. In this case the mechanism is connected to the weathering of the shale and the formation of an impermeable layer in the soil at depths up to 30 cm. In Mediterranean surface-mined lands, rills and gullies develop on steep and overburden-covered spoil banks, damaging the first herbaceous mixture and preventing succession evolution (Nicolau, 1996). A model for the 'competition' between soil erosion and plant establishment and growth in these spoil banks is developed later in this chapter.

However, exploitation is not the only ecological aspect of erosion on reclaimed lands. Sediments and water coming from mined/reclaimed landscapes produce one of the most important external environmental impacts of mining (Curtis, 1979; Clotet et al., 1983). Sediments and water exported from reclaimed areas can disturb fluvial ecosystems. Halverson and Sidle (1992) describe such disturbance: 'As sediment load increases, channels become shallower and wider; thus the size and distribution of pools—important habitat features for fish—can change. Increase in the fines in stream-bed gravels, which result from accelerated sediment load, can restrict oxygenation of incubating fish eggs and impede fry emergence'.

EROSION PROCESSES IN LANDSCAPES DERIVED FROM OPENCAST COAL MINING AND THEIR INFLUENCE ON ECOSYSTEM DEVELOPMENT

In the last two decades, the geomorphological controls of soil erosion on mined lands have been intensively explored. Findings, related to temporal and spatial dynamics of the processes involved and factors that control them, indicate how ecological succession is affected.

Characteristics of Soil Erosion in Reclaimed Areas

Accelerated erosion on reclaimed lands results as a consequence of the disequilibrium between forces—slope gradient, overland flow—and resistance—soil erodibility, vegetation protection (Toy and Hadley, 1987). This erosion can be very active in the early stages of the new landscapes, when the disequilibrium is great. These landscapes evolve to a more stable state in two different ways. Erosion may reshape the slope profile to a less erosive form. The development of vegetation and soils may increase the resistance and diminish the force of erosion. Some authors suggest that erosion is more weathering-limited than transport-limited (Haigh, 1985). Erosion may also instigate increased erosion through a positive feedback mechanism, whereby deeper channels and steeper slopes maintain the instability. As the new system evolves—towards ecosystem development or towards soil removal—various rainfall erosion processes take place, and at different rates. The characteristics of rainfall erosion in any reclaimed area are conditioned by climate, lithology-soil properties and ecosystem development.

The characteristic erosion processes in reclaimed-surface-mined lands are associated with overland flow—especially in subhumid to arid regions. Soil surface compaction with low plant cover and biomass in the period immediately after reclamation favour an overland flow regime that frequently leads to the generation of a rill-interrill system. However, subsurface flow may also play an important role in humid regions in some substrata and under some management practices. Erosion processes on recently reclaimed lands occur in unbalanced systems that are moving towards a new equilibrium state. Here the processes of plant colonisation, soil formation, evolution of landforms, weathering etc. can be very active and so condition the trajectory of the erosion.

Vertical and lateral spatial variations of substrate properties affect runoff, sediment yield and transport. Two types of factors which control erosion on reclaimed lands can be identified: those dependent on the original physical and chemical characteristics of the redistributed materials and those dependent on the methods and equipment used in stripping and contouring operations, as well as time of year when these operations take place (Groenewold and Rehm, 1982).

The development of erosion is also affected by changes caused by the settlement and autocompaction of the spoils. Groenewold and Rhem (1982) describe area-wide settling in surface-mined landscapes in the Northern Great Plains (USA). The volume of redistributed overburden materials is greater than in the undisturbed state (25% greater in W. North America). With time, the contoured spoils settle and their density increases, sometimes (in the upper most layer at least) reaching levels greater than those that existed before mining. This process is influenced by the textural characteristics of the original overburden and by the mechanical equipment used in reclamation. The influence of this variation on the erodibility of the substrata and of the shape of the relief on soil erosion remains to be clarified.

The main agencies of soil erosion are rainsplash and sheet erosion. Additional agencies are rill and gully erosion as well as piping.

Rainsplash

Downslope sediment transport may be initiated by raindrop impact, with storm intensity and duration controlling soil detachment. The role and magnitude of rainsplash in reclaimed lands are not known but ought to be significant on sparsely vegetated surfaces, especially in the early stages of ecological succession and near the top of spoil banks (Porta et al., 1989). The authors applied the Mirstskhulava model to spoil banks in Catalonia, Spain, estimating the splash erosion at 2.5 $Mg \cdot ha \cdot yr^{-1}$ on non-vegetated mine soils. Rainsplash conditions sheet-wash and rill formation (Thornes, 1984) and contributes to the formation of soil crusts and superficial compacted layers. These are very common on unvegetated reclaimed surfaces. They favour overland flow generation and limit water availability, soil formation, seed germination and plant community development.

Sheet Erosion

The main modes of infiltration/overland flow generation are all observed on reclaimed soils. Saturated overflow takes place on topsoiled spoil banks under temperate climate, where the material placed under the topsoil has a high bulk density (Haigh, 1992). Surface characteristics—related to vegetation cover and to microtopography—and bulk density control the infiltration on topsoil in semi-arid environments (Sanchez and Wood, 1989). A complex infiltration process, similar to that in many agricultural soils, is described by Wells et al. (1986) for topsoil materials (on reclaimed sites): initial infiltration through macropores (large cracks), with surface sealing and infiltration rate later decreasing through micropores.

Sheetflood and Vegetation

Sheetflood and sheet erosion decrease over time as a plant community develops. Sanchez and Wood (1989) conducted some rainfall experiments in reclaimed coal-mine spoils in West Central New Mexico, where a plant community composed of grasses, forbs and shrubs had been successfully established. They found that sites of three and five growing seasons displayed greater infiltration rates and yielded less sediments than sites reclaimed one year earlier. The reclaimed area with the most established vegetation allowed the most infiltration. This was the three-year reclaimed area because grazing—biotic exploitation—on the five-year reclaimed area delayed ecological succession. The biological control of runoff and sediment release was pointed out by the variables related to them: foliar grass cover, grass production, bulk density, foliar shrub cover, shrub production, horizontal roughness, litter cover, vertical roughness, rock cover and organic matter content.

Similar trends in greater stability of the geomorphic system as the plant establishment progresses have been recorded by other authors. Nicolau (1996) conducted rainfall experiments in a Mediterranean environment on topsoiled and grassed spoil slopes and found that the final infiltration rate climbed from 12.0 $mm \cdot h^{-1}$ the second year after revegetation to 50.4 $mm \cdot h^{-1}$ four years later. Concomitantly, sediment concentration dropped from 1.87 $g \cdot l^{-1}$ to 0.75 $g \cdot l^{-1}$.

Toy (1989) recorded sheet erosion rates in a reclaimed community of grasses and shrubs in Wyoming. The magnitude to the sheet erosion rates in the reclaimed area was similar to those recorded for natural communities, indicating successful revegetation and plant community development. Seasonal variations of soil density were also identified. During the months of freezing temperature, there was in increase in ground surface elevation and during the months of rainfall and runoff, a decrease.

Nevertheless, there are many places where plant colonisation and growth have failed and overland flow and sheet erosion have maintained high rates over many years, and where rill-interrill and gully systems have developed.

A third response has been described in reclaimed lands. It consists of a marked change in the trajectory of the reclaimed ecosystem. After a period of ecosystem development and a greater control of water and sediments by the plant community, a modification of soil conditions can lead to a new soil hydric regime that disturbs the biocoenosis, moving back the ecological succession. One case is described by Haigh (1992) in Wales where, as a consequence of accelerated weathering of mine stones, an impermeable layer formed in the soil at depths of 30 cm. Thus water infiltration declined, accelerating erosion and causing progressive deterioration of the reclaimed grasslands. Elsewhere, deterioration may result from chemical changes in the spoil, especially the development of acid conditions.

Slope profile evolution

One of the consequences of overland flow action is the reshaping of the slope profile. Haigh studied the evolution of the slope profile of spoil banks in Wales (UK), Illinois and Oklahoma (USA). These works show that slope evolution is conditioned by the location of two erosion maxima, situated at either extreme of the rectilinear main slope segment. These are indicative of mainslope replacement by the extension of both the upper convexity and lower concavity, a process which appeared to be more advanced on the more active non-vegetated profiles (Haigh, 1979, 1980).

Three factors controlling slope evolution were considered: basal control, vegetation cover and slope aspect, and three models of slope evolution described (Haigh, 1985):

1. Base level constant (Milfraen, Wales): growth of the upper convexity, growth of slope-foot concavity and retreat and slight steepening of the main slope. A migration of the peaks of erosion towards the midslope is evidenced.

2. Basal deposition (Illinois): extension of the upper convexity, growth of slope-foot concavity inhibited by burial and retreat and slight steepening of the main slope. The zone of peak erosion migrates towards the midslope.

3. Basal undercutting (Blaenavon, Wales): extension of upper convexity, retreat and steepening of the main slope:; a slope-foot bench is created.

In each case there is a general tendency for initially rectilinear technogenetic slope profiles to evolve towards a sigmoid shape. The processes that shape these slopes are mainly those related to rain water (soil creep is much less obvious). Vegetation had no influence on the pattern of slope evolution but reduced the magnitude of erosion.

One ecological consequence of the erosion working on the initially rectilinear profile was the creation of new microhabitats with different slope angles, moisture conditions, fine particles and organic matter distribution.

Rainfall erosion can be limited by weathering (sediment availability) or by the capacity of transport of the runoff (Thornes, 1984). Goodman and Haigh (1981) studied the long-term evolution of modern, 30-year-old and 60-year-old dumps in Oklahoma. They found that the long-term evolution of slope profiles responds to the balance between the processes of weathering, generation of fine spoil particles and suffosion. Modern mine spoils were composed of large, coarse, blocky fragments and few fines. The 30-year-old mine spoils contained a large proportion of fines, which had been created by weathering. The 60-year-old dumps were relatively coarse, indicating that a substantial proportion of the fines in these subsurface samples had moved out of the spoil material. As a particular characteristic of those spoils, Haigh (this volume, chapter 7) states that natural weathering processes work more rapidly in mine spoils than in undisturbed rock strata. The rock disruption creates a larger surface area for weathering while the mixing of rock fragments creates a greater range of chemical and moisture conditions within the topsoil.

Rill and Gully Erosion

There are essentially two general modes of rill and gully formation: (a) the Hortonian mode of formation involving concentration of overland flow and abstraction of larger by smaller gullies in the network to produce a regular channel system, and (b) the non-Hortonian mode involving headward extension and bifurcation. In the second type the dominant mechanisms appear to be overfall, mechanical collapse and piping (Thornes, 1994).

Hortonian rilling produced by incision occurs frequently in spoil banks. Rilling is common on freshly disturbed areas (e.g. roadside cuttings and ploughed fields) and often represents the early stages of establishment of equilibrium landforms on disturbed slopes (Riley, 1995). Extreme mete-orological events may further accelerate rilling while development of vegetation cover may cause some to heal (Soulliere and Toy, 1986).

Porta et al. (1989) described a typical pattern of Hortonian rills from spoil banks in Catalonia, Spain, formed at 1.0–1.5 m from the top. Rill erosion was estimated as about 200 $Mg \cdot h \cdot yr^{-1}$. The ubiquity of the rill phenomena in the spoil banks was attributed to the high erodibility of the substrate. This was caused by the poor soil structural conditions leading to crusting and slaking, which promoted high runoff. These authors observed that the first few millimetres quickly became saturated when the rains started. Sealing occurred and a crust formed as it dried. This mechanism is very common on fine overburden materials. In such conditions, plant colonisation and development are prevented.

Causative factors of Hortonian-rill erosion are related to mine-soil properties: bulk density (Soulliere and Toy, 1986), variable compaction due to the mechanical effects of machinery during overburden replacement, presence of vegetation and surface stoniness (Porta et al., 1989). Topo-graphy also influences rill formation: Soulliere and Toy (1986) found a positive correlation with slope length but not slope angle. Nicolau and Puigdefá-bregas (1990) found a positive correlation between slope angle and distance of rill initiation from the top of the spoil bank.

A non-Hortonian mechanism of rill formation has been described for overburden-covered spoil banks in a semi-arid environment in Teruel, Spain (Nicolau, 1996) as well as in badlands and road cuttings (Gerits et al., 1987). Subsuperficial hydric flow runs along cracks and macropores in late spring, leading to pipe and rill generation on sodic soils.

Rills can dissipate if soil erodibility and runoff diminish over time (Riley, 1995) or develop into gullies (Porta et al., 1989; Soulliere and Toy, 1986). When the latter occurs, ecological succession cannot progress. Porta et al. (1989) describe two types of gully network, both very active in the Catalonia study site. The first type received water from adjacent upstream areas, so gully formation was mainly from rills. In the other type, gully growth was promoted by breakdown of berms and also by piping. The growth of gullies took place irrespective of the materials, being controlled by external factors: water inflow and base level. Gullies developed in slopes well below 20%.

Haigh (1980) recorded soil retreat in two gullies—Hortonian and non-Hortonian—in artificial slopes in Wales. The former—developed on a non-vegetated profile—was continuous, with broad shallow channels which became narrower and deeper downslope. It evolved actively by incision. The latter—on a vegetated slope—was discontinuous, with a rectangular cross-section, starting as an abrupt headcut and receiving water and sediments from pipes. All the recorded incision values were negative, implying that this gully was actively healing.

It is interesting to note a particular type of gully erosion described by Osterkamp and Toy (1994) in constructed road banks, but probably occurring in surface-mining reclaimed areas also (Toy and Osterkamp, 1996). This is the gully gravure, explained as follows: 'A capping of coarse, erosion-resistant rock debris concentrates in channels or depressions formed by erosion on a hillslope of less resistant geological materials. The coarse rock debris gradually entraps interstitial finer grained erosion products, either derived from surrounding material or weathered from the deposited debris. The permeability and porosity of the channel filling are thus reduced and runoff is diverted to the periphery of the channel fill. The fill initially inhibits channel expansion but later expedites it. During

UNIVERSITY OF HERTFORDSHIRE LRC

high-intensity runoff events, the loci of incision and gully expansion are the margins, thereby promoting continued incision into the less resistant material. The sequence is completed by the filling of the newly enlarged channel with fresh debris on the sides of the older fill'.

The most remarkable aspect of this process is that gully gravure results in stable, planar surfaces, over which minimal transport of sediment occurs. Processes that cause and maintain planation appear to provide landscape stability. We can interpret this as a negative feedback mechanism in which vegetation is not involved. Applying (R) USLE, Toy and Osterkamp (1996) report that veneering with coarse material can reduce soil loss by 99% and, if native vegetation grows amongst a coarse veneer, predictions indicate soil loss reductions > 99%, regardless of slope.

EROSION MODELS TO BE USED IN OPENCAST MINING RECLAMATION

Models allow predictions to be made without repetitive monitoring and experimental measurements. Qualitative and quantitative descriptions also gain if they are provided with a theoretical framework of the phenomena being studied. If it is necessary to take restorative action, models provide a guide to the variables that must be manipulated. Models fulfil these and other tasks because, not only do they permit planning for the prevention of erosion, they also help in understanding the problems.

In general, models have undergone a notable change, linked in part to the development in the power of calculation and in part to the growing interest in the environment. The first of these factors has caused empirical models, using relatively easy calculations, to be replaced by 'physically based' models, which inculcate the complexity of the processes involved in the phenomenon of erosion (Arranz Gonzalez et al., 1993). The emphasis on the environment, for its part, has favoured the change from emphasis on technology to one on naturally occurring interrelationships and their importance in erosion. Puigdefábregas (1996), for example, points out that the role of vegetation is more important than previously supposed.

Types of Models

As there are many models of different types, it might be useful to classify them and give some examples. Those which are more commonly employed or especially useful in the reclamation of damaged lands are particularly detailed here. Erosion models are classified according to the type of processes and subprocesses they consider (Table 1) and their mode of operation (Table 2).

Table 1: Characteristics of erosion models according to the processes they consider

Type of Processes	USLE	RUS-LE	MUS-LE	MUS-LE (Rojia-ni et al.)	Khan-vilba-rdi & Rogo-wski	WEPP	EURO SEM	Thor-nes	SIBE-RIA	MED-ALUS
1. EROSION										
1a. Rill erosion		X	X	X	X	X	X		X	X
1b. Inter-rill	X	X	X	X	X	X	X	X	X	X
1c. Channels						X	X		X	
1d. Sediment yield		X	X	X	X	X	X		X	X
1e. Transport					X	X	X		X	X
1f. Sedimentation					X	X	X		X	X
2. HYDROLOGY: CLIMATE										
2a. Rainfall	X	X	X	X	X	X	X		X	X
2b. Interception						X	X		X	X
2c. ETP						X	X		X	X
2d. Moisture						X	X		X	X
2e. Temperature						X	X		X	X
2f. Wind						X	X		X	X
2g. Radiation						X			X	X
3. HYDROLOGY: INFILTRATION										
3a. Infiltration						X			X	X
3b. Seepage						X				X
3c. Runoff						X			X	X
4. MOVEMENT OF WATER IN SOIL										
4a. Soil moisture content						X	X		X	X
4b. Water flow in saturated zone									X	X
4c. Water flow in unsatu rated zone									X	X
4d. Superficial flow					X	X	X		X	X
4e. Piping										
5. SOIL										
5a. Soil surface	X	X	X	X	X	X	X	X	X	X

Contd.

Contd. **Table 1:**

Type of Processes	USLE	RUS-LE	MUS-LE	MUS-LE (Rojia-ni et al.)	Khan-vilba-rdi & Rogo-wski	WEPP	EURO SEM	Thor-nes	SIBE-RIA	MED-ALUS
5b. Particle size distribution						X	X		X	X
5c. Structure						X	X		X	X
5d. Erodibility	X	X	X	X	X	X	X	X	X	X
5e. Roughness						X	X		X	X
5f. Organic matter						X			X	X
5g. Nutrients cycle										
6. PLANT DYNAMICS										
6a. Plant cover	X	X	X	X	X	X	X	X	X	X
6b. Above ground growth						X		X		X
6c. Below ground growth						X				X
6d. Reproduction										
6e. Colonisation										
6f. Succession										
7. LAND USE	X	X	X	X	X	X				
8. POLLUTANTS										
9. INTERACTION								X		X
10. STABILITY								X		

USLE and Models based on USLE

Toy (1989) analyses the applicability of the (R) USLE (Revised Universal Soil Loss Equation) to geomorphological studies and describes the advantages as: (1) simplicity, (2) inclusion of a large database, (3) ease of obtaining parameters, (4) widely used by government agencies and (5) adaptability to uniform areas where sedimentation does not exist. The disadvantages he mentions are: (1) inability to estimate sedimentation, production of sediments and erosion in channels and badlands and (2) lack of precision in estimates of soil loss for a single event. Another serious limitation is the difficulty of incorporating new advances in research by calculating them as a product of factors.

In general, however, while many managers are satisfied with the ease and level of accuracy of its estimates, many scientists remain unsatisfied

Table 2: Characteristics of erosion models according to the operational procedure

OPERATIONAL PROCEDURE	USLE	RUS-LE	MUS-LE	MUS-LE (Rojiani et al.)	Khan vibardi & Rogowski	WEPP	EURO SEM	Thornes	SIBE-RIA	MED-ALUS
METHOD										
Empirical	X	X	X	X				X		X
Physically determined	X	X	X		X	X	X	X	X	X
Stochastic				X						
Simulation					X			X	X	X
VARIABLES										
Parametric	X		X	X	X	X		X	X	X
Analytical					X	X	X	X	X	X
RELATIONSHIPS BETWEEN PROCESSES										
Aggregated (arithmetic link)	X	X	X	X	X		X		X	
Modular (linear link)						X				
Synthetic (no linear link)								X		X
DISTRIBUTION										
Non-distributed	X	X	X	X				X		
Spatially distributed			X	X	X		X		X	X
Temporally distributed									X	X
SPATIAL SCALE										
Uniform slope	X							X		
Slope		X	X	X	X	X	X			X
Watershed			X	X	X	X	X		X	
Regional						X				
Global										
TEMPORAL SCALE										
Real time							X			
Hour										X
Day						X				
Event			X	X	X				X	
Period	X	X	X	X	X			X	X	

by the empirical approximation to physical facts. Toy and Ostercamp (1995) also point out that the (R) USLE will now become a more attractive research tool with the incorporation of an easier-to-use program and a database.

For artificial slopes, Arranz Gonzalez et al. (1993) warn of the need to investigate, above all locally, the USLE 'K' factor of erodibility and the 'S' factor, especially in slopes of over 20 degrees, as well as the C and P factors. They also point out the importance of paying attention to coarse particles, incorporating their effect into one of the K or CP factors.

Barfield et al. (1979) studied the application of the modified USLE, the MUSLE model (Modified Universal Soil Loss Equation) (Williams, 1975), estimating the production of sediments over a period of time of one event in a unitary form or distributed over a basin. They found that the MUSLE, together with a route function, is adequate for this purpose and that it is better to use it in a distributed way, although it can be quite difficult to subdivide the basin. An interesting adaptation and extension of the MUSLE is found in Rojiani et al. (1984). Using the randomness inherent in runoff (Q), peak flow (q_p) and erodibility factor (K), the MUSLE can be converted into a probability model. The SEDIMOT (Simple Distributed Parameter Approach to Surface Mine Sedimentology) (Wells et al., 1980) uses the USLE parameters and the runoff curve number CN (concentration time and travelling time). This can be calculated using the data from the United States Soil Conservation Service. The methods most frequently used for erosion production on reclaimed lands are based on MUSLE and runoff Curve Number method (Fifield, 1994).

Schroeder (1995) used rainfall simulation to calibrate the (R) USLE with soil losses from various slopes and degree of vegetation on spoil banks in North Dakota.

Model of Khanbilvardi and Rogowski

Khanbilvardi and Rogowski (1986) propose a sediment transport model, which is used to calculate erosion in each cell of a net that covers the drainage area and which can go from a plot to a basin. The model predicts the routes followed by sediments through the flow network, as well as soil losses due to the growth of rills and the contributing areas between the rills. It allows alterations in the values of the parameters between one event and another.

Rogowski (1985) also presents a model to evaluate the potential productivity of the land in reclaimed or non-reclaimed areas. This is based on the climatic index (Lieth's Miami Model) and the introduced soil index, which is a product of factors such as available water, apparent density, porosity, pH, electrical conductivity etc. Productivity is expressed as the product of the climatic index multiplied by the sum of the soil index for each soil horizon, balanced by a factor which takes into account the influence of root distribution in each horizon (Kiniry et al., 1983). In 1987, Rogowski and Weinrich (1987) offered a modified version of this soil index that uses Leibig's minimum law instead of the balance factor based on the relative distribution of roots in the soil layers. They combine it with the erosion and transport model (Khanbilvardi and Rogowski, 1984) to compare, by simulation, the effects on productivity of both mining and erosion. Mining proves to have the greater impact.

WEPP

In 1986, four United States federal agencies—Agricultural Research Service (ARS), Soil Conservation Service (SCS), US Department of Agriculture (USDA) and Bureau of Land Management—agreed to develop a new generation of technologies for erosion prediction. This was to include knowledge accumulated since the initiation of use of USLE and would enable the USLE to be applied to a wider range of situations than those for which it had been developed. The result of the project, developed in three phases, is the WEPP (Water Erosion Prediction Project). WEPP is a process-based erosion model including many of the processes that accompany erosion and possessing qualities that make it our best hope for providing an instrument as versatile as the USLE (Laflen et al., 1991). Much of the knowledge incorporated into WEPP is more than technical; it is also strategic. Its design is open and modular, thereby avoiding the limitations found in USLE with its factorial formulas that make change difficult. It is also evolutionary in the sense that it does not attempt to change the USLE method radically. Instead, it incorporates the method of working with single and uniform areas and extends this to work in a basin by subdividing the latter into reasonably uniform areas. It includes many more details, such as how to take a rectangular distribution of rills and regular spacing to calculate erosion in rills, which, although not totally realistic (Ibañez-Marti, 1991), simplifies the entry of parameters. It develops means of introducing land erodibility using its own properties and thereby reducing the problem of determining K through the USLE. Included amongst these processes are vegetation growth, evolution of organic matter, soil moisture and even freezing and snowmelt.

EUROSEM

EUROSEM (European Sediment Model) (Morgan et al., 1993) is a more precise erosion model, demanding the entry of parameters of isolated events. It accepts the entry of automated data tables, which means it functions almost in real time, calculating erosion along with the variations in precipitation. So, if rainfall volume is used as an entry, it considers the intensity. In conjunction with the KINEROS (a kinematic runoff and erosion model) (Woolhiser et al., 1990), it can be applied to both a singular and relatively uniform surface as in the USLE, and to a mesh of cells, as mentioned above, to any surface up to the size of a small basin. It takes into account the partition of precipitation, erosion by impact and by flow, and calculates the sediment yield, transport and sedimentation. The main disadvantage with respect to land reclamation, is the need for considerable data entry.

SIBERIA

SIBERIA is a land-form evolution model that was developed by Willgoose and others in 1989, to predict the long-term stability and erosion potential of the proposed post-mining land-form for the Ranger uranium mine, in the Alligator River region, Australia. It is a physically based model that models the runoff and erosion process that occur over the land-form and that adjusts the land-form (and thus the runoff and erosion) in response to the erosion that occurs. It can be considered as "new technology" in the field of mining landform design. SIBERIA can model both transport-limited and detachment-limited sediment transport. However, its primary use is for transport-limited environments. Willgoose and Riley (1994) used SIBERIA to model the evolution of a final landform design. Later, work has been undertaken to include the impacts of vegetation. (Evans, 1997 Evans et al. 1996).

Models Based on Ecological Relationships

Vegetation is poorly represented in erosion models despite recognition of the important role it plays in erosion control and land recovery. Only in the last few years have the ecological effects of vegetation—interrelationships with the soil and therefore with erosion—been incorporated. Kirkby and Neale (1987) offer a model which explicitly incorporates this interaction in an empirical manner (Puigdefábregas, 1996). Thornes (1987, 1990) offers an interactive model of a stable vegetation-erosion system, assimilating two competing species. This model has been tested on a sample of 18 slopes on spoil banks in the province of Teruel, Spain. A modification for the expression of the maximum biomass, which can change with the ecological succession, and a Liapunov function to determine the radius of stability around the critical point were introduced. The adjustment was in line with the data at $p = 0.05$ (95% confidence level) (Asensio, 1995).

A notable effort has recently been made with groups of ecological scientists and plant physiologists to produce erosion models such as the MEDALUS (Mediterranean Desertification and Land Use) (Kirkby et al., 1996), which integrates processes involving the atmosphere, vegetation, the land and its surface (Puigdefábregas, 1996). Although MEDALUS is

more than an erosion model, it is also a desertification mode
of successive change due to a wide range of processes, it ha
used in mining contexts.

SUMMARY AND CONCLUSIONS

1. Land reclamation strategies have to be holistic. The self-sustainability of reclaimed landscapes derived from opencast coal mining depends on stable landforms, a functional soil, and productive plant communities controlling water and nutrient cycles and capable of self-replacement of species.

2. Two aspects of erosion must be considered from the reclamation point of view. On the one hand, erosion is an abiotic mechanism of exploitation that delays ecological succession, limiting the on-site development of reclaimed ecosystems. On the other hand, runoff and sediments exported by mined areas represent a disturbance to fluvial ecosystems. This on-site influence of erosion accumulates in off-site effects on other land uses.

3. The following characteristics of erosion in reclaimed lands may be emphasised:

— Erosion processes in recently reclaimed lands occur in unbalanced systems that are moving towards a new equilibrium state. Here, the dynamics of plant colonisation, soil formation, landform evolution, weathering etc. can be very active.

— Erosion is controlled by both 'natural' and 'artificial' factors: those dependent on the original physical and chemical characteristics of the redistributed materials and those dependent on methods and equipment used in stripping and contouring operations, as well as the time of year when these operations occurred.

— Especially in subhumid to arid regions, soil surface compaction and low plant cover and biomass (in the earlier phases after reclamation) favour a Hortonian overland flow regime that often leads to the generation of a rill-interrill system. However, subsurface flows play an important role in humid regions and in permeable substrata.

— Runoff generation, sediment yield and transport are often conditioned by strong spatial heterogeneity—both vertical and lateral—of substratum properties.

— Erosion is more weathering-limited than transport limited.

—Area-wide settling is a typical phenomenon that can influence landform evolution.

— The geomorphic stability of the whole reclaimed system can be menaced by some critical areas in which infiltration is very low and/or sediment yield very high. Such areas can be: roads, berms, platforms, slope banks or ditches.

4. The use of erosion models for designing reclaimed areas is limited. WEPP, USLE and (R) USLE are the most frequently used models. However, from the conceptual point of view, the new 'ecological' models help greatly in understanding reclamation problems.

References

Aldon, E.F. and Oaks, W.R. (eds.) 1982. *Reclamation of Mined Lands in the Southwest: a Symposium.* Albuquerque, NM. Soil Cons. Soc. Amer., New Mexico Chapter, 218 pp.

Arranz Gonzalez, J.C., Almorox Alonso, J. and Antonio Garcia, R. 1993. Análisis critico de modelos de predicción de la erosión hidrica en mineria y obra civil. *Bol. Geol. y Minero* 104–4: 422–430.

Asensio, E. 1995. Modelización de las relaciones interactivas entre vegetación y transporte de sedimentos. Ph.D. thesis, Univ. Polit. de Valencia, Spain (unpubl.).

Barfield, B.J., Moore, I.D. and Williams, R.G. 1979. Prediction of sediment yield from surface mined watershed. *Symp. Surface Mining Hydrology, Sedimentology and Reclamation.* Univ. Kentucky, Lexington, Kentucky, pp. 83–91.

Beeby, A. 1993. *Applying Ecology.* Chapman & Hall, London, 441 pp.

Bradshaw, A. 1983. The reconstruction of ecosystems. *J. Appl. Ecol.,* 20: 1–17.

Bradshaw, A. 1988. Alternative endpoints for reclamation pp. 69–85. In: J. Cairns, Jr. (ed.). *Rehabilitating Damaged Ecosystems.* CRC Press, Boca Raton, FLA.

Chambers, J. and Wade, G. (eds.) 1992. Evaluating reclamation success using ecological principles, pp. 105–107. In J. Chambers and J. Wade (eds.). *Evaluating Reclamation Success: The Ecological Consideration.* USDA, General Technical Report NE-164.

Clotet, N., Gallart, F. and Calvet, J. 1983. Estudio del impacto en la dinámica del medio fisico de las explotaciones de lignito a cielo abierto en el valle del torrente de Valicebre. Alto Lobregat. *II Reunión del Grupo Español de Geologia Ambiental y Ordenación del Territorio.* Lleida, Spain, pp. 3.19–3.37.

Curtis, W.R. 1979. Surface mining and the hydrologic balance. *Mining Cong. J.* 65 (7): 35–40.

Diaz-Pineda, F. 1989. *Ecologia. Ambiente fisico y organismos vivos.* Editorial Sintesis, Madrid.

Evans, K.G. 1997. Runoff and erosion characteristics of a post-mining rehabilitated landform at Ranger Uranium Mine, Northern Territory, Australia and the implications for its topographic evolution. Unpublished Doctoral Thesis. University of Newscastle. Dep. of Civil, Surveying and Environmental Engineering.

Evans, K.G., Saynor, M.J., Willgoose, G.R. and Unger, C.J. 1996. Land-form erosion assessment. *Min. Envir. Man.* 4 (4) 23–25.

Fifield, J.S. 1994. Evaluating the effectiveness of erosion and sediment control plans. Int. Land Reclamation and Mine Drainage Conf., Pittsburgh. Technical Workshop, Course Notes.

Gentcheva-Kostadinova, Sv., Zheleva, E., Petrova, R. and Haigh, M. 1994. Soil constraints affecting the forest-biological recultivation of coal-mine spoil banks in Bulgaria. *Int. J. Surface Mining, Reclamation and Environment,* 8: 47–53.

Gerits, J., Imeson, A.C. and Verstraten, J.M. 1987. Rill development and badland regolith properties. *Catena Suppl.,* 8: 141–159.

Goodman, J.M. and Haigh, M.J. 1981. Slope evolution on abandoned spoil banks in eastern Oklahoma. *Phys. Geog.* 2 (2): 160–173.

Groenewold, G.H. and Rehm, B.H. 1982. Instability of contoured surface-mined landscapes in the Northern Great Plains: causes and implications. *Recl. and Reveg. Res.* 1: 161–176.

Gu, H., Chen, S., Qian, X., Ai, N. and Zan, T. 1986. River geomorphologic processes and dissipative structure, pp. 211–224 In: V. Gardiner (ed.). *Int. Geomorphology: Proc. 1st Int. Conf. Geomorphology.* John Wiley & Sons, New York.

Haigh, M.J. 1979. Ground retreat and slope evolution on regraded surface-mine dumps, Waunafon, Gwent. *Earth Surface Processes*, 4: 183–189.

Haigh, M.J. 1980. Slope retreat and gullying on revegetated surface-mine dumps, Waun Hoscyn, Gwent. *Earth Surface Processes*, 5: 77–79.

Haigh, M.J. 1985. The experimental examination of hill-slope evolution and the reclamation of land disturbed by coal mining, pp. 123–138. In: J.H. Johnson (ed.). *Geography Applied to Practical Problems*. Geo Books, Norwich, U.K.

Haigh, M.J. 1992. Problems in the reclamation of coal-mine-disturbed lands in Wales. *Int. J. Surface Mining and Reclamation* 6: 31–37.

Haigh, M.J. 1995. Soil quality standards for reclaimed coal-mine-disturbed lands. A discussion paper. *Int. J. Surface Mining, Reclamation and Environment* 9: 187–202.

Haigh, M.J. 1996. Towards the management of soil structure for erosion control on reconstructed coal lands. *Lecture Book, 1st European Conf. & Trade Exposition on Erosion Control*. IECA, Barcelona, Spain, pp. 125–134.

Halverson, H. and Sidle, R. 1992. Cumulative effects of mining on hydrology and water quality, pp. 99–104. In: J. Chambers and J. Wade (eds.). *Evaluating Reclamation Success: The Ecological Consideration*. USDA, General Technical Report NE-164.

Ibañez Marti, J.J. 1991. Evolución cuaternaria de las cuencas de drenaje y procesos de desertización en ambientes mediterráneos. *Curso de la U.I.M.P. sobre "Procesos de desertificación"*. Valencia, Spain, tomo I: 704–4.

Khanbilvardi, R.M. and Rogowski, A.S. 1984. Mathematical model of erosion and deposition on a watershed. *Amer. Soc. Agric. Eng., Trans.*, 27: 73–79.

Khanbilvardi, R.M. and Rogowski, A.S. 1986. Modeling soil erosion, transport and deposition. *Ecol. Modelling* 33: 255–268.

Kiniry, L.N., Scrivner, C.L. and Keener, M.E. 1983. A soil productivity index based upon predicted water depletion and root growth. *Univ. Missouri, Columbia, MO, Res. Bull.* 105, 26 pp.

Kirkby, J.J. and Neale, R.H. 1987. A soil erosion model incorporating seasonal factors. *Int.* pp. 189–210. In: V. Gardiner (ed.). *Int. Geomorphology: Proc. 1st Int. Conf. Geomorphology.* John Wiley & Sons, New York.

Kirkby, M.J., Baird, A.J., Diamond, S.M., Lockwood, J.G., McMahon, M.D., Mitchell P.L., Shao, J., Sheehy, J.E., Thornes, J.B. and Woodward, F.I. 1996. The MEDALUS slope Catena model: a physically based process model for hydrology, ecology and land degradation interactions, pp. 303–354. In: J. Brandt and J. Thornes (eds.). *Mediterranean Desertification and Land Use.* John Wiley & Sons, New York.

Laflen, J.M., Lane, L.J. and Foster, G.R. 1991. WEPP: a new generation of erosion prediction technology. *J. Soil and Water Conservation* 46 (1): 34–38.

Margalef, R. 1968. *Perspectives of the Ecological Theory.* Univ. Chicago, Chicago, ILL., 109 pp.

Margalef, R. 1980. *Ecologia.* Omega, Barcelona, 951 pp.

Morgan, R.P.C., Quinton, J.N. and Rickson, R.J. 1993. *EUROSEM: A User Guide.* Silsoe College, Cranfield Univ., UK.

Nicolau, J.M. 1992. Evolución geomorfológica de taludes de escombreras en ambiente mediterráneo continental (Teruel). Ph.D. thesis, Univ. Autonoma, Madrid (unpubl.).

Nicolau, J.M. 1996. Effects of topsoiling on erosion rates and processes in coal-mine spoil banks in Utrillas, Teruel, Spain. *Int. J. Surface Mining Reclamation and Environment* 10: 73–78.

Nicolau, J.M. and Puigdefábregas, J. 1990. Erosión hidrica superficial en los paisajes restaurados de la mineria a cielo abierto turolense, vol. II: pp. 123–133. In: M. Gutiérrez Elorza et al. (eds.). *I Reunión Nacional de Geomorfologia* Zaragoza, Spain.

Osterkamp, W.R. and Toy, T.J. 1994. The healing of disturbed hillslopes by gully gravure. *Geol. Soc. Amer. Bull.*, 106: 1233–1241.

Palmer, J. 1992. Nutrient cycling: Key to reclamation success, pp. 27–36. In: J. Chambers and J. Wade (eds.). *Evaluating Reclamation Success: The Ecological Consideration USDA, General Technical Report, North East-164.*

Perez-Trejo, F. and Clark, N. 1996. Dynamic modelling of complex systems, pp. 431–446. In: J. Brandt and J. Thornes (eds.). *Mediterranean Desertification and Land Use*. John Wiley & Sons, New York.

Porta, J., Poch, R. and Boixadera, J. 1989. Land evaluation and erosion practices on mined soils in NE Spain. *Soil Tech. Series* 1: 189–206.

Puigdefábregas, J. 1996. El papel de la vegetación en la conservación del suelo en ambientes semiáridos, pp. 79–88. In: T. Lasanta and J.M. Garcia-Ruiz (eds.). *Erosióny recuperación de tierras en áreas marginales*. Instituto de Estudios Riojanos, Sociedad Española de Geomorfologia, Logroño, Spain.

Riley, S.J. 1995. Geomorphic estimates of the stability of a uranium mill tailings containment cover: Narbalek, Northern Territory, Australia. *Land Degradation & Rehabilitation* 6: 1–16.

Rogowski, A.S. 1985. Evaluation of potential topsoil productivity. *Environmental Geochemistry and Health* 7 (3): 87–97.

Rogowski, A.S. and Weinrich, B.E. 1987. Modeling the effects of mining and erosion on biomass production. *Ecol. Modelling* 35: 85–112.

Rojiani, K.B., Tarbell, K.A., Shanholtz, F., and Woeste, F.E. 1984. Probabilistic modeling of soil loss from surface mined regions. *Amer. Soc. Agric. Eng., Trans.*, 27 (6): 1798–1804.

Sabey, B., Herron, J., Scholl, D. and Bokich, J. 1987. Particle size distribution, pp. 59–74. In: R. Williams and E. Schuman (eds.). *Reclaiming Mine Soils and Overburden in the Western United States. Analytic Parameters and Procedures*. Soil Cons. Soc. Amer., Ankeny, Iowa (USA).

Sanchez, Ch. E. and Wood, M.K. 1989. Infiltration rates and erosion associated with reclaimed coal mine spoils in West Central New Mexico. *Landscape and Urban Planning* 17: 151–168.

Schroeder, S.A. 1995. First estimation of cover for reclaimed mineland erosion control. *J. Soil and Water Conservation* 50 (6): 668–671.

Soulliere, E.J. and Toy, T.J. 1986. Rilling of hillslopes reclaimed before 1977 surface mining law, Dave Johnston Mine, Wyoming. *Earth Surface Processes and Landforms* 11: 293–305.

Thornes, J.B. 1984. Erosional processes of running water and their spatial and temporal controls: A theoretical viewpoint, pp. 129–182. In: M.J. Kirkby (ed.). *Soil Erosion*. John Wiley & Sons, New York.

Thornes, J.B. 1987. Nonlinear dynamics in relation to soil erosion-vegetation cover and grazing and forecasting their behaviour. Symp. IGU Commission on Geographical Monitoring and Forecasting '*Dynamics of Geosystems: Monitoring Control and Forecasting*. Moscow, Russia.

Thornes, J.B. 1990. The interaction of erosional and vegetational dynamics in land degradation: Spatial Outcomes, pp. 41–54. In: J. Thornes (ed.). *Vegetation and Erosion*. John Wiley & Sons, New York.

Thornes, J.B. 1994. Channel processes: evolution and history, pp. 288–314. In: A.D. Abrahams and A. Parsons (eds.). *Geomorphology of Desert Environments*. Chapman & Hall, London-New York.

Toy, T.J. 1989. An assessment of surface-mine reclamation based upon sheetwash erosion rates at the Glenrock Coal Company, Glenrock, Wyoming. *Earth Surface Processes and Landforms* 14: 289–302.

Toy, T.J. and Hadley, R. 1987. *Geomorphology and Reclamation of Disturbed Lands*. Academic Press, New York, 480 pp.

Toy, T.J. and Osterkamp, W.R. 1995. The applicability of RUSLE to geomorphic studies. *J. Soil and Water Conservation* 50 (5): 498–503.

Toy, T.J. and Osterkamp, W.R. 1996. The prospect of gully gravure on reclaimed hillslopes. In: *Proc. Conf. Amer. Soc. Surface Mining and Reclamation, Knoxville, USA*, pp. 781–786.

Visser, S. 1985. Management of microbial processes in surface mined land reclamation in western Canada, pp. 203–241. In: R.L. Tate and D.A. Klein (eds.). *Soil Reclamation Processes*. Marcel Dekker, New York.

Wade, G. and Chambers, J. 1992. Introduction pp. 1–2. In: J. Chambers and J. Wade (eds.). *Evaluating Reclamation Success: The Ecological Consideration*. USDA, General Technical Report NE-164.

Wells, L.G., Ward, A.D., Moore, R.E. and Phillips, R.E. 1986. Comparison of four infiltration models in characterizing infiltration through surface mine profiles. *Amer. Soc. Agric. Eng. Trans.* 29 (3): 785–793.

Wells, L.G., Barfield, B.J., Moore, I. D., Benock, G.T. and Uhl, R.A. 1980. *SEDIMOT: A Simple Distributed Parameter Approach to Surface Mine Sedimentology: Final Report,* Project S80-2681-20, US EPA, Cincinnati, Ohio.

Willgoose, G.R. and Riley, S.R. 1994. Long term erosional stability of mine spoils. *Australian Institute of Mining and Metallurgy, Annual Conference (Darwin); Proceedings* pp. 423–427.

Williams, J.R. 1975. Sediment yield prediction with universal equation using runoff energy factor. In: *Present and Prospective Technology for Predicting Sediment Yields and Sources.* Publ. ARS S-40, USDA, Washington, D.C.

Whitford, W.G. and Elkins, N.Z. 1986. The importance of soil ecology and the ecosystem perspective in surface-mine reclamation, pp. 151–188. In: C. Reith and L. Potter (eds.). *Principles and Methods in Reclamation Science.* Univ. New Mexico Press, Albuquerque, New Mexico (USA).

Woolhiser, D.A. Smith, R.E. and Goodrich, D.C. 1990. KINEROS: *A Kinematic Runoff and Erosion Model: Documentation and User Manual* USDA-ARS, ARS-77.

Young, J. 1992. Population-level processes: Seed and seedbed ecology, pp. 37–46. In: J. Chambers and J. Wade (eds.). *Evaluating Reclamation Success: The Ecological Consideration.* USDA, General Technical Report NE-164.

5

Erosion Control: Principles and Some Technical Options

Martin J. Haigh

Abstract

Erosion control engineering aims to control running water, minimise the erodible surface and trap mobilised sediments. Several techniques for determining design-runoff for drainage systems are described. In each case, the process resolves as a series of compromises and estimations, each involving a calculated risk. Three kinds of erosion control technology are available for reclaimed lands: conventional mechanical, biotechnical and ecological. The most reliable and predictable are the techniques of conventional erosion control and drainage engineering. However, these rigid structures are artifacts that have a finite lifespan and require regular maintenance. This is not ideal if mine-site closure is the main objective and the ambition is to quit without leaving behind residual obligations. The promise of ecological engineering is that it creates living structures that are naturally self-sustaining and require no maintenance. Bioengineering strategies are compromises that combine conventional engineering to provide secure structures in the short term and include the possibility of achieving self-sustainability when the technique's biological component is fully fledged. The best route to erosion control is to design reclaimed land that will not suffer accelerated runoff and not require erosion control structures. If erosion control is required, then the aim should be to make these measures as self-sustainable as possible. Since there is no single route to success in erosion control, it remains best to use an array of measures and to be prepared to support these measures until self-sustaining control is fully achieved.

Land Reconstruction and Management Vol. 1, 2000, pp 75–110.
ISSN 1389-2541
ISBN 90 5410 793 6
A.A. Balkema, Rotterdam, The Netherlands

INTRODUCTION

Erosion control engineering shares an agreed list of priorities. These are to control running water, minimise the erodible surface and trap any mobilised sediments. These have been elaborated by Goldman et al. (1988) into ten, and here into twelve, principles of erosion control (see inset) and are incorporated with water-quality considerations in USA training manuals for land reclamation inspectors (Curtis et al., 1988). Across China's 2×10^6 ha of coal-mine damaged lands, erosion control is regarded as an essential part of land reclamation (Zhenqi Hu et al., 1994). This chapter lists some of the techniques available for runoff prediction and erosion control. Many of the key technologies are detailed in the next chapter by Kostadinov and Stanojevic.

Newly reclaimed and immature reclaimed lands often exhibit temporary and unusual vulnerability to runoff and erosion. Engineering is conducted as a temporary measure and as part of the process of restitution. However, by and large, erosion control structures should not be necessary on fully reclaimed lands wherein the land should be restored to tolerances that lie within the natural system. Indeed, resorting to the creation of structures that will require perpetual care and maintenance must be regarded as an admission of failure. They are needed only because the reclaimed land has been improperly planned or designed, so its surfaces remain in a condition whereby they are unable to cope with the normal processes of runoff and erosion.

Engineered structures may be sustainable but they do not repair themselves. Sustained investment in the repair and maintenance of erosion control structures may be an expense that future land-users do not wish to have thrust upon them. Increasingly, the evolving legal frameworks for land reclamation aim to ensure that the duty of care is one that mining companies find harder to relinquish (AGRA, 1995).

In a different context, between 1969 and 1990, the Government of India invested around Rs 16 billion in soil and water conservation. Evaluating the result, the Planning Commission's SWC Working Group agreed that land-users had neither willingly adopted conservation measures nor maintained those installed for them (Kerr and Sanghi, 1992: 3). Similar problems beset the erosion control structures constructed on reclaimed coal lands that have been returned to community control in Wales (Haigh, 1992). These issues are the driving force behind a revolution in erosion control engineering. In the past, this was dominated by conventional civil and agricultural engineering. Its future seems to lie in the new, evolving strategies of biotechnical and ecological engineering. The problem with the rigid engineering structures of the past is that they require care, maintenance and eventual replacement. In other words, while they may be, or have been, sustainable not always have they been, or will they be, sustained. The promise of the new biotechnical and ecological engineering strategies is that they lead to once-and-for-all solutions. In other words, these new technologies are designed to be self-sustainable. They aim to hand the duty of care and maintenance back to Nature. Self-sustaining behaviour is a particular property of living systems. Biotechnical and ecological engineering involves using living organisms, not only plants, in engineering contexts and so creating erosion control structures that, to some degree, self-regenerate and self-repair (Barker, 1994).

Twelve Targets for Erosion Control (after Goldman et al., 1988)

   1.   Fit land-use to terrain
   2.   Minimise soil exposure
   3.   Retain existing vegetation cover where possible
   4.   Protect denuded surfaces
   5.   Divert surface runoff from denuded areas
   6.   Minimise length and steepness of slopes
   7.   Keep runoff velocities low
   8.   Provide adequate drainage
   9.   Trap sediment on site
 10.   Maintain control measures
            Plus two more for reclaimed lands:
 11.   Attend to access roads and receiver watercourses
 12.   Monitor site margins

TWELVE TARGETS FOR EROSION CONTROL ON RECLAIMED LAND

Whatever the strategy invoked, erosion control work shares the same ambitions, as summarised by Goldman et al. (1988) (see insert). This advice may be expanded as given below (cf IECA, 1994).

1. Fit Land-use to Terrain

In site development, the best practice is to change natural, stable conditions as little as possible. If the area is forest, then trees should feature in the recommended land-use. If the land is steep or the soils unstable, then disturbance by tillage and traffic should be avoided. Do not create a system that will cause sediment or chemical pollution in other parts of the environment. The soil loss tolerance should not be exceeded. Do create a land system capable of remaining high quality permanently (Goldman, 1988: 2.1).

2. Minimise Soil Exposure

The less soil is exposed, the less it is disturbed and the less the threat of erosion. Long ago, Curtis (1971) showed that the sediment load generated by surface coal mines depends more on the area disturbed than the rainfall, even though heavy rainstorms yield more sediments. The timing of disturbance can be of critical importance. Construction work should not coincide with erosive conditions. If the surface must be exposed temporarily, then try and provide a protective layer. This could be a chemical sealant such as an emulsion or tackifier, a (hydraulic) surface mulch, a geotextile blanket or a layer of quick-growing vegetation (IECA, 1994; Table 9).

3. Retain Existing Vegetation Cover

The less the existing vegetation canopy is disturbed, the less the soil is exposed to erosion. In surface mining and reclamation, dust is often a problem. Wind erosion may be reduced by creating or preserving a shelter-belt of trees on the site margin. Where the problem is water erosion, preserving the natural vegetation along site margins, especially in the riparian zone of streams, may reduce off-site sediment release. In the case of reclaimed land, vegetation may also need to be protected by the treatment of acid mine drainage (chapter 9, 230 *et seq*).

4. Protect Denuded Surfaces

Try to minimise the area and time of soil exposure but, where this cannot be done, protect the soil by means of vegetation, a cover crop or a mulch of biodegradable debris, or if this is not possible, protect the surface chemically or mechanically. Dust can be controlled by wetting, or better yet by the spraying of chemical tackifiers or crusting agents. As in agriculture, roughening the surface by tillage may reduce wind erosion and foster the infiltration of rainfall. Construction sites, channels and embankments may be defended or protected mechanically by surface and subsurface 'geotextile' mats, through which plants may grow. In severe environments, perforated concrete lattices may serve the same function (IECA, 1994; Table 9).

5. Divert Surface Runoff

Erosion by overland flow can be minimised by drainage and by diverting runoff from vulnerable areas. Erosion rills and gullies thrive only where there is surface runoff that can collect and concentrate in their channels. If there is no runoff, then these features can not grow and may heal. In many nations, one of the first steps in the reclamation of major gully is to cut off its source of water by diversion. On reclaimed sites, excess runoff may be collected in channels, ideally grassed waterways, or in underground drains or harmlessly dispersed into buffer zones of vegetation or stilling ponds (Schwab et al., 1993; Table 9).

6. Minimise Length and Steepness of Slopes

The steeper the slope, the greater the energy of the flowing water and the greater the capacity of that water to erode and transport sediment. The longer the slope, the greater the likelihood that running water will accelerate to an erosive velocity. Equally, the longer the slope, the more chance there is for the water to collect into rills and gullies. One solution is to break up slopes at regular intervals with terraces, ditches, or other obstacles such as cross-slope strips of vegetation. Research from Oklahoma has shown how reclaiming the dogtooth ridges of orphan strip-mine spoils as long smooth slopes can dramatically increase off-site sediment release (Meleen 1986). The hollows between the dogtooth ridges prove excellent sediment traps. Vegetation and bacteria developing in ponds in these hollows help trap sediment and, like any reed bed, neutralise some of the adverse chemical characteristics (Kalin et al., 1991). The long smooth slopes of many reclaimed sites also foster gully evolution (Meleen, 1986). Smaller and more gentle slopes are also, by and large, more likely to be geotechnically stable. However, slope failure can cause huge sediment problems, so attention should be paid to slope stabilisation. The creation of buttress or retaining structures can reinforce the toe-zone of vulnerable sites (Coates, 1977). These structures can be preserved, reinforced or replaced by vegetation (Gray and Leiser, 1982).

7. Keep Runoff Velocities Low

It is said that the energy of flowing water increases as around the square of velocity (e.g. 4~8 vs. 4~16). In channels, the velocity of runoff can be controlled by increasing the roughness (the frictional resistance of the surface over which it flows) through lining the channel with riprap (cobbles), or with vegetation, or by providing other barriers such as check dams and stilling or settlement ponds (Little and Mayer, 1976). The advantage of vegetation is that it has the capacity to grow and repair itself. Channel armouring with riprap can be made more sustainable if the cobbles are part of a thick valley floor pad of mixed coarse grade sediments. Natural stream processes will then produce the most stable armoured channel floor (Sawatsky, 1995; Sawatsky et al., this book). Moving drainage water down a gradient can also be accomplished without causing erosion if channel fall can be concentrated into armoured waterfalls called drop structures, which use up the energy of the water in armoured plunge pools and baffles (Table 9). In extreme cases, engineers have designed concrete energy dissipation structures to reduce the velocity and erosivity in watercourses.

8. Provide Adequate Drainage

Underdrainage can remove excess water from the surface and from saturated areas so that surface runoff does not develop and thus erosion rates are reduced. However, it can be difficult to deliver water to field drains on those former surface mine sites that have suffered compaction or which contain impermeable traffic-pan layers from their construction.

9. Trap Sediment on Site

In many instances it is impossible to prevent some sediment being mobilised. It then becomes important to stop that sediment moving into the environment. If the problem is temporary, straw bales, gravel bags, gravel filters and fabric silt containment fences may suffice (IECA, 1994; Table 9). Ditches and buffer strips of vegetation can help prevent the sediment travelling far. Check dams, bunds and some types of crop ridges may be designed to hold sediment in the body of the site, while on its margins, stilling ponds and retainment basins may be employed to slow up runoff and encourage the deposition of excess sediments (Table 9). All too often off-site drains, ditches, ponds and reservoirs inadvertently serve a similar function.

10. Maintain Control Measures

Erosion control measures may work fine while in a good state of repair. However, most erosion control measures have a finite lifespan and require replacement and/or regular repair. Indeed, many can contribute to the acceleration of erosion and sediment pollution problems if they are allowed to break down and degrade. The cost and probability of long-term maintenance and survival must be considered as a major factor in any erosion and sediment control plan. (Goldman et al., 1988). It is vitally important to consider the needs, capabilities and resources of the end-user of the site (AGRA, 1995).

11. Control Erosion in Access Roads and Receiver Watercourses

Surface mines are often provided with large distances of access and haul roads. In some cases, especially in mountains, the areas disturbed by these roads may be as great as that for the mine itself. Many of these roadways are abandoned when mining is completed. The degraded remnants of haul roads are a common feature of the margins on many reclaimed lands. The road surfaces can concentrate runoff and trigger landslides, gully erosion or the trenching of minor stream channels. Road drains can become blocked and overflow or fail. In sum, they may become major sources of erosion and sediment pollution. One survey in the Kentucky mountains found that soil losses from haul roads amounted to 1300 and 2530 m^3 per km of road-bed (Weigle, 1966). Studies are underway to develop soil-loss models for these sites (Ulman and Lopes, 1995). Meanwhile, these structures need careful monitoring and focussed erosion control work if extension of environmental damage associated with the mine into large peripheral areas is to be precluded (Curtis et al., 1988).

Whether accelerated, or merely redirected, the runoff and drainage from reclaimed lands is returned at some juncture to a natural watercourse (Kilmartin 1989, 1999—this book). If the change in the hydrological regime caused by this circumstance is substantial, and it is not easy to assess what may constitute a substantial change, then the river channel will change shape and plan through bank erosion and/or the sedimentation of riparian lands. Following the analysis of Schumm (1969, 1971) a channel might be expected to adjust as follows to increase in water volume or sediment load (Table 1).

Table 1: Channel response to increments in water and/or sediment from new reclaimed lands in catchment area (after Schumn 1969, 1971; Santos-Cayade and Simons, 1971)

Channel characteristic	Increase in water discharge or mean annual flood	Increase in sediment load	Increase in water and sediment discharge
Width (Bankfull)	Increase	Increase	Increase
Maximum Depth	Increase	Decrease	Not known
Width/Depth Ratio	Decrease	Increase	Decrease probable
Channel Slope	Decrease	Increase	Not known
Sinuosity (Ratio of channel length to valley length)	Decrease	Decrease	Decrease
Average Size (D_{50}) of Sediment Load	Increase	Depends on source	Not known

12. Monitor Site Margins

In the inspection of any parcel of reclaimed land, the best place to start is at its downstream edge. This is where runoff and sediment problems, created on site, begin to affect the landscape off site. It is also the point at which the designer of the reclaimed area relinquishes control. The systems used to protect the land from erosion finish. If the new land is not perfectly integrated into the natural landscape, this is the point at which the consequences will most quickly emerge. Site edges are notorious for problems in watercourses, either incision, sedimentation or increased flood hazard. In mountains, reclaimed site edges are more likely to exhibit slope failure and emerging gully erosion problems than the site's heartland. If there are water-quality problems, the impacts will be most obvious in the immediate downstream and seepage zones of the site (Curtis et al., 1988).

Several nations have adopted strict limitation standards for the quality of water discharged by surface mined areas. The United States Public Law 95-87 of the 'Surface Mining Control and Regulation Act of 1977' calls for control of effluent standards. In practice, these include that the loading of total suspended solids should not exceed 7.0 $mg·l^{-1}$ for any one day nor an average value of 3.5 $mg·l^{-1}$ through any 30 consecutive days (Albert et al., 1988).

DRAINAGE OF RECLAIMED LANDS

The first rule of erosion control is to control running water. Drainage structures are required when reclamation designers or land managers seek to exploit the land beyond its natural or present limits, for example, cultivating land to pasture where the degree of runoff could only be mitigated by forest. On land reclamation sites, where soil compaction is a common problem, there is often a surfeit of erosive runoff caused by low soil-water infiltration rates and excessive concentration of waters in channels and drains. Three of the most basic principles of successful erosion and sediment control are: divert runoff from erodible areas, keep runoff velocities low and provide adequate drainage (Goldman et al., 1988). The problem remains: how to forecast the amount of runoff for which drainage has to be provided?

The problems of managing runoff on reclaimed land are fundamental. Prior to mining, many mine sites were underlain by bedrock. Streams flowed in rock-floored channels across landforms that were adjusted to their flow regimes. Others were well vegetated, perhaps by forest, which included amongst its capacities, the ability to return up to 40% of rainfall to the atmosphere as evapotranspiration. Others had deep well-textured soils. By contrast, reclaimed sites have, on the whole, an erosive substratum composed of rock fragments and fine spoils, and an artificial topography that is not adapted to minimise erosion (Haigh, 1992). Their biological systems are young and relatively inactive in hydrological terms. Finally, despite the low density of the spoils that fill the former mine void, the surface layers of many former coal-mine sites often become highly compacted and impermeable. Hence, these lands tend to have poorly vegetated surfaces and more than usual surface water runoff.

In sum, newly reclaimed lands suffer from the twin problems of having more erodible land and more erosive runoff. Hence it becomes hard to avoid the necessity of using runoff regulation engineering. (see Sawatsky et al., 1999—this book).

Design Runoff Estimation

This section does not aim to provide details for engineers seeking to predict runoff and to design drainage on reclaimed land. It attempts only to indicate some of the approaches that are available and some of the arguments used. The following topics are mentioned: the rational method, the ADAS method, SCS Runoff Curve Method and British Flood Studies Report method for predicting flow from ungauged catchments. Some issues of channel capacity and design of field drainage (ADAS method) are also discussed.

Designing effective drainage and water erosion control works depends on the availability of accurate empirical data or, failing that, of reliable derived criteria. Frequently, empirical data are not available. So, the problem resolves to knowing how much runoff may be generated by the most severe storm that may affect the site during an arbitrarily agreed period. Since progressively greater runoff-generating storms occur with progressively lower frequency, it is possible to speak of designing for the worst storm that might be expected in 5 years, 20 years, or 100 years and for the maximum intensities of rainfall that might be sustained for 5, 15, 30, or 60 minutes in this period. Engineers calculate the risk of the design runoff being exceeded within the life of a project as:

$$r = 1 - (1 - 1/T)^L.$$

where : r is the risk, T the return period in years of the design event and L the life of the project.

The problem is that as one designs structures that are safer because they cater for larger and larger storms, these structures concomitantly grow larger, more expensive and become redundant for longer and longer periods. One classic example is the Del Prado Dam of Orange County, California, created to cope with a 100- to 200-year event; it stood empty for decades. Mature forest regenerated on its floor. Inevitably, its designers were accused of overcautiousness during these years. However, when a near-design event occurred, because of changes in its catchment, the reservoir was quickly filled and overwhelmed. Such problems magnify in designing for the field-scale, since smaller areas respond more directly to rainfalls of shorter duration. In most nations, there are only a few stations where records of rainfall intensity have been kept for more than a couple of decades. Relatively little is known about the kind of rare rainfall events that could cause erosion and runoff disasters since such may occur only once in a hundred years or more.

Rational Method

Most of the drainage design calculations performed for reclaimed lands are based on the inaccurate, antique, but simple 'Rational Method'. This method is inappropriate for large areas because it assumes continuous rain at uniform intensity. It also ignores the fact that, during runoff, depth of flow also increases and hence a certain amount of water is stored (see Linsley and Franzini, 1972: 59). It tends to seriously overestimate maximum flood and is best employed for a first approximation. However, the rational method is routinely employed for designing storm drains and other runoff conveying structures on small areas.

The Rational Method expresses a design peak runoff rate in terms of the equation:

$$Q = 0.0028\, C\, i\, A\,.$$

where: Q is the design peak runoff rate in cubic metres per second, C the runoff coefficient, i the rainfall intensity in $mm \cdot h^{-1}$ for the design return period (and for a duration equal to: T_c—the 'time to concentration' for the watershed) and A the watershed area in hectares.

In the original formulation of this equation, the consequence of rain falling at one inch per hour on one acre was 1.008 cubic feet per second of runoff... functionally unity. The proportion of rain that resulted in runoff could then be expressed as a dimensionless ratio: C—the runoff coefficient. A conversion constant is needed to convert the old imperial to SI units and the result has to be divided by 360 (i.e. multiplied by 0.0028) to reconcile the units with time.

T_c, the 'time to concentration', is the length of time a particular rainfall has to be sustained in order for its full impact to be felt in a drain or channel. It assumes that the affected area is already saturated and capable of contributing runoff. T_c is often calculated from the Kirpich formula which expresses T_c as a function of slope angle and slope length:

$$T_c = 0.0195\ L^{0.77}\ S^{-0.385}$$

where: T_c is the time to concentration in minutes, L the maximum length of flow (m), S the watershed gradient (metres per metre). 'S' is calculated as the difference in elevation between source and outlet divided by L.

The runoff coefficient (C) represents the proportion of any given rainstorm that is not eliminated by infiltration into the soil or other factors, which might prevent the water moving as runoff. It is calculated for particular land-uses and soil types (see Schwab et al., 1981: 74–75; 1993). Soils are classified according to their hydrological characteristics. Coal-mine spoils tend to classify at either extreme of the range, either as Group A or Group D (Table 2). There are many examples of lignite spoils that drain too freely, say in Central Europe. However, hard coals of Pennsylvanian/Carboniferous age, tend to have spoils with high runoff potential. In many cases, the rates of water infiltration into the former surface coal-mine spoils at Blaenavon, Wales are less than 0.1 $mm \cdot h^{-1}$

Table 2: Soil characteristics for rational method calculations (adapted from USDA after Schwab et al. 1981: 75)

Soil Group	Description	Final infiltration rate $(mm \cdot h^{-1})$
A	Low Runoff Potential—deep sands with little silt and clay, deep permeable loess and some lignite spoils	8–12
B	Moderate—Low Runoff Potential—light, loessic, organic soils and some deep-mine coal spoils and very deep applied topsoils	4–8
C	Moderate-High Runoff Potential—shallow soils, including thin applied topsoils on reclaimed lands, some silty spoils—especially deep-mine spoils, soils with a significant clay and colloidal content	1–4
D	High Runoff Potential—most weathered and compacted clayey mine spoils, clays and soils with impermeable layers near the soil surface	0–1

Table 3: Runoff coefficient 'C' for different land-uses and soils (after Schwab et al., 1981)

Cover	Runoff coefficient 'C' for soils of group B at different rainfall intensities			Conversion factors for other soil groups		
	25 mm. h^{-1}	100 mm.h^{-1}	200 mm.h^{-1}	A	C	D
Mature forest	0.02	0.10	0.15	0.45	1.27	1.40
Pasture	0.20	0.17	0.23	0.64	1.21	1.31
Arable-grain	0.18–0.38	0.21–0.38	0.22–0.38	0.85	0.12	0.17
Arable-row crop	0.47—0.63	0.56–0.65	0.62–0.66	0.88	1.09	1.13

(Haigh, 1992; Table 2). C also varies with current land-use (see Schwab et al., 1981: 73–75; 1993).

ADAS Method

The British Ministry of Agriculture (MAFF) and Agricultural Advisory Service (ADAS, 1982: 9) propose a slight variation on this technique. They suggest that peak flood flow (Q_o: $1 \cdot s^{-1}$) may be estimated as:

$$Q_o = St * F * A$$

where: A is catchment area in hectares, St a soil type factor including permeability and soil texture (Table 4), and the F coefficient is calculated through a nomograph, which also requires selection of land-use type: a choice of grass, arable or horticulture (Figure 1). The solution also requires the input of a 'Catchment characteristic' C where: C = 0.0001 (L/S) and L and S are catchment length and slope as in the Rational Method procedure.

U.S. Soil Conservation Service Model

The American Soil Conservation Method for runoff estimation was developed from many years of storm flow records for agricultural

Table 4: Determination of Soil Type Factor (St)—ADAS Method

Permeability class	Range (m.day)	Soil texture ranges	St
	—	Peat	1.3
Very slow	<0.001–0.1	Clay	1.0
Slow—moderate	0.1–0.3	Silty clay loam	0.8
Moderate	0.3–1.0	Clay loam/sandy clay	0.5
Moderate-rapid	1.0–10.0	Medium to light loam	0.3
Very rapid	> 10	Peat/sand-S. loam	0.1

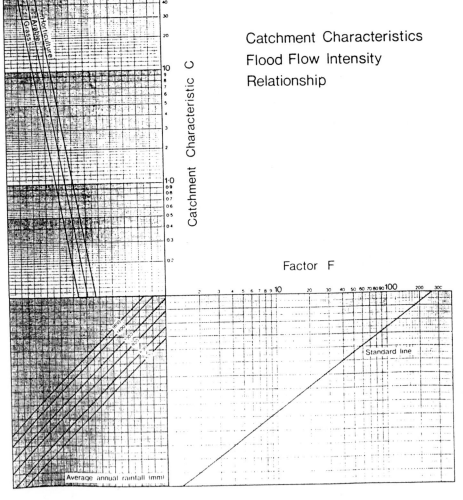

Catchment Characteristics
Flood Flow Intensity
Relationship

Fig. 1: Catchment characteristics flood flow intensity relationship (ADAS, 1982, Courtesy of HMSO, London)

catchments all over the United States. The SCS curves were developed for small agricultural catchments where conditions are likely to be homogeneous (Boughton, 1989). The percentage runoff is based, equally, on land-use/vegetation cover and soil type. Schwab et al. (1981, 1993) include a classification of the effects of vegetation on runoff curve numbers.

The SCS method is sensitive to the influence of storm events. The method consists of adding values representing the extent to which the runoff is influenced by catchment characteristics. The result is a set of rainfall/runoff curve numbers for the prediction of flow from ungauged catchments (Hudson, 1995; Schwab et al., 1993; USDA Soil Conservation Service, 1972, 1973). The technique has produced satisfactory results on semi-arid reclaimed sites in North Dakota (Schroeder, 1994).

British Flood Studies Report (FSR) method

This technique was developed for predicting Mean Annual Flood (Q_{ma}) on small, rural, ungauged catchments. Britain's National Flood Studies Report estimated flood discharge for ungauged basins through their catchment characteristics (NERC, 1975). Arguing from correlation and principal component analyses, the report finds that 4 major components account for 92% of the variance in the mean annual flood of large basins. These are size, slope, lithology and drainage density. Adding soil storage and climatic indices, a 6-variable equation is proposed (NERC, 1975 and seq.; Newson, 1994) and later adjusted for application to small (< 20 km^2) rural catchments (Poots and Cochrane, 1979; Poots 1979). Further examination concluded that, for very small catchments, little divides the results of the full equation and those from a 3-variable equation derived directly from the small catchment data, especially for catchments with heavy soils. In fact, Poots and Cochrane (1979) and Poots (1979) found that slightly better predictions were produced by an equation that employs stream area and the two minor variable to predict (Q_{ma}) the mean annual flood. Hence

$$Q_{ma} = 0.015 \ AREA^{0.866} \ RSMD^{1.462} \ SOIL^{1.904}$$

where: AREA is the basin area in square kilometres, RSMD is computed by taking the net 1 day maximum rainfall of a 5-year return period less the soil moisture deficit, and SOIL is a composite index determined from soil survey. This is calculated by the formula:

$$SOIL = \frac{(0.15S1 + 0.30S2 + 0.40S3 + 0.45S4 + 0.5S5)}{S1 + S2 + S3 + S4 + S5}$$

where S1.....S5 are the proportions of the catchment covered by each of the soil classes 1–5. Soil class 1 has the highest infiltration rate so the lowest runoff potential and class 5 has the lowest infiltration rate and the highest runoff potential. Reclaimed lands tend to either extreme but especially to class 5.

Channel Capacity

Once runoff is estimated, the next problem is to design drainage channels that will remove it (Lane, 1955; Chow, 1959; Schwab et al., 1993). The basic relation for estimating the capacity of flow in any channel is:

$$Q = A \cdot V$$

where: Q is discharge, V the velocity, A the cross-sectional area.

In drainage engineering, all three variables may be adjusted. Discharge may be adjusted by providing additional surface or subsurface channels; cross-sectional area by adjusting the channel geometry and the surface area for frictional resistance to flow; and velocity by making the channel bed more or less straight, thus altering its slope, or by adjusting the channel bed's surface roughness and, hence, its resistance to flow. The objective is to find a design that removes the design runoff at less than erosive velocities. This is especially critical when the channel is not to be armoured by concrete, geotextile or wire mattress, but managed as earth, grassed, or with riprap as its bed. On reclaimed land, the problem can often be complicated by active subsidence that can distort channel gradients and cause ponding (Kilmartin, 1989). However, in sum, channels must be designed such that: $V = Q/A$, where V does not exceed the threshold where erosion is triggered (Schwab et al., 1981: 148; 1993). Table 5 indicates some typical maximum non-erosive velocities for open channels.

Table 5: Typical maximum non-erosive velocities in channels (after Linsley and Franzini, 1972: 286)

Channel bed material	Maximum velocity: Clear water (m/s)	Maximum velocity: Sediment laden (m/s)
Fine sand	0.46	0.46
Silt loam	0.61	0.61
Fine gravel	0.76	1.07
Coarse gravel	1.22	1.83
Stiff clay	1.22	0.92
Concrete	12.20	3.66

When cohesive materials are being engineered, discovering the maximum sustainable velocity is achieved with the help of the Manning equation. This formula considers the surface roughness (capacity for flow retardance) of different channel linings and channel shape. The formula employs the Manning coefficient of channel roughness or resistance to flow (Manning's 'n') and the hydraulic radius (R).

$$V = (R^{0.66} \, S^{0.5}) \, 1/n$$

where: R is the hydraulic radius (metres), S the slope (per cent), and V the velocity in metres per second.

A nomographic solution for this equation has been devised but, for most purposes, the value of the roughness coefficient, Manning's 'n', still has to be estimated from tables (Schwab et al., 1981: 491–493; 1993). Some typical values for 'n' are listed in Table 6.

Table 6: Typical values for Manning's 'n'

Channel type	Variants	Manning's 'n'
Earth ditches	R < 1.5 m	0.030–0.025
	1.5 > R > 0.8 m	0.040–0.030,
	R > 0.8 m	0.040 (+/- 0.005),
Grassed waterway	grass > 250 mm	0.230–0.049
	grass < 60 mm	0.160–0.023
Concrete lined channel		0.015 (+/- 0.003),

For wide shallow channels, like most grassed waterways (Temple et al. 1987), in which width exceeds depth by a factor of 5 or more, it is possible to substitute channel width (W: metres) and average depth of flow (D: metres) for hydraulic radius.

$$V = (1/n) \, (W.D)^{1.66} \, S^{0.5}$$

Field Pipe Drains

If it is not possible to deal with the water on the surface, one alternative is to remove it to subterranean channels. This can be difficult on reclaimed lands that include impermeable layers created by trafficking or soil autocompaction (Rands and Bragg, 1988) and on reclaimed lands undergoing accelerated weathering, where voids become clogged (Haigh, 1992).

Several types of pipe drains are employed in field drainage operations. A lateral pipe drain is a subsurface pipe that collects water continuously over its length, either through seepage or slots in the pipe wall or through gaps at the pipe joints. It feeds into a restricted-inlet main pipe, which collects the flow from several laterals and takes the flow to an outfall. The main may also act as a collector through gaps or slots in its walls. Sometimes pipe drains are used in replacement of either parts of ditches, when they are called culverts, or because they are ditches converted into open-inlet pipes. The British ADAS method for the design of field drainage systems (ADAS, 1982) involves solving the equation:

$$Q_L = 0.13 \; r \; . \; f \; . \; A_L$$

where: Q_L is the design flow in $l \cdot s^{-1}$, r the design rainfall (daily $mm \cdot day^{-1}$) scaled to the drain type and proposed land-use, f the drainflow factor, which is affected by the future land-use, the type of drainage system and especially by whether or not the pipes are supplemented by subsoil shattering or the creation of mole drains, A_L the catchment area in hectares for each lateral drain.

The design rainfall (r) is determined by reference to an ADAS map of the agroclimatic areas of England and Wales. Map units are linked to tables that list the average annual rainfall and daily design rainfalls for both mole drainage and piped drainage schemes under different land-uses. Design rates for grassland range from 6 $mm \cdot day^{-1}$ for areas with 600 $mm.yr^{-1}$ rainfall up to 21 for areas with more than 1500 $mm.yr^{-1}$. Design rates for arable land range from 7 $mm.day^{-1}$ up to 25 $mm.day^{-1}$ and those for horticulture from 9 to 28 $mm.day^{-1}$ respectively.

This is multiplied by the drain-flow factor (f), a fraction which reflects the hydraulic conductivity of the subsoil, and the slope angle (Table 7). The factor varies depending on whether the subsoil has been shattered by subsoiling, or whether a subsoiler has been employed to create a discrete channel, or mole drain.

Table 7: ADAS drain-flow factors (for subsoil permeability classes rapid-moderate-slow)

Drain-flow factor (f)		Slope		
Drainage system	Land-use	<1%	≈3%	>3%
Pipes only	Arable	1.0-0.8-0.7	0.9-0.7-0.6	0.7-0.6-0.5
Pipes only	Grass	0.9-0.7-0.6	0.8-0.6-0.5	0.6-0.5-0.4
Pipes with subsoiling	Arable	–0.9-0.8	0.8-0.7	0.7-0.6
Pipes with subsoiling	Grass	–0.9-0.7	0.7-0.6	0.6-0.5
Pipes with moling	Arable/Grass	0.8	0.8	0.7

Once the design flow is determined, the pipe gradient is determined by reference to the specific site, the pipe type selected and the necessary pipe size read from charts.

The procedure for calculating the design flow for the inlet main (Q_M), which intercepts flow from the pipe drains, is identical. However, the catchment area (A_M) is the entire area to be drained by the whole pipe system measured to its out-fall.

$$Q_M = 0.13 \; . \; r \; . \; f \; . \; A_M$$

Table 8: Inflow to pipe drains in different soils

Inflow rate per 100 m of drain (l.s^{-1})	Slope			
Soil texture class	< 2%	2–5%	5–12%	> 12%
Coarse sand/gravel	1.3–9.0	1.4–10.0	1.6–11.0	1.7–12.0
Sandy loam	0.6–2.2	0.7–2.4	0.7–2.6	0.8–2.9
Silt loam	0.4–0.9	0.4–1.0	0.4–1.1	0.5–1.2
Clay/clay loam	0.2–0.6	0.2–0.7	0.2–0.7	0.2–0.8

Sometimes, it is not possible to determine the catchment area because the area to be drained is fed from indeterminate sources: seepage, springs and so on. In this case, it is possible to frame an approximation of the pipe size needs from an approximation of soil hydraulic conductivity scaled to slope angle and expressed as water inflow rates per 100 m of pipe (Table 8). ADAS (1982: 2) adds that for flows greater than 1.5 m·s, pipe joints should be sealed.

TECHNOLOGIES FOR EROSION CONTROL

Three kinds of erosion control technology are available for reclaimed lands: conventional mechanical, biotechnical and ecological (see Table 9). The most reliable and predictable methods are the tried and trusted techniques of conventional erosion control and drainage engineering. However, rigid engineering structures have a finite lifespan and require maintenance. This is not ideal if mine-site closure is the main objective and the ambition is to quit without leaving behind residual obligations (AGRA, 1995). The promise of ecological engineering is that it creates living structures that are naturally self-sustaining. They last forever and require little maintenance. Bioengineering strategies are compromises that use the certainties of conventional engineering to provide secure structures in the short term, but which include the possibility of achieving self-sustainability when the associated biological component is fully fledged (see Barker, 1994).

Ecological engineering offers the promise of low cost, more ecologically sensitive, self-sustainable techniques for landscape reconstruction and erosion control. Its main problem is that few accurate and reliable values are available to predict the resistance of ecologically engineered slopes (Barker, 1994; Howell 1999, Hewlett et al., 1987; Gray and Leiser, 1982) and river banks to disturbance (Florineth, 1982; Lfu, 1996). Thus, while Howell et al. (1991, Howell 1999) recommend several eco- and bioengineering measures for stabilisation of road-cuts in the Nepal Himalaya, they also find several widely used techniques to be unsatisfactory. These include turfing—which does not counter soil

Table 9: Some current erosion control technology options

Technique	Description	Advantages/Disadvantages
Ecological Engineering	*Leaves provide protection from rainfall and wind erosion, add organic matter to soil and improve drainage; roots strengthen and loosen soil. New developments include using animals, such as earthworms to relieve soil compaction, algae, fungi and microbes for soil surface stabilisation, enzymes and biostimulants.*	*Vegetative covers provide self-repairing and potentially self-sustaining erosion control but they can be slow to establish, show unpredictable behaviour under stress, and prove prone to the development of weak points.*
Seeding	Application of seeds, often in a mixture including fertiliser and bitumen (Florineth, 1994)	Normal seeding by hand or truck is restricted to fairly level ground. On steeper slopes, hydroseeding from a truck or helicopter mounted truck. A common mixture might include: 1000 l water, 12.5 kg seed, 50 kg organic fertiliser, 30 kg cellulose, 50 kg algal binder pumped at around 2 l.min^{-1}. Hydroseeding from helicopter is accomplished about 5 m above the surface. In the Alps, bitumen straw seeding is commonly used; a hand-laid straw layer 4 cm deep (650 g.m^2) including seeds and an organic slow release fertiliser protected by an unstable black bitumen emulsion. This warms the ground and, along with the straw, conserves moisture and protects seedlings from frost damage (Florineth, 1994).
Hydraulic Mulches	Mulching the surface is a traditional method of soil and moisture conservation. Hydraulic mulches are applied directly by hosepipe from a specially designed machine (e.g. IECA, 1994).	Hydraulic mulching, combined with seed and fertiliser application, can be an effective way of establishing vegetation cover on degraded or reconstructed lands. Of course, if the labour force is available, it may be cheaper to do this work by hand, using local materials.
Turf	The application of a layer of living grass and roots to unvegetated area (cf. Gray, 1998).	This achieves a more rapid ground cover than seeding but turf mats are not initially well connected with the soil beneath. They are prone to slippage and rupture. Sourcing the turf can be a problem (Howell et al. 1991: 35).

Contd.

Table 9: (*Contd.*)

Grass lines	Grasses are planted along the contour on steep slopes that are prone to erosion.	An inexpensive but temporary solution to sediment release problems, sediment accumulates rapidly behind the grass wall and, in time, a small, not necessarily stable, terrace is formed. Sometimes the grass lines are overwhelmed. Some grasses are notable for their invasion of adjacent lands.
Woody live barriers: fences, hedges, sprouting brushwood mattresses and wattles	Lines of shrubs or trees are established as hedges along the contour. When established, these trap sediment and may convert hillsides into series of small accumulation terraces. Sprouting brushwood mattresses, wattles and fascines have also proven valuable in protecting stream banks from scour (Gerstgrasser, 1998).	While non-live brushwood fences and wattles lack strength, durability and are prone to collapse and theft (Haigh et al., 1995), live fences provide a sturdy and self-sustaining hedge against erosion for both river banks and slopes. On slopes, in time, the accumulation of sediment may convert hedges into terrace-risers, in extreme cases, some metres high. In streams, they may help mitigate acid mine drainage (Tuttle et al., 1969). Disadvantages include their slow rate of establishment, problems from the failure of parts of plantings to establish and, on slopes, their tendency to balance sediment accumulation up-slope with soil impoverishment immediately downslope.
Forest trees	Trees shelter and bind the soil, reducing erosion and increasing slope stability but, because they are expensive and take a long time to become established (Gray, 1994), they are often used in conjunction with other biotechnical erosion control structures. For their application, see technical details in Gray and Leiser (1982).	Trees rebuild soils, increase soil permeability and infiltration capacity, reduce runoff and erosion, and also strengthen slopes with their roots. However, they also dry out, which reduces destabilising pore-water pressure but increases stresses due to desiccation cracking. In addition, trees add extra weight to destabilise a slope and trap the wind, thus transmitting stresses into the soil (Greenway, 1989). The growth of young trees can be inhibited by competition with grass turf.
Vegetated Rip-rap	Rip-rap is coarse cobble used to face and protect erosion-prone sites such as stream channels and road embankments. It is also the fill used in gabions. Adding vegetation aims to achieve the same effect as gabions using the structural strength of roots as the binder.	Creates a permanent and self-sustaining stabilising structure. However, vegetated riprap may not have the structural strength of either gabions or masonry (Howells et al. 1991). New design criteria for stream banks have been developed by the US Army Corps of Engineers (1991) and Teal (1995).

Contd.

Table 9: (*Contd.*)

Grassed Waterways	These are shallow watercourses (width:depth ratio is usually greater than 5:1) used to carry intermittent flow and protect headwaters, especially agricultural, from erosion. They are often used in conjunction with subterranean pipe drains. See full technical design specifications in Temple et al. (1987).	Grassed waterways provide an inexpensive and self-sustaining drainage that fosters infiltration and traps mobilised sediment. Their disadvantage is a proneness to selective degradation and scour and the grass's, capacity to invade nearby land. There is a large technical literature that tries to calculate the greatest velocity the channel can stand without erosion (Schwab et al., 1993). For grass mixtures, this ranges from 1.5 to 1.2 m·s^{-1} on erosion-resistant soils and channel slopes of 0–5 and 5–10% respectively, or 1.2 and 0.9 m·s^{-1} on easily eroded soils. For tough grasses, such as Bermuda Grass, higher velocities may be sustained: 2.4, 2.1 or 1.8 m·s^{-1} for channel slopes of 0–5, 5–10% and over 10% on erosion-resistant soils, decreasing to 1.8, 1.5, and 1.2 m·s^{-1} respectively on easily eroded soils. Newly seeded grasses offer less resistance to erosion and velocities should be kept below 1.2 m·s^{-1} (Hudson, 1995).
Bio-filters and sediment traps	Grass strips, riparian vegetation and reedbeds rank among the techniques used to prevent sediment pollution. Reedbeds can also reduce acid drainage problems (Brooks and Gardner 1994, 1995, Kalin et al., 1991)	The techniques benefit from their low cost and self-sustainability but have the disadvantage of involving a relatively high land-take and low capacity compared to equivalent, conventional engineering measures.
Bioengineering	*These measures combine strength, immediacy and reliability of mechanical structures with self-sustainability and regenerative capacity of biological structures*	*A compromise between ecological and mechanical engineering, bioengineering provides long-lasting structures and includes the potential to return a site to the control of nature. Techniques are relatively new and unproven over the long term. The US Natural Resources Conservation Service is building a database on plants of value in bioengineering (Dickerson et al., 1997).*
Soil conditioners	Chemical additives that encourage soil aggregation, hence a soil's capacity to resist erosion, while fostering plant growth and improving water-holding capacity of the soil (cf. O'Malley, 1998).	Commercially available conditioners, often designed for agricultural use, soil conditioners remain expensive in cost-benefit terms and do not work reliably on all soils. Their use remains in the process of active research and development (De Boodt, 1990; Levy, 1996).

Contd.

Table 9: *(Contd.)*

Erosion control blankets	Permanent or biodegradable mats, some impregnated with fertiliser and seeds, designed to protect erosion-prone steep land and allow vegetation to become established (e.g. IECA, 1994; Rickson, 1995).	The provision of effective erosion control and the reliable establishment of vegetation cover is offset by expense. Costs are as high as domestic fitted carpet and hence blankets are used only to protect slopes in the context of expensive engineering projects. The relative efficacy of different systems is not yet fully proven (e.g. Sutherland, 1998).
In-soil structural blankets	Mattresses, typically 100–200 mm deep, of plastic or organic fibres that are embedded in the soil to reinforce the root mats of young vegetation covers on steep slopes (e.g. IECA, 1994) or reinforced grass waterways (Hewlett et al., 1987; Temple et al., 1987). Coir and copra are the fibres of choice in S. Asia (see Joseph and Sarma, 1997).	This technique provides an effective insurance against problems emerging during the slow re-establishment of vegetation on steep embankments. The reinforcement holds the slope until woody vegetation gains a grip. Once again the blankets are expensive and prohibitively so outside the context of major engineering or landscaping works. Problems often develop when the vegetation fails to become established, although the exposed blanket does provide erosion control. Used in waterways, reinforced grass has been called an aesthetically pleasing, lower cost, alternative to riprap in channels requiring high-performance protection (Lancaster, 1997).
Gabions with trees	Baskets or mattresses of wire that are filled with local rubble and used for the structural reinforcement of steep slopes or stream channels. Gabion bolsters may be buried on steep slopes to inhibit gully evolution. Gabions are permeable to water and useful for trapping sediment (Howell et al., 1991). See technical specifications listed in Gray and Lieser (1982: 231–235)	An inexpensive, reliable, well-researched technique, gabions have a lifespan measured in terms of a few years. However, they can be planted with woody vegetation, which with luck, will extend the lifespan of the structure. Their most frequent use is in road-bank and river engineering. Along the Pokhara Highway, Nepal, where landslides from road-cuts and road-waste material impact on adjacent rivers, gabion barriers are used as the main form of slope and river-bank protection. However, they are less effective than masonry structures because of their lower strength, tendency to corrode and vulnerability to scour in river-beds (Schuster and Hubl, 1995).

Contd.

Table 9: (*Contd.*)

Mechanical	*Engineering structures*	
		These have the advantage that they are easily designed and behave predictably. Unfortunately, they do not repair themselves and are sustainable only through perpetual care and maintenance.
Emulsions, tackifiers and binders	Chemical sprays that provide the soil surface with a shield against water or wind erosion (e.g. IECA, 1994).	Easily applied but also easily disturbed, these offer a temporary solution for dust problems on active construction and surface mine sites. Sometimes bitumen and asphalt emulsions are used with seeds and hydromulching as part of a programme of revegetation (Florineth, 1994; Waller, 1995).
Silt fences	These are permeable fabric curtains that are established across the course of flowing waters and used to trap sediments (e.g. IECA, 1994).	An expensive and temporary solution to a sedimentation crisis, they work quite well, but are quickly clogged or breached. They are used during construction and mining operations to prevent sediment pollution. In the long term, a stilling pond and reed-bed is far more effective.
Sedimentation ponds	Sedimentation ponds serve to detain surface runoff in order to allow suspended solids to settle. They are normally temporary structures that are created before mining and removed when reclamation is successful.	Sedimentation ponds may be constructed by excavation or by containment within an embankment. Their function is to delay surface runoff not storage and cannot be constructed in stream channels. Their size is scaled to the mining area. In the USA, PL-95-87, supplement to the 'Surface Mining Control and Reclamation Act of 1977', details regulations for sediment ponds. In general, storage volume should be at least the area disturbed times 125%; ponds should be cleaned when they fill to 60% capacity. Maximum project sizes allowed by different states run from 2 to 8×10^6 m^3 (Albert et al., 1988).
Gravel filters/French drains	Layers of coarse material that allow the rapid infiltration of surface runoff to groundwater or buried pipe or mole drains	These systems are common features on reclaimed mines and reconstructed steep lands. They are inexpensive to create but their lifespan is limited by a tendency to become choked with fine sediments.

Contd.

Table 9: (*Contd.*)

Pipe drains and inlets	Subterranean pipes of clay or plastic (e.g. ADAS, 1982). Inlet and technical capacities are assessed by Waterhouse (1982: 228–247, 261).	A traditional, expensive, but well-researched and potentially effective way of draining lands and preventing excess surface water. Pipe and mole drains have suffered rapid loss of efficiency due to siltation on clayey mine-spoil sites that are prone to accelerated weathering (see Bragg et al., 1984). In one Welsh case study, mole drains silted to closure within weeks of establishment (Rands and Bragg, 1988). Problems occur in establishing pipe drains in thin or stony soils and of getting water into them in soils of low permeability. Surface drain inlet structures come in many forms and are best known for their capacity to malfunction or, in the case of French drains, to silt up.
Contour Bunds	Bunds are small embankments constructed across a slope. Contour bunds run along the contour and are most appropriate in arid areas. In areas of heavy to moderate rainfall, graded bunds have a low (0.2–0.3%) gradient against their upslope side to channel runoff. Bunds are usually protected by vegetation. Recently grasses like vetiver have become popular. Several principles are suggested for the spacing of contour bunds. 1. The seepage zone below the upper bund should merge with the saturation zone held up by the lower bund. 2. The bunds should check flowing water before it attains erosive velocity. 3. The bunds should cause the smallest possible inconvenience to agricultural operations and trafficking. 4. They should consume as little land as possible. Bund spacing is calculated as: $VI = s/a+b$ [where: VI is the vertical interval in cm, s is land slope in per cent and $a + b$ constants for soil and rainfall type]. Murthy and Sachan (1990) commend values of 10 and 60 in medium to heavy rainfall zones and 15 and 60 in low rainfall areas.	Bunding is appropriate for lands of up to 6%, or possibly 10% grade. Contour bunds cannot be employed in areas with very shallow soils (Bono and Seiler, 1984). In India, bunds have been applied successfully in many locations but not in areas with deep, cracking clay soils (Vertisols) (Murthy and Sachan, 1990). The reason is that water channels through these cracks and destroys the bund. This kind of soil is common on reclaimed land. Field studies of bunds find that a very high percentage fail to perform effectively. The most common causes are damage by animals or vehicles, siltation and overflow, breaching by animal burrows, and poor or ineffective design.

Contd.

Table 9: (*Contd.*)

| Riprap | Riprap is made from cobbles of rock deposited to armour a channel bank or to foster infiltration above a (French) drain. [Very often, riprap is held in place by vegetation or wire mesh. It is the core of most gabions.]. Limestone riprap can also help treat acid mine drainage in armoured open channels (Ziemkiewicz et al., 1997). | The effectiveness of riprap depends upon it remaining where it is dumped and neither slumping nor being washed away. The fact that this is hard to ensure is the reason many workers prefer to add further stability to the deposits by adding wire or vegetation. Failures commonly involve the flowing water bypassing the riprap mat, the elimination of the riprap in pockets of especially intense scouring or where the cover is breached by undermining.

However, for what they are worth, here follow typical considerations for estimating the stability of riprap. This depends on the 'tractive force' of the flowing water (Chow, '1959: 168 and seq.): $T_{cw} = 19.5\ D_{50}\ W$[here D_{50} is the mean particle size of the riprap]. The tractive force at the water is $T_{cw} = f[w\ R\ S]$ where: w is the unit weight of water (1 gm/cm^3 for pure water but more for sediment-laden water), R is the hydraulic radius of the channel, S is the slope (hydraulic gradient) of the water]. However, the tractive force required for erosion is different for a particle on the sides of a channel to one on its floor. On the channel walls the tractive force is supplemented by gravity, which also tends to start that particle rolling down the channel wall. So, sometimes the equation is refined as follows: $T_{cw} = w\ d\ S\ K$ [where: d is the depth of flow, and K is the ratio of the tractive force needed to move a particle resting on the floor to the necessary to start motion of the same particle on the channel floor]. This ratio is 1 for an infinitely wide channel and less than 1 for all others. The US Geological Survey simplifies this relation as $D_{50} = 0.01V^{2.44}$ where V is velocity in feet per second B.L edgett; Mc Conough H5, 1995. The US Army Corps of Engineers (1991) apply a more complex formula based on D_{30}. |

Contd.

Table 9: (*Contd.*)

Sandbags, bales, log-wood cribs	Canvas bags filled with sand, gravel or concrete and used for the structural reinforcement of slopes and watercourse (e.g. IECA, 1994). Straw bales are useful for temporary emergency erosion control (Miles et al; 1989). Cribs are meshworks of timber and other materials used for slope reinforcement as an alternative to gabions (Coates, 1977; Gray and Leiser, 1982: 235 and *seq.*).	Inexpensive, usually ad hoc solutions, used in the temporary control of small-scale erosion and sediment problems. Jute and hessian bolsters have proved inferior to wire (Howell et al., 1991). Cribs are best used to buttress cut slopes and are better as a preventive measure than a control. They may be made of a variety of materials including logs, timber, metal concrete etc. (Coates, 1997).
Hydraulic matrices	Polygonal cellular matrices of plastic or similar material into which soil or concrete may be delivered by a hydraulic sprayer (e.g. IECA, 1994)	An expensive new technology, most frequently employed on road-banks, erosion and slope failure prevented initially by the strength of the cellular matrix and later by the concrete or vegetation which is grown within the matrix.
Contain-ment structures	Typically, these are 200–400 mm deep meshes of plastic (geogrid) or precast concrete. The grid confinement decreases the lateral movement of soil particles when wetted or loaded (O'Grady, 1984; Oliver, 1984). They may be backfilled with more concrete or with soil for cosmetic revegetation (e.g. IECA, 1994).	An up-to-date variation on the theme of poured concrete. The lifespan of these structures depends on the durability of their materials and the stress placed upon them. The Armater system has proved useful against mudslides on embankments (Ward, 1995).
Formed concrete	Concrete is pumped into woven fibre bags that are laid over the area to be defended (e.g. IECA, 1994).	Inexpensive, provides a sterile and not especially durable, low-grade impermeable concrete surface that is deployed mainly on road and river banks.
Concrete and masonry	Traditional cut stone and formed and set concrete, provided with weep-holes to prevent water pressure disturbance, used to line channels and to construct retaining toe-walls for steep slope (e.g. IECA, 1994; Coates, 1977).	Fairly inexpensive, easily designed and predictable material of high durability. Provides sterile surfaces that have the disadvantage that they are relatively impermeable to water, thus tend to create high water pressures in slopes, and they tend to be smooth—hence accelerating water flow. A modern variation on this theme, the use of articulated concrete blocks for revetments and bank protection has the advantage that the systems are inherently permeable (e.g. IECA, 1994; Scott Queen and Clopper, 1992).

Contd.

Table 9: (*Contd.*)

Check dams	Barriers of various materials ranging from brushwood, through masonry to formed concrete, set across channels to trap sediment and slow runoff. (Kostadinov and Stanojevic, 1999).	Traditional textbook solution applied widely in most land reclamation work. Check-dams have a limited capacity to store sediment but can be employed to create flat areas for agriculture when silted. Problems are associated with failure or bypassing of the spillway, piping through the dam and scour downstream. Requires regular maintenance.
Drop structures	Drop structures are used to conduct watercourses down steep grades where there is a danger of scour or incision. They are also used to reduce the slope of an oversteep or incising channel by breaking the long profile into a series of steep defended steps, separated by long reaches of low slope. Miniature waterfalls reduce runoff velocity and control the erosive power of the running water by dissipating it in on an armoured plunge pool and apron (e.g. Blaisdell, 1981; Blaisdell and Moratz, 1961, Fig. 2). Hydraulic engineers recognise two classes of drop structure 'Low' and 'High'. Only high drop structures are common on reclaimed lands. However, a quick distinction is that, in high drop structures, conditions downstream do not affect the pattern of flow upstream. There is always an abrupt fall in the water surface. There are many types of high drop structure. Blaisdell (1981) describes three: the straight drop-spillway, the dissipation-bar drop structure and the box inlet drop structure. The discharge capacity of a drop structure weir is calculated from the formula: $$Q = 0.552\ M\ (B.H)^{1.5}$$ where: Q is discharge in cumecs, B the width of inlet, H the full head over the weir (metres), 0.552 acceleration due to gravity, and M a coefficient which	A standard textbook measure, often used in land reclamation, agricultural drains, river engineering and gully plugging work, the purpose of a (high) drop structure is to reduce the excess energy produced by the loss of channel elevation. Drop structures may be constructed out of local materials or precast concrete. Drop structures are inexpensive but require repair and maintenance to avoid scour of the spillway and plunge-pool apron. All drop structures include three design features. 1. They have an inlet or weir that may be straight, curved, notched, flanged, or a box-type feature. This collects and focuses the flow of water to the spillway and discourages bank erosion due to the drawdown of the water surface at the fall. Defective inlet structures allow bank scour, fail to collect all of the channelled flow, or fail to focus the flow into the centre of the spillway/weir. 2. They have a headwall or spillway that represents an abrupt increase in the gradient of the channel where the structure is placed. The purpose of the spillway is to control the fall of the water. Common failings include scour of the weir or spillway, which means that it was not built of appropriate materials, or failure to direct the accelerated water jet onto the appropriate armoured structure in the drop apron. It is not unusual to find badly constructed drop structure weirs where, at high flows, the nappe of falling water completely overshoots the apron area causing serious erosion downchannel.

Contd.

Table 9: *(Contd.)*

varies with the character of the weir, its inlet, depth, type, reservoir and outlet.

Schwab et al. (1993) recommend an M-value of 3.2 for all but curved inlets (metres). This is provided that the ratio of head to inlet width is at least 0.2, the drop at least 0.25 metres above the height of the water, and there are no obstacles in the approach channel within 1.5 times the width of the inlet or 3 times the height of the water surface. Grassed spillways have M values between 0.34–0.37 with the smoother surfaces having the higher number. Earthen spillways and broad crested weir M-values commonly range between 0.32–0.39. More complete specifications are provided by Schwab et al. (1993; Blaisdell 1981; Waterhouse, 1982: 212–228).

3. They have an armoured apron and/or stilling pond down channel that is usually protected by armoured channel banks. The purpose of the apron or stilling basin is to receive the nappe of the falling water. The force of the falling water is expended either on an armoured surface or in a pool of water.

Some aprons also include roughness elements called baffles, which are included to impede the accelerated flows of water moving through the structure. The baffles create flow resistance and turbulence, which reduces the energy of the flowing water and so helps prevent scour down channel. Some also include downslope wing-walls and toe-walls, which prevent erosion of the structure by back currents and aid energy dissipation by allowing a controlled widening of the channel below the structure.

Typical failings of drop-structure aprons include their construction of inappropriate materials, which are not capable of resisting erosion, and their failure to consume a sufficient proportion of the energy created in the wier or spillway. The result is scour. Drop-structure aprons are often not extended enough down channel and the downstream edge of the structure is commonly a zone of incision. Baffles are very vulnerable to clogging by trash or floating debris. They are also prone to wear and damage and require a relatively high level of maintenance. Broken baffles may also help divert the channel flow towards the undercutting of a bank downstream.

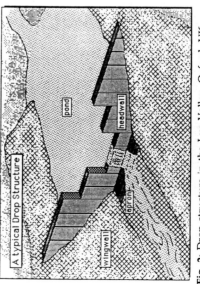

Fig. 2: Drop structure in the Lye Valley, Oxford, UK.

Contd.

Table 9: (*Contd.*)

Drop chutes and flumes	Used where the slope is too steep for drop structures, drop chutes or flumes are short sections of armoured channel way (e.g. Blaisdell, 1981; Blaisdell and Moratz, 1961). The formless flume, a low-cost chute, is employed where the fall does not exceed 2 m or its width 7 m. It consists of shaped soil armoured with wire-reinforced concrete or protected by a gabion. Schwab et al. (1981) suggest that the capacity of a formless flume is: $$Q = 2.13 \, L.h^{1.5}$$ where: Q is discharge in cumecs, L notch width in metres and h flow depth in metres.	Relative to conventional drop structures, these are expensive to design and ineffective. Chutes are to control heads up to 6 metres. They have the advantage that they require less armouring than an equivalent drop structure and are less prone to sedimentation at the outlet (Schwab et al., 1981). However, they are more prone to undermining by seepage and erosion. Their main problem is that they cause acceleration of the water they transport and lead to scour downstream (Waterhouse, 1982).
Energy dissipators	Formed concrete weirs designed to reduce the velocity and erosivity of flowing waters.	Expensive, specially designed, structures that are prone to damage and failure if not properly installed and maintained. Used mainly in badly engineered watercourses as a last-ditch measure to prevent excessive runoff.
Terraces	Terraces are benches created either by accumulation or cut- and -fill processes. Their function is to break a steep slope into small, relatively horizontal benches that may be used for agriculture etc. Terrace benches are separated by risers. These are defended from erosion and failure by vegetation or masonry (Hudson, 1995; Moldenhauer and Hudson, 1988; Sheng, 1989; Curtis, 1974; Kostadinov and Stanojevic, 1999).	This classic solution to the problems of sustainable farming of steep lands is both very expensive and very labour intensive. Terraces come in many designs. However, all require special drainage systems and regular repair to the riser. Today, poor farmability of many terrace benches on steep slopes is leading to terrace abandonment and collapse. Terrace risers can be major sources of sediment (Critchley and Bruijnzeel, 1995). The problem of maintenance is the reason terraces are no longer recommended in the reclamation of surface-mine-disturbed lands. (*Sawatsky et al. this volume*).

Contd.

slippage, live fences—which require many source cuttings that suffer a high failure rate, and algal mats—which are slow growing and difficult to establish. This team found grass lines to be the most effective bioengineering measure for the steepest slopes (>50 degrees). These lines were variously oriented to trap or shed water according to local conditions. On more gentle slopes, tree plantings, perhaps with grass lines, wire herringbone bolsters on wet and jute geotextile mats on dry slopes, proved useful.

In river engineering, Varma and Sivaramakrishnan (1995) found that natural jute and coir fibre geotextile had value in river-bank stabilisation. The mats had 3 times as many plants, and the plants had 3 times the root extension (6 cm vs. 2 cm), of untreated banks. Gerstgraser (1998), experimenting with a 170-metre riverside test-bed, found that a 3-month-old brush mattress with sprouting willows provided the best resistance to bank erosion. Wattle fences and fascines also performed well on shallow sloping banks. Forms of bank protection that do not defend the whole surface perform poorly. The problem remains to turn such promising experimental results into textbook prescriptions and formulae than can be applied by a run-of-the-mill working engineer (see Schwab et al., 1993).

Meanwhile, in India, there are promising results from the application of biotechnical techniques to the reclamation of steep, loose, spoil banks and degraded slopes. In the Krol Limestone Mining belt above Dehra Dun, U.P. India, initial sediment losses can be as high as 550 $t \cdot ha \cdot yr^{-1}$ (Juyal et al., 1995), compared with just 3 $t \cdot ha \cdot yr^{-1}$ from local forest. Sediment pollution used to damage 100 ha of forestland each year. Here, erosion controllers have also used geotextiles. At Lambidhar Mines, Mussoorie, U.P. India, steep banks protected by coir became covered with natural vegetation within one year (Jaganatha Rao et al., 1995). In the 64-ha Sahastradhara basin, geojute geotextile mats helped revegetate steep slopes with grasses (including *Thysanolaena maxima* and hybrid Napier). The geotextile mats were used as part of a larger suite of measures. These included conventional engineering—gabion check dams, silt detention basins, gabion channel spurs to divert flow from unstable slopes, and gabion toe-walls to support such slopes. Bioengineering and revegetation strategies were reserved for more gentle slopes. Here, 0.3 m by 0.3-m contour ditches, backfilled with good soil and farmyard manure were planted with local species. Vegetation cover increased from 10% to 17%. Small gullies were controlled with brushwood check dams and steep landslides stabilised with logwood crib structures. Forestation species employed include: *Lucaena leucocephala*, *Pueraria hirsuta* and *Acacia catechu*.

In 6 years, soil losses declined to 8 $t \cdot ha \cdot yr^{-1}$, surface runoff from 57–37% and water quality improved. The structural measures retained a huge volume of debris (62,000 m^3). Juyal's team (Sastry et al., 1995) also detail a similar approach to the restoration of landslide affected slopes. In the Nalota Nala reclamation project, integrated watershed management supported by a large investment in erosion control structures raised vegetation cover from 0% to 95% and reduced stream sediment loads from 320 to 5.5 t.ha.yr^{-1}.

CONCLUSIONS

The best erosion control solutions are land-use solutions. They involve discovering a system of land husbandry that is inherently self-sustainable because the amount of soil lost and damaged through human use is less than the amount that is replaced by natural regeneration. In sum, the best erosion control strategies are those that work within nature's limits, rather than against it, and which require no erosion control engineering or drainage engineering work at all (see Bradshaw, 1997).

Erosion control structures are often required on land reclamation sites simply because the mining contractors treated reclamation as an afterthought. Erosion control structures are employed because there is no feasible alternative. The reasons can be several, including cost, culture, existing circumstances and an inability to influence the wider picture. However, the main reason is that erosion control workers are rarely employed to plan ahead and preclude a problem. They are employed, instead, to deal with a problem that has already emerged.

Great advances have been made in recent years to prevent mining companies from creating impossible land reclamation problems. However, lack of foresight and poor planning remain the major cause of erosion problems on reclaimed and to-be-reclaimed lands.

The following cases are typical of the problems that appear when land reclamation planning is left too late. In the Kumaun Himalaya, the early operators of the Jhiroli Magnesite mine elected to create their spoil pile by tipping down a long steep Himalayan hillside that, coincidentally, overlooked a local village. After debris flows damaged some of the village's fields, the company was called on to reclaim a scree of several hundred metres of loose-tipped, rapidly weathering, spoil and debris cones plus several kilometres of badly constructed haul roads, on some steep, not very stable mountain slopes. It is very difficult to stabilise a loose scree slope that is perched halfway up a hillside and the work involved dwarfed the resources available to the mining company. Again, in Bulgaria's Maritsa-Iztok Economic Association some early post-mine sites were returned to the land reclamation agencies with toxic clay exposed at the surface. These sites still resist vegetation regrowth even after 20 years of treatment. Later, the mining agency was persuaded to bury these clays at the bottom of the open pit and leave more favourable materials near the surface. These new materials are so fertile that they quickly regenerate natural vegetation, to the degree that this can be used as a green manure to give the desired crop or forest species a head start. In sum, many problems are not beyond solution through the application of a little foresight.

The final advice may be that the erosion control methods described here should not be considered 'alternatives'. Most successful projects employ an array of techniques. Agnew and Humphries (1991: 155) advise: 'The combination of appropriate conservation measures and an aggressive repair program add up to successful treatment of reconstructed drainage channels and consequently successful reclamation, as at Trapper Mine, Colorado, while permanent vegetation establishes'.

In sum, the best route to erosion control on reclaimed lands is to design landscapes that will not suffer accelerated runoff or erosion and not require erosion control structures. If erosion control is required, perhaps temporarily in the early stages of reconstruction, then the aim should be to make these measures as self-sustainable as possible. Finally, since there is no single route to success in erosion control, it remains best to use an array of measures and be prepared to support these measures actively until self-sustaining control is fully achieved.

References

ADAS, 1982. *The Design of Field Drainage Pipe Systems*. HMSO (Ministry of Agriculture, Fisheries and Food, Agricultural Development Service, Land and Water Service, Reference Book 345), London, 20 pp.

Agnew, W. and Humphries, H.B. 1991. A systems approach to permanent drainageway stabilisation on reclaimed lands. *Int. Erosion Control Assoc. Proc.*, 22: 149–155.

AGRA. 1995. *Minesite Reclamation Planning: Issues and Solutions*. AGRA Earth and Environmental, Calgary, 300 pp.

Albert, E.K., Conrad, P. and Phelps, L.B. 1988. Sedimentation ponds: a review of United States regulations affecting design. *Int. J. Surface Mining* 2(1): 7–18.

Barker, D. 1994. The way ahead: continuing and future developments in vegetative slope engineering or ecoengineering. Keynote paper, 1–17. In D. Barker et al. (eds.). *Vegetation and Slopes: Stabilisation, Protection and Ecology*. Institution of Civil Engineers, London, 300 pp.

Batajoo, A.K. Maskey, D.R. and J.H. Howell. 1995. Bioengineering as a low cost technology for erosion control in the road sector of Nepal, pp. 551–560. In: R.B. Singh and M.J. Haigh (eds). *Sustainable Reconstruction in Highland and Headwater Regions*. Balkema, Rotterdam/Oxford and IBH, New Delhi. 692 pp.

Blaisdell, F.W. 1981. Engineering structures for erosion control, pp. 325–355. In: R. Lal and E.W. Russell (eds.). *Tropical Agricultural Hydrology*. J. Wiley Chichester, 450 pp.

Blaisdell, F.W. and Moratz, A.F., 1961. Erosion control structures, pp. 426–491. In: C.B. Richey, P. Jacobson and C.W. Hall (eds.). *Agricultural Engineers Handbook*. McGraw Hill, New York.

Blodgett, J.C. and McConaughty, C.E. 1995. Rock Riprap Design for Protection of Stream Channels. Vol. 2: Evaluation of Riprap Design Procedures. US Geol. Surv., Water Res. Invest. Rep. 86–4128.

Bono, R. and Seiler, W. 1984. *Suitability of Soils in the Suke-Harenge and Abdit Tid Research Units (Ethiopia) for Contour Bunding.* Provisional Military Government of Socialist Ethiopia, Ministry of Agriculture, Soil Conservation Research Project, Research Report 4: 80 pp.

Boughton, W.C. 1989. A review of the USDA SCS curve number method. *Australian. J. Soil Research* 27: 511–523.

Bradshaw, A. 1997. Restoration of mined lands—using natural processes. *Ecol. Eng.*, 8: 255–269.

Bragg, N.C., Griffiths C.W., Jones, A.O. and Bell, S.J. 1984. A study of the problems and implications of land drainage on reinstated opencast coal sites. *Proc. North England Soils Disc. Group* 19: 37–59.

Brooks, R.P. and Gardner, T.W. 1994. *Optimizing Wetland Creation on Coal-Mined Lands* US. Dept. Interior, Office of Surface Mining, Final Report. ER 9404.

Brooks, R.P. and Gardner, T.W. 1995. *Handbook for Wetland Creation on Reclaimed Surface Mines.* US. Dep. Interior, Office of Surface Mining, Washington. Final Report: ER 9503.

Chow, V.T. 1959. *Open Channel Hydraulics.* McGraw Hill, New York, 680 pp.

CIRIA 1990. *Use of Vegetation in Civil Engineering.* Constr. Ind. Res. Assoc/Butterworths, London 200 pp.

Coates, D.R. 1977. Pit-slope Manual Supplement 6-1: Butresses Retaining Walls. *CANMET (Canadian Centre for Mineral and Energy Technology) Report* 77–4: 79 pp.

Curtis, W.R. 1971. Strip-mining, erosion and sedimentation. *Amer. Assoc. Agric. Eng., Trans.,*14: 434–436.

Curtis, W.R. 1974. Terraces reduce runoff and erosion on surface mine benches. *J. Soil Water Cons,* 26 (5): 198–199.

Curtis, W.R., Dyer, K.L. and Williams, G.P. 1988. *A Manual for the Training of Reclamation Inspectors in the Fundamentals of Hydrology.* Scil and Water Conservation Society, Ankeny Iowa 178 pp.

Critchley, W.J. and Bruijnzeel, L. 1995. Terrace risers: erosion control or sediment source. pp. 529–544. in: R.B. Singh, and M.J. Haigh (eds). *Sustainable Reconstruction in Highland and Headwater Regions.* Rotterdam: Balkema: 692 pp.

De Boodt, M. 1990. Applications of polymeric substances as physical soil conditioners, pp. 517–556. In: M. De Boodt et al. (eds.). *Soil Colloids and their Associations in Soil Aggregates.* NATO ASI Series B, vol. 215. Plenum Press, New York.

Dickerson, J.A., Miller, C.F., Burgdorf, D.W. and Van der Grinten, M. 1997. A critical analysis of plant materials needed for soil bioengineering. *Int. Erosion Control Assoc., Proc.* 28: 293–300.

Florineth, F. 1982. Erfahrungen mit Ingenieurbiologischen Massnahmen bei Fleissgewassern in Gebirge. *Landschaftswasserbau* (Techn. Univ. Wien) 3: 243–260.

Florineth, F. 1994. Erosion control above timberline in South Tyrol Italy, pp. 11/1–16 In: Barker et al. *Vegetation and Slopes: Stabilisation, Protection and Ecology.* Inst. Civil Engineers, London, 300 pp.

Gerstgaser, C. 1998. Soil bioengineering methods for bank protection, pp. 373–380. In: M.J. Haigh, J. Krecek, G.S. Rajwar and M.P. Kilmartin (eds.). *Headwaters: Hydrology and Soil Conservation.* Balkema, 460 pp.

Goldman, S.J., Jackson, K. and Bursztynsky, T.A. 1988. *Erosion and Sediment Control Handbook.* McGraw Hill, New York, 454 pp.

Gray, D.H. 1994. Influence of vegetation on the stability of slopes. Keynote Address, pp. 1: 1–17. In: D. Barker et al. (eds.). *Vegetation and Slopes: Stabilisation, Protection and Ecology.* Institution of Civil Engineers, London 300 pp.

Gray, D.H. and Leiser, A.T. 1982. *Biotechnical Slope Protection and Erosion Control.* Van Nostrand Reinhold, New York, 270 pp.

Greenway, D.R. 1989. Biotechnical slope protection in Hong Kong. *Int. Erosion Control Assoc. Proc.* 20: 401–411.

Grey, J. 1998. Tough Turf. *Erosion Control* 5(6): 68–76 [http://www.ces.ncsu.edu/Turf Files].

Haigh, M.J. 1992. Problems in the reclamation of coal-mine disturbed lands in Wales. *Int. J. Surface Mining and Reclamation* 6: 31–37.

Haigh, M.J., Gentcheva-Kostadinova, Sv. and Zheleva, E. 1995. Forest-biological erosion control on coal-mine spoil banks in Bulgaria. *International Erosion Control Association, Proceedings* 26: 383–394.

Hewlett, H.W.M. et al. 1987. *Design of Reinforced Grass Waterways*. Const. Ind. Res. Inform. Assoc. Rep., London 116 pp.

Howell, J. 1999. *Roadside Bioengineering: Site Handbook* Department of Roads, Kathmandu, HMG, Nepal, 160 pp.

Howell, J.H. Clark, J.E., Lawrence, C.J. and I. Sunwar. 1991. *Vegetation Structures for Stabilising Highway Slopes: Manual for Nepal*. Department of Roads, Kathmandu, HMG Nepal, 182 pp.

Hudson, N. 1995. *Soil Conservation* (3e). Batsford, London, 391 pp.

IECA 1994. *Soil Erosion and Sediment Control*. Amer. Soc. Civil Eng. and Int. Erosion Control Assoc., Steamboat Springs, Co. Video: 32 minutes.

Jaganatha Rao, P., Bhagwan, P. and Arun, U. 1995. Environmental degradation from slope erosion and its mitigation, pp. 597–606. In: R.B. Singh and M.J. Haigh (eds.). *Sustainable Reconstruction in Highland and Headwater Regions*. Balkema, Rotterdam/Oxford and IBH, New Delhi, 692 pp.

Joseph, K.G. and Sarma, U. 1997. Retted (white) coir fibre nettings—the ideal choice as geotextiles for soil erosion control. *Int. Erosion Control Assoc., Proc.*, 28: 67–76.

Juyal, G.P., Katiyar, K.S., Joshie, P. and Arya, R.K. (1995). Reclamation of mine spoils on steep Himalayan hill, pp. 441–450. In: R.B. Singh, and M.J. Haigh (eds.). *Sustainable Reconstruction in Highland and Headwater Regions*. Balkema, Rotterdam/Oxford and IBH, N. Delhi, 692 pp.

Kalin, M., Cairns, J. and McCready, R. 1991. Ecological engineering methods for acid mine drainage treatment of coal wastes. *Resource Conservation and Recycling* 5: 265–275.

Kerr, J. and Sanghi, N.K. 1992. Indigenous soil and water conservation in India's semi-arid tropics. *Int. Inst. Environ. Develop. Gatekeeper Series* 34: 30 pp.

Kilmartin, M.P. 1989. Hydrology of reclaimed surface coal-mined land: a review. *Int. J., Surface Mining* 3: 71–83 (see also updated version in this book).

Kostadinov, S. and Stanojevic, G. 1999. Design of technical erosion control measures for reconstruction of degraded steep lands. In: M.J. Haigh (ed.) *Reclaimed Land: Erosion Control, Soils and Ecology*. Oxford and IBH, New Delhi. (see also Kostadinov in: *IECA, Proc.* 28 (1997): 385–396 and 26 (1995): 111–124).

Lancaster, T. 1997. Geosynthetically reinforced vegetation vs riprap: a collection of case studies and cost analysis *Int. Erosion Control Assoc., Proc.*, 28: 47–54.

Lane, E.W. 1955. Design of stable channels. *Amer. Soc. Civil Engineers, Trans.*, 120: 1234–1260.

Levy. G.J. 1996. Soil stabilisers, pp. 267–299. In: M. Agassi (ed.). *Soil Erosion, Conservation and Rehabilitation*. Marcel Dekker, New York.

Lfu (Landesanstalt for Umweltschutz) 1996. *Naturnahe Bauweisen im Wasserbu Badenwurtenberg*, Karlsruhe (in Gerstgaser 1998).

Linsley, R.K. and Franzini, J.B. 1972: *Water Resources Engineering* (2e). McGraw Hill, New York 690 pp.

Little, W.C. and Mayer, P.G. 1976. Stability of channel beds by armouring. *Amer. Soc. Civil Engineers, J. Hydraulics Div.*, 102 (HY1): 101–135.

Meleen, N.H. 1986. Fluvial sediment problems from strip-mining in northeast Oklahoma. pp. 3–11 In: M.J. Haigh (ed.). Geomorphological perspectives on land reclamation. *Oxford Polytechnic Discussion Paper in Geography* 22: 78 pp.

Miles, T.R. Burt, J., Hales, K. and Lofton, J. 1989. Emergency watershed protection using straw bales. *Int. Erosion Control, Proc.*, 20: 381–386.

Mitchell, M.P. 1998. Erosion control at the watershed scale. *Erosion Control* 5 (2): 68–78.

Moldenhauer, W. and Hudson, N.W. (eds.). 1988. *Conservation Farming on Steep Lands*. WASWC/SWCS, Ankeny, Iowa, 296 pp.

Murthy, V.V.N. and Sachan, R. 1990. Conservation measures for sloping agricultural lands in the semi-arid tropical regions of India. *Topics in Applied Resource Management* (DITSL, Germany) 2: 39–60.

NERC 1975. *Flood Studies Report: I—Hydrological Studies; II—Meteorological Studies; III: Flood Routing Studies; IV: Hydrological Data; V—Maps.* Natural Environment Research Council, London, 570 + 81 + 76 + 594 + 24 pp.

NERC 1978. Flood Prediction for Small Catchment. *Flood Studies Supplementary Report 6.*

NERC 1978. A Comparison between the Rational Formula and the Flood Studies Unit Hydrograph Procedure. *Flood Studies Supplementary Report 8.*

NERC 1985. The Flood Studies Report. Rainfall-runoff Parameter Estimation Equation Updated. *Flood Studies Supplementary Report 13.*

Newson, M.D. 1994. *Hydrology and the River Environment.* Clarendon Press, Oxford.

O'Grady, P. 1984. Three dimensional geogrid soil stabilisation, 16: 1–17. In: A.K.M. Rainbow (ed.). *Symposium on the Reclamation, Treatment and Utilisation of Coal-mining Wastes Proceedings.* National Coal Board, London.

Oliver, T.L. H. 1984. The use of geogrid in reinforced minestone structures, 11:1–15. In: A.K.M. Rainbow (ed.). *Symposium on the Reclamation, Treatment and Utilization of Coal-mining Wastes Proceedings.* National Coal Board, London.

O'Malley, P. 1998. Stimulants and amendments. *Erosion Control* 5(6): 80–94.

Poots, A.D. 1979. A floods prediction study for small rural catchments. M.Sc. Thesis, (unpubl.). Queen's University, Belfast,

Poots, A.D. and Cochrane, S.R. 1979. Design flood estimation for bridges, culverts and channel improvement works on small rural catchments. *Proc. Civil Engineers.* (London) 66 (TN229): 663–666.

Rands J.G. and Bragg, N.C. 1988. The design of subsurface drainage systems in England and Wales on fine textured soils on which opencast mining operations have created extremely impermeable barriers at a shallow depth. ADAS, Field Drainage Experimental Unit, Cambridge, Internal Report, pp. 2–37.

Rickson R.J., 1995. Simulated vegetation and geotextiles, pp. 95–132. In: R.P.C. Morgan and R.J. Rickson (eds.). *Slope Stabilisation and Erosion Control: A Bioengineering Approach.* E. & F.N. Spon., 274 pp London.

Santos-Cayade, J. and Simons, D.B. 1971. River response, 1:1–1:25. In: H.W. Shen (ed.). *River Mechanics,* I. Water Resources Publ., Fort Collins, Co.,

Sastry, G., Juyal, G.P. and Samra, J.S. 1995. Mass erosion in the Himalaya, pp. 431–440. In: R.B. Singh and M.J. Haigh (eds.). *Sustainable Reconstruction in Highland and Headwater Regions.* Balkema, Rotterdam/Oxford and IBH, New Delhi, 692 pp.

Sawatsky, L. 1995. Sustainable landscape design, pp. S2: 1–11. *In: Minesite Reclamation Planning: Issues and Solutions.* AGRA Earth and Environmental, Calgary, 300 pp.

Sawatsky, L., McKenna, G., Keys, M-J., and Long, D. 1999. Towards minimising the long-term liability of reclaimed mine-sites. In: M.J. Haigh (ed.). *Reclaimed Land: Erosion Control, Soils and Ecology.* Oxford and IBH, New Delhi.

Schroeder, S.A. 1994. Reliability of SCS curve number method on semi-arid reclaimed minelands. *Int. J. Surface Mining, Reclamation and Environment* 8(2): 41–45.

Schumm, S.A. 1969. River metamorphosis. *Amer. Soc. Civil Engineers, Proc., J. Hydraulics., Div.,* 95: 255–273.

Schumm, S.A. 1971. Fluvial geomorphology, pp. 4: 1–4:27. In: H.W. Shen (ed.). *River Mechanics, I.* Water Resources Publ., *Fort Collins, Co.*

Schuster, M.J. and Hubl, J. 1995. Impact of road construction on the Pokhara-Baglung Highway, Kashi, Nepal, pp. 175–182. In: R.B. Singh and M.J. Haigh (eds.). *Sustainable Reconstruction in Highland and Headwater Regions.* Balkema Rotterdam/Oxford and IBH, New Delhi, 692 pp.

Schwab, G.O., Frevert, R.K., Edminster, T.W., and Barnes, K.K. 1981. *Soil and Water Conservation Engineering* (3rd ed.) Wiley International, New York, 525 pp.

Schwab, G.O., Elliot, W., Fangmeier, D. and Frevert, R.K. 1993. *Soil and Water Conservation Engineering* (4th ed). Wiley International, New York, 585 pp.

Scott Queen, B. and Clopper, P.E. 1992. Case Study: the use of articulated concrete block revetments for grade control and bank protection, Temecula Creek, Riverside County, California. *Int. Erosion Control, Proc.,* 23: 191–205.

Sheng, T.C. 1989. Soil Conservation for Small Farmers in the Humid Tropics. *FAO Soils Bulletin* 60: 80 pp.

Sutherland, R.A. 1998. A critical assessment of the research conducted at the hydraulics and erosion control laboratory—a focus on rolled erosion control systems applied to hillslopes. *Geotextiles and Geomembranes* 16: 87–118.

Teal, M. 1995. Computer-aided design of rip rap bank protection. *Int. Erosion Control Assoc., Proc.,* 26: 127–134.

Temple, D.M., Robinson, K.M., Ahring, R.M. and Davis, A.G. 1987. *Stability Design of Grass-lined Open Channels.* USDA Agriculture Handbook 667: 167 pp.

Tuttle, J.H., Dugan, P.R. and Randles, C.I. 1969. Microbial dissimilatory sulphur cycle in acid mine water. *J. Bacter.* 97: 594–602.

Ulman, P. and Lopes, V. 1995. Determining inter-rill soil erodibility for forest roads. *Int. Erosion Control Assoc., Proc.,* 26: 345–357.

US Army Corps of Engineers. 1991. *Hydraulic Design of Flood Control Structures [and] Additional Guidance for Riprap Channel Protection.* Department of the Army, Engineer Manual 1110-2-1601 and Technical Letter 1110-2-120, Washington D.C.

USDA Soil Conservation Service, 1973. *A Method for Estimating Volume and Rate of Runoff in Small Watersheds.* Soil Conservation Service, Technical Paper 149, Washington DC.

USDA Soil Conservation Service, 1972. *Hydrology.* National Engineering Handbook Section 4. Washington, DC.

Van Molle, M. and Van Ghelue, P. 1988. Soil erosion and river sedimentation, pp. 221–228. In: *Int. Symp. Water Erosion, Varna, Bulgaria.* UNESCO:MAB/IHP, Sofia. 376 pp.

Varma, C.R.R. and Sivaramakrishnan, R. 1995. Potentials of natural fibre geotextiles in bioengineering applications, pp. 561–570. *In:* R.B. Singh and M.J. Haigh (eds.). *Sustainable Reconstruction in Highland and Headwater Regions.* Balkema, Rotterdam/Oxford and IBH, New Delhi, 692 pp. (see also *International Erosion Control Association, Proc.,* 28 (1997): 419–426).

Waller, H.F. 1995. The use of emulsified asphalt for mulching. *Int. Erosion Control Proc.,* 26: 279–284.

Ward, L.E. 1995. Kolb Road project design and installation of a cellular containment system. *Int. Erosion Control, Proc.,* 26: 53–61.

Waterhouse, J. 1982. *Water Engineering for Agriculture* Batsford Academic, London, 395 pp.

Weigle, W.K. 1966. Erosion from abandoned coal-haul roads. *J. Soil and Water Conservation* 21 (3).

Zhenqi Hu, Longian Chen and Haibin Liu. 1994. Erosion problems in Chinese coal-mining area. *Int. Erosion Control, Proc.,* 25: 477–482.

Ziemkiewicz P.F., Skousen, J.G., Brant, D.L., Sterner, P.L. and Lovett, R.J. 1997. Acid mine drainage treatment with armoured limestone in open limestone channels. *J. Environ. Quality* 26, 1017–1024.

6

Design of Technical Erosion Control Measures for the Reconstruction of Degraded Steep Lands

Stanimir Kostadinov and Gradimir Stanojević

Abstract

Erosion control of degraded steep land should be carried out by combined biological and engineering works. The soil over steep land may be very shallow or even absent. If biological measures of erosion control are to be applied successfully—afforestation, grassing, or the establishment of perennial agricultural crops (orchards)—it may be necessary to undertake mechanical erosion control works first to ensure a stable environment for the forest, orchard crops or grass, to develop their root systems and grow. Another objective is to prevent adverse and uncontrolled runoff from the slopes after heavy rains and thereby prevent further soil erosion. This chapter reviews the most significant classical structures for erosion control on degraded steep lands, such as various kinds of terraces, walls, contour ditches, followed by more recently introduced prefabricated erosion control elements made of concrete, briquettes, nets etc.

INTRODUCTION

The reconstruction of damaged steep lands without protection measures often causes serious watershed problems. The results include land degradation 'on site' by water erosion and aggravation of silting and flood damage 'off site', i.e., in the downstream area (Sheng, 1977, 1989).

Erosion control on all degraded steep lands demands combined application of biological and mechanical engineering works. The soil on steep land is often very shallow or even removed by intensive erosion. If

Land Reconstruction and Management Vol. 1, 2000, pp 111–136
ISSN 1389-2541
ISBN 90 5410 793 6
A.A. Balkema, Rotterdam, The Netherlands

biological measures of erosion control are to be applied successfully—afforestation, grassing, or the establishment of perennial agricultural crops (orchards)—it may be necessary to undertake technical erosion control works first, to ensure a stable environment for development of the root system and plant growth.

The choice of erosion control measures for vulnerable lands is based on the following principles (Rosić, 1960).

1. Overland water flow should be prevented by stimulating infiltration into the soil. This will help control erosion caused by water running down-slope.
2. Wetting should be as uniform as possible throughout the slope. There should be no excessive accumulation of moisture downslope.
3. Overland flow should be dispersed throughout the slope and with the least possible velocity, so that as much water as possible is absorbed in the soil.
4. The applied erosion control works and measures should release the greatest possible land area for productive use.

The basic technical erosion control measures and works for the reconstruction of degraded and vulnerable lands include:

1. Contouring
2. Contour bunds
3. Terraces
4. Contour ditches
5. Waterways
6. Stabilisation structures.

In addition to these traditional erosion control measures, there are more modern technical (mechanical) measures such as:

7. Nets for erosion control
8. Geotextiles for erosion control
9. Erosion control blankets
10. Cellular containment systems
11. Prefabricated concrete blocks and formed concrete structures.

CONTOURING

Conducting ploughing, planting and cultivation operations along the contour can reduce soil loss from sloping land by up to 50% compared with cultivation up and down the slope. The effectiveness of contouring varies with slope steepness and slope length. It is ineffective for slope lengths greater than 180 m at 1.8% (ca. 1°) steepness. The allowable length declines with increasing steepness to 30 m at 5.5° and 20 m at 9% (5°09′). Moreover, the technique is only effective during storms of low rainfall intensity. Protection against more extreme storms is improved by supplementing contour farming with strip-cropping. The soil loss from contour strip-cropped fields is 25 to 45% of that from fields managed by up-down tillage, depending on the slope steepness. Using contour strip-cropping, with strips 50 to 100 m wide and alternating five fields of grain with three of grass and one of legume, effected a 50% reduction in erosion on the Kamennaya steppe near Voronezh, erstwhile USSR (Tregubov, 1981, in Morgan, 1986).

CONTOUR BUNDS

Contour bunds are earth banks, 1.5 to 2 m wide, thrown across the slope to act as a barrier against runoff. They form a water storage area on their upslope side and break up a slope into segments shorter in length than required to generate overland flow. They are suitable for slopes of 1° to 7° (2–12%) and are frequently used on small holdings in the tropics, where they form permanent buffers in a strip-cropping system involving grasses or trees (Roose, 1966, in Morgan, 1986). The banks, spaced at 10 to 20 m intervals, are generally hand constructed. There are no precise specifications for their design and deviations in their alignment of up to 10% from the contour are permissible. Humi (1984) calculated the effectiveness of contour bunds to control erosion in Wallo Province, Ethiopia, and showed that they could only reduce soil loss sufficiently on the lowest of the slopes examined.

TERRACES

Terraces are earth embankments constructed across the slope to reduce slope length, to intercept surface runoff and to convey it to a stable outlet at a non-erosive velocity. Thus they perform functions similar to those of contour bunds. They differ from the latter in being larger and designed under more stringent specifications. Decisions are required on the spacing and length of terraces, location of terrace outlets, gradient and dimensions of the terrace channel and layout of the terrace system.

Types of Terraces

Three basic types of terraces are used for erosion control on steep slopes in arid regions:
— Infiltration bunds—Algerian terraces
— Gradons
— Terraces with small walls
    In humid countries, however, six types of terraces are used (Sheng, 1977, 1989):
— Bench terraces
— Hillside ditches

— Individual basins
— Orchard terraces
— Mini-convertible terraces
— Hexagons

Terraces for Arid Regions

Infiltration bunds—Algerian terraces

An Algerian terrace, in transverse profile, resembles a roadside cut with a longitudinal slope of 0.5% (0°17′), but has along its outer side a somewhat elevated fill, resembling a wide channel. Algerian terraces can be laid

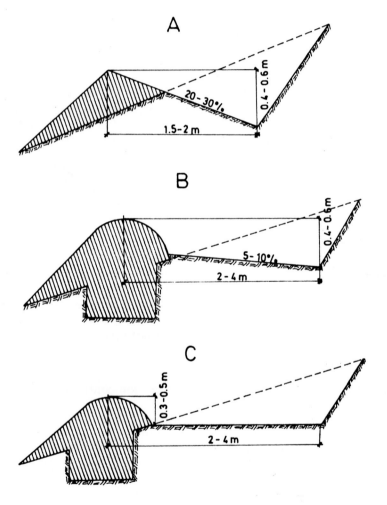

Fig. 1: Algerian terraces.

horizontally as well as with a mild longitudinal slope. Horizontal terraces are reserved for places with gentle slopes (3–4%, 2°) and very permeable soils. This type of terrace is not used on poorly permeable soils. Since these terraces should absorb all the water input, their ends are closed. They are rather short (200–300 m) to avoid overflow, i.e., bursting due to too great quantity of water.

Terraces with a longitudinal slope, on the other hand, while checking overland flow and supporting infiltration, are nonetheless capable of conducting surplus water, with lower velocity, to a catchment (drain). This type of terrace is used on slopes of 5–35% (3–19°).

Terraces should not be regarded in isolation. They should always constitute part of a system that extends from the divide to the bottom of a slope. Terraces are distinguished by four characteristics: distance between terraces, terrace length, longitudinal slope and transverse slope (Vučićević, 1995).

Distance between terraces. The Algerian Soil Conservation Service has adopted a formula for calculating the distance between terraces. The formula shows the dependence of terrace distance in the vertical direction (H) and slope (P):

$$H^3 = 260 \times P \pm 10,$$

where P is the value of the slope tangent (for 25% (14°), P = 0.25) and the value of H is in m.

In the USA, Ramser's formula is applied:

$$H = 7.5 \times P + 0.6 \pm 0.15.$$

In this formula, H and P have the same meaning as in the previous one. However, terraces are denser in the USA than in Algeria.

Terrace length depends on soil cohesion and subgrade conditions. A greater quantity of water can be conducted over grassed subgrade surfaces than over ploughed subgrades, so the former terrace can be longer. In the USA, it is recommended that terrace length not exceed 450 m in one direction. Furthermore, for terraces exceeding 300 m in length, the width should be increased and the fill elevated. In Algeria, the permissible terrace length is 400 m in one direction.

In the USA, the longitudinal slope of terraces ranges between 0.25 and 0.50% (0°08′–0°17′). In Algeria, the longitudinal slope is generally 0.50% (0°17′) and need not be uniform throughout the length of the terrace. On the contrary, it is considered better for soil and water conservation if the slope is low initially, gradually increasing towards the outlet.

Cross-section of a terrace. In Algeria, calculations of cross-section terrace profiles are based on the maximum storm of prolonged duration with an intensity of $i = 3$ mm·min^{-1}, assuming that the coefficient of overland flow is equal to 1 and that the terrace can take all the rainfall (Saccardy, 1951).

Drainage area above the Algerian terrace (A) is calculated:

$$A = L \cdot Z$$

where: L is the average horizontal interval between terraces and Z the length of Algerian terrace.

$$V = C \cdot \sqrt{R \cdot J}$$

where: V is the average velocity of overland flow in m·s^{-1}, R the hydraulic radius, J the longitudinal slope of terrace (J – 0.5% (0°17′)), $C = \dfrac{87 \cdot \sqrt{R}}{\rho + R}$ the coefficient of velocity, ρ the specific gravity of water($\rho = 1.0$ t · m^{-3}).

$$V = 3.2 \cdot R,$$

$$D = \frac{A \cdot i}{60 \cdot 1000}$$

where: D is the total quantity of water in m^3·s^{-1}, i the intensity of rainstorm in mm · min^{-1}.

The cross-section of an Algerian terrace is σ in m^2

$$\sigma = \frac{D}{V_d}$$

where: V_d is the allowable velocity of water in m·s^{-1} ($V_d = 0.65 -$ 0.80 m·s^{-1})

$$\sigma = \frac{A \cdot i}{60 \cdot V_d \cdot 1000}$$

$$\sigma = \frac{L \cdot Z \cdot i}{60 \cdot V_d \cdot 1000}$$

1. An inclined profile (Fig. 1A) is recommended for steeper slopes and should be made manually. The disadvantage of this type of terrace is that it concentrates the water on a hillside, which leads to decreased infiltration.

2. The normal profile (Fig. 1B and C), with a flat bench, can be constructed by machine. It is used on slopes greater than 10% (6°). The bench of the normal terrace type can be cultivated, which results in increased infiltration, so that overland flow is made possible. In the construction of this type of terrace, the bench can be inclined towards the uphill rise as much as 5–10% (3–6°) (type B) so as to protect edge fill. Later on, subsidence results in another profile (type C). All types of Algerian

terraces (except horizontal) have a longitudinal slope in the direction of its receptor, usually a drainage channel or a forest.

Gradons

Gradons have two functions:
1) erosion control in interspaces,
2) soil preparation for afforestation.

The first is the primary function while the second is supplementary. Recent investigations have proven that erosion can be successfully controlled by gradons distributed less densely than previously advised; thus gradons will be a more economical option in future.

Gradons are terraces 70–90 cm wide, with a subgrade (Fig. 2) inclined 30% (17°) towards the uphill. They are made along the contour with a longitudinal slope of 0.5% (0°17′).

Gradons are created on erosive steep lands, downhill in a series. The spacing depends on slope gradient. In practice, they are usually spaced with a vertical distance of 4–5 m (70%, 35°). In gradons, the slope of the fill should be 1:1.5 and the slope of the cut 1:1 (100%, 45°). Distances and cross-sections are calculated by the same equations used for Algerian terraces but instead of 400 m, the length of gradons is 200 m because they are made on steeper and more erosive terrain. The maximum storm is $i = 2$ mm·min^{-1}. Gradons are implemented on extremely vulnerable steep sites. In the Balkans they are often deployed on steep south, south-west and south-east slopes, which are arid because of too fast overland flow

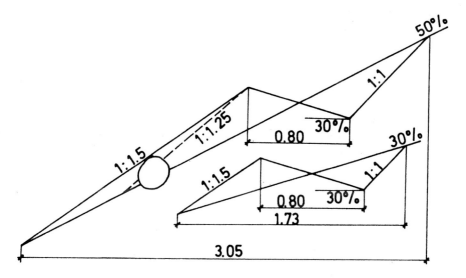

Fig. 2: Gradons.

and extremely intense insolation throughout the year. The rule applied to all retention works—terraces, waterways and ditches as well as gradons—is that the sites should not be vulnerable to landslides. Gradon terraces are water-holding structures that can increase the stress on a treated slope. So, as soon as sliding is suspected, gradons should be immediately withdrawn from the system and grassing and shrub planting in pits undertaken.

In general, very steep slopes should not be a feature of reconstructed land. However, on steeper slopes (near to 40%, 22°) it is difficult to form the fill with a gradient 1:1.5 (70%, 35°), so in such places 1:1.25 (80%, 39°) is formed instead. On very steep slopes (above 60%, 31°), the soil cannot remain on the fill; in such cases the fill must be supported and fixed by small dry-laid masonry walls and by wattles.

Terraces with small walls

Step-wise terraces with small walls (Fig. 3) are the longest lasting structures and a reliable means of erosion control. They can be used on slopes up to 84% (40°). Such terraces are characterised by near-vertical risers and the coefficient of effective use of slope is almost one. Despite many advantages, these terraces are rarely used in practice due to high construction and maintenance costs. They can be used on slopes with abundant stone.

The evaluation of these terraces can be performed with formulae used for cut-and-fill terraces. However, in this case static evaluation of a small wall should be done. If the effective height of a wall is less than 0.5 m, such evaluation is not necessary and the following guidelines are adopted: width of the spillway 0.3 m, gradient of downstream slope 4:1 (25%, 14°). If the height is greater, evaluation is done by Thiery's method (Gavrilović,

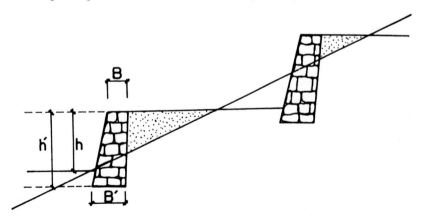

Fig. 3: Terraces with small walls.

1982). As for the cross-section, the terraces can be horizontal or with a counter-slope. The latter type is more economical.

Terraces for Humid Regions

Bench terraces

These are, essentially, a series of level or nearly level strips running across the slope and supported by steep risers. The risers are made of earth protected with grass banks or they are rock walls, if rocks are available. Bench terraces can be built and cultivated by manual labour, animal draught tools or by machines. A detailed cross-sectional view of this type of bench terrace is shown in Fig. 4.

Hillside ditches

The hillside ditch is a discontinuous narrow, reverse-sloped, terrace built across the land. The purpose is to break long slopes into a number of short sections so that runoff can be intercepted and drained safely before causing erosion. The cross-section of this kind of narrow bench is more convenient for maintenance than a conventional ditch. The ditches can also be used,

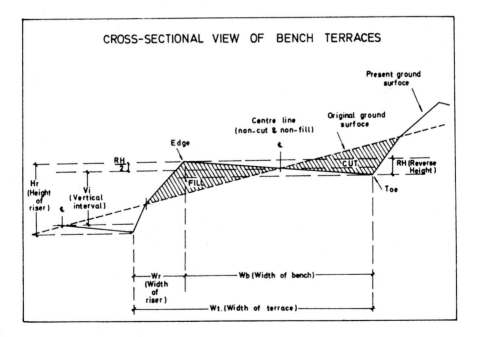

Fig. 4: Cross-sectional view of bench terraces (after Sheng, 1977).

simultaneously, as roads. The distance between two ditches is determined by the degree or per cent of the corresponding slope. The strip between two ditches should be protected with soil conservation measures.

Individual basins

Individual basins are small round benches for planting individual plants. They are particularly useful for establishing semi-permanent or permanent crops on slopes. Their form helps control erosion while conserving fertilisers and moisture, especially if mulched, and it helps keep weeds out. They can be used on steep and dissected lands. They also provide

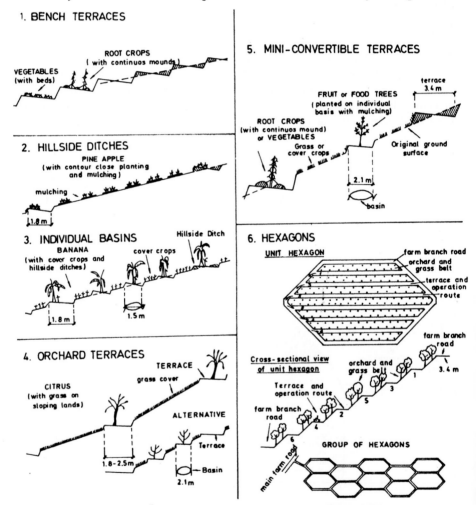

Fig. 5: Types of terraces for humid regions (after Sheng, 1977).

sites where fertilisers productives can be concentrated. They may be used in conjunction with hillside ditches or orchard terraces.

Orchard terraces

Orchard terraces are narrow bench terraces built on steep slopes of 45–60% (25–30°). Their spacing is determined by the planting distance of the trees. Because of steepness, these spaces should be kept under permanent grass. The trees can be planted either on the terraces or in individual basins in the grass strips.

'Mini-convertible' terraces

'Mini-convertible' terraces are of medium width and built according to the distances used for hillside ditches. Field crops may be planted on the terraces whereas trees are planted in-between.

Hexagons

Used in reclamation for agriculture, a unit hexagon is a special arrangement of farm road on a slope that envelops a piece of land readily accessible to four-wheel tractors. The enveloped road or branch road goes around the slope to connect with each operation route or terrace, which is entered at an obtuse angle. A group of hexagons forms a honeycomb with no land wasted. This land treatment is primarily for the mechanisation of orchards on a large block of uniform terrain. It can also be applied on steep slopes (above 35%, 20°) for a small farm (quarter hectare). In the latter case, the operation routes can be cultivated to produce cash crops until the food or fruit trees in the grass strip have grown.

Specifications and applications

Width: For bench terraces, the proper width should be determined by the crop needs, tools to be used, soil depth and slope, as well as the land-user's interest and financial position. A critical consideration in the survival of bench terrace systems is their 'farmability' from the perspective of the land user.

For hillside ditches and orchard terraces, 1.8 m width is usually sufficient although the latter can be wider when the soil is deep and the slope is around 25°. A width of 3.4 m is minimal for machine-built bench terraces and for mechanisation.

Vertical intervals and spacing: The vertical interval (V_i) is actually the elevation difference between two succeeding bench terraces. It is determined by the slope of the land and the width of the benches, using the formula (Sheng, 1977):

$$V_i = \frac{S \cdot W_b}{100 - S \cdot U}$$

where: V_i is the vertical interval in feet or metres, S the slope in per cent (%), W_b the width of bench in feet or metres, U the slope of riser (ratio or horizontal distance to vertical rise using value 1 for machine-built terraces and 0.75 for manually constructed ones).

For hillside ditches and mini-convertible terraces the following equation is employed (Sheng, 1977):

$$V_i = \frac{S + 4}{10} \text{ or } V_i = \frac{S + 6}{10}$$

where : V_i is the vertical interval in feet or metres and S the slope in per cent.

Length: Based on current experience, a maximum length of 100 m is recommended for humid countries.

Slope limit: Hand-made terraces can be used over a slope range of 7 to 25° (12–47%) and machine-built terraces from 7 to 20° (12–36%). For tree crops, 1.8 m discontinuous orchard terraces can be employed up to 30° slope if the soil permits. Thirty degrees is a practical limit for all kinds of terraces. Beyond this the riser will be too high and wide and the bench will be too narrow. However, steeper slopes are terraced in places like the Himalaya.

CONTOUR DITCHES

Contour ditches have long been used to control overland flow and for water retention (Fig. 6). In theory, these ditches are designed to retain all water they receive. In practice, this is not possible to achieve because such ditches (in areas where rainfall intensity and quantity is great) would have to be too large and too densely distributed.

Spacing of the ditches is very important. This should be such that runoff water cannot reach a speed sufficient to erode the soil. Spacing should also be harmonious with ditch capacity.

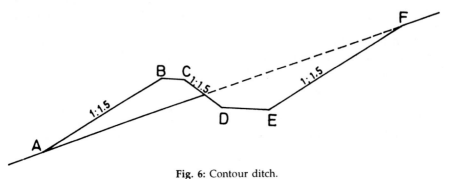

Fig. 6: Contour ditch.

The distance between ditches is calculated by the formula of input volume, by which ditch volume per 1 m (in m^3) should correspond to the quantity of water that flows from interspaces, x metres long and 1 m wide. The formula is as follows:

$$W = f \cdot x \cdot a \cdot \cos \alpha$$

where: W is the volume of a 1 m ditch in m^3; f the coefficient of overland flow; x the length of overland flow, i.e., spacing between two ditches in m; a the maximum storm in m per hour and cos α the slope gradient.
From this formula, the value of x is:

$$x = \frac{W}{f \cdot a \cdot \cos \alpha}$$

American Types of Contour Ditches

According to the American concept, contour ditches are constructed from the slope crest downhill, with definite intervals, according to the capacity and expected quantity of water. Contour ditches do not accept all the water from the maximum storm: they take, for example 75% and the remaining 25% overflows longitudinally sideways, without spilling over the fill. Each ditch is cut without interruption along the contour, from one receptor drain to another. At each 6–12 m the ditch is divided into sections by small transverse equalisers. Equalisers (barriers in a ditch) are 21 cm wide (at the spillway) and 9 cm lower than the fill. Ditches are divided into sections for two reasons:
1. Since a ditch is not uniformly loaded with water throughout its length, the parts above which a larger quantity of snow melts, or those situated in depressions are more charged with water. When the ditch is divided into sections, water can flow into less loaded sections and be absorbed there.
2. If a ditch fails in one section (due to a burrowing animal or other disturbance), water outflows only from that section into a lower ditch, whereas all its other sections remain full of water.
On more gentle slopes, ditches are cut at an interval of 7.62 m and on steeper slopes at 9.14 m. To protect one hectare with contour ditches, theoretically, it is necessary to cut 1300 m of ditches. Practically (due to spacing resulting from lesser slopes etc.), the length of contour ditches is almost always less than 1300 m.
Evaluation of standard contour ditches: Figure 7 illustrates three styles of American contour ditches. To adjust a standard contour ditch to the climate of a region, its size is evaluated in terms of the most dangerous storm possible, i.e., its intensity and duration.

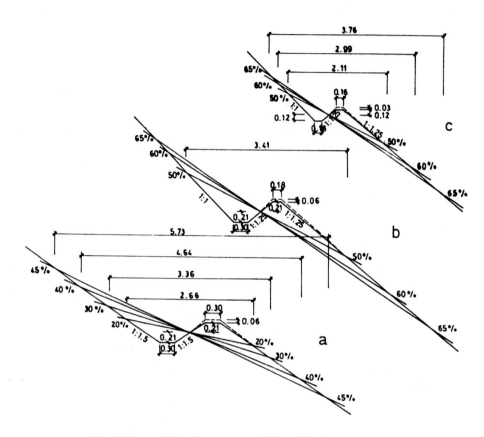

Fig. 7: Types of American contour ditches (after Lujic, 1973)

WATERWAYS

A waterway is an integral part of terracing in humid regions. In many cases, a natural depression without shaping and protection cannot safely accommodate the extra runoff concentrated by terracing. The purpose of waterways in a conservation system is to convey runoff at non-erosive velocity to a suitable disposal point. To achieve this, a waterway must be carefully designed. Normally, its dimensions must provide sufficient capacity to confine the peak runoff from a storm with at least a ten-year return period. On average, a hectare needs 100 m of waterway. For larger blocks of land, the same waterway could serve up to two hectares (Sheng, 1989).

The design procedures are based on the principles of open-channel hydraulics (Goldman et al., 1986). The cross-section of the waterway may

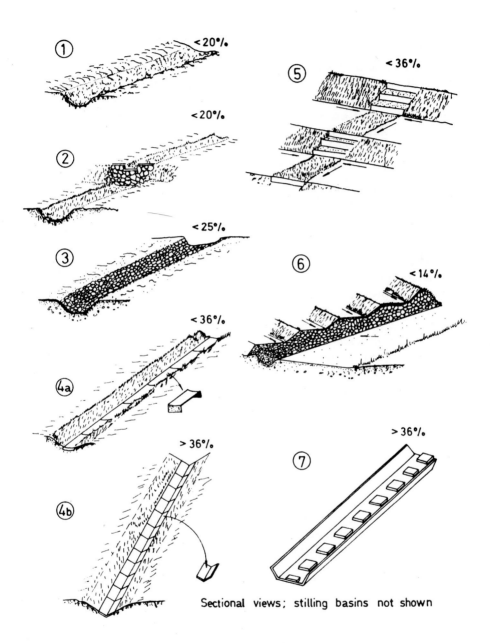

Sectional views; stilling basins not shown

Fig. 8: Major types of waterways (after Sheng, 1977, 1989).

be triangular, trapezoidal or parabolic. Triangular sections are not recommended because of the risk of scour at the lowest point.

Waterway Planning

The site and kind of waterway depend on the slope, velocity and amount of runoff and land-use. It is always desirable to find a gentle depressed area for the site of a waterway plus shaping and revegetation. When the velocity of runoff exceeds $1.8 \text{ m} \cdot \text{s}^{-1}$, engineering structures are usually needed for additional protection. A grassed waterway alone is seldom secure on slopes greater than 11° or 20% slope.

The size of the waterway is determined by the peak flows of the area; its estimation is not covered here. However, waterways wider than 3 m are not desirable on small farms because too much land is taken out of production.

Types of Waterways and Structures

Sheng describes many types of waterways depending on material available, shape of the channel, purposes and structural needs, as well as uses and approximate limits (Sheng, 1977; Fig. 8). These include:
1. Grassed waterway
2. Grassed waterway with drop structures
3. Ballasted waterway
4. Prefabricated concrete waterway
5. Stepped waterway
6. Waterway and road ditch complexes
7. Footpath and chute complexes

STABILISATION STRUCTURES

Small Walls along the Contours-Rosić Walls

These structures of low height are built along the contours to prevent quick flow of surface water and to check the movement of eroded sediment. The purpose is to create stable conditions for the establishment of vegetation (grass or forest). Several types of small walls are used and they are built of different types of materials. The most well known are the horizontal walls of S. Rosić (1960). They are made of prefabricated concrete elements and there are two types.

Type I (Fig. 9) is constructed along the contour two-thirds of the way down a slope. The wall foundation may be cast in situ or made of prefabricated concrete elements. A channel runs throughout the length of the wall. On the uphill side (through on opening left between the elements), it accepts water. On its downhill side, there is a water outlet at

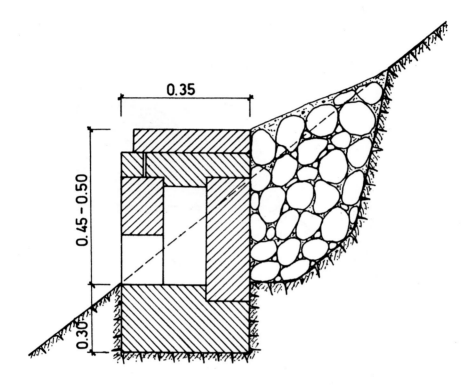

Fig. 9: Horizontal wall—Type I (after S. Rosić, 1960).

each 2 m. On the uphill side behind the wall, a pile is made of stones collected on the slope. Water passes through the pile and sediment is left behind. In this way, a new layer of soil is formed, with decreased grade. This can easily be covered with vegetation. Thus, these small walls function to reclaim soil and regulate surface runoff.

Type II is also made of concrete elements (Fig. 10). These walls are smaller and serve as supplements to type I. They consist of just one row of concrete elements placed in a series on the cut along the contour (horizontal). On the uphill side, fascines are placed throughout the wall length.

Rosić walls (both types) are suitable for bare land with parent rock liable to weathering, e.g. serpentine, limestone and gneiss. For pastures, Rosić recommends that this horizontal wall is supplemented by the addition of two rows of hedges, parallel to the wall, on its downhill side.

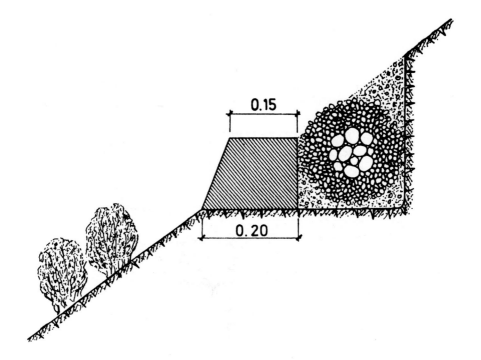

Fig. 10: Horizontal wall—Type II (after S. Rosić 1960).

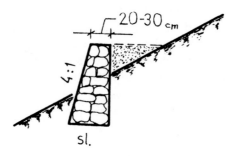

Fig. 11: Dry-laid masonry wall.

Dry-laid Masonry Walls

These are used on slopes that are devoid of vegetation, which are subject to intensive erosion. They are built of dry (mortarless) stone, collected on the spot, with a shallow footing (20–30 cm) on the downhill side; the footing on the uphill side will be somewhat deeper due to slope grade (Fig. 11). Spillway thickness is 20–30 cm. Slope grade is 4:1 (25%, 14°). Total height of the wall is 60 cm but may be higher in some cases. Effective height on the uphill side ranges between 20 and 50 cm, depending on the hill slope. Spacing between walls ranges between 10 and 50 m. Check dams of dry-laid masonry are also constructed in gullies.

Wattles

Wattles are made of vertical sticks (often of sprouting willow) and interlaced rods. Both materials can usually be found on the spot or in its vicinity. The advantage of wattles is they can be made simply and quickly. Their disadvantage is low durability, which restricts their usability. They last a maximum of 5 years.

a) Single wattles are placed along the contours at intervals of 5–10 m. Their height is 0.3–0.7 m (Fig. 12). Their purpose, along with afforestation, is to fix erosive soil or to support the revegetation of steep slopes where that vegetation has been destroyed. They are necessary when erosion is so vigorous that vegetation cannot be restored without prior erosion control. Sometimes, criss-cross wattle are made at the angle of 45° to fix the terrain. In addition to single wattles, fascines are sometimes placed horizontally on the soil and fixed with sticks. To protect the eroded slope from overland flow, branches are laid between the fascines.

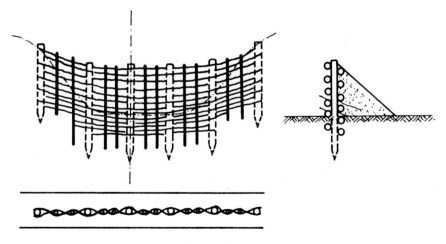

Fig. 12: Single wattle.

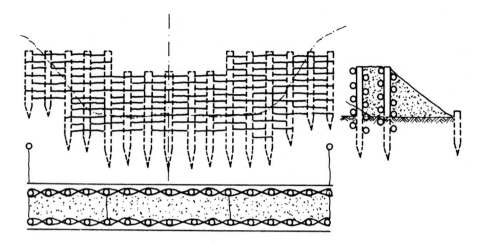

Fig. 13: Double wattle.

b) Double wattles are applied where wattles should be higher (usually in gullies) and where water flow is more intense. Their effective height can be 1 m (Fig. 13). The first row of sticks is driven into the soil and interlaced with rods. Then, at the interval of 0.8 to 1.0 m, another row of wattles is made, and the space between the rows is filled with large gravel or stone. The rows of wattles are fastened by ties, which are nailed to the sticks. An apron is usually made on the downstream side. Wattles should be well driven into the bank.

MODERN MATERIALS AND WORKS FOR EROSION CONTROL

A great number of technical, biological or combined systems are used for control of erosion on slopes, both of natural and man-made origin. Today, an expanding number of systems result from a wide choice of new materials and industrially designed products for erosion control. Pride of place is given to light, porous and flexible geotextiles.

Reinforced Turf

Turf provides a natural aesthetic and long-lasting finish to earth ditches, channel slopes and many other earthen structures. However, when erosive forces are high, it is difficult for turf to establish and it is prone to washout.

One solution for this problem is use of metal nets (metal, plastic or fibre) to reinforce the soil and turf. Turf reinforcement allows natural benefits of grass, wild flowers and other vegetation cover on sites where it was never practical before. Turf reinforces the vegetation canopy so that it can stand up to high erosive forces that would rip up unreinforced sod.

Reinforced turf forms a 'soft armour' that provides permanent protection and natural beauty. The net serves as a tensile reinforcement and anchorage layer that protects the soil against washout while maintaining a natural appearance.

Geotextiles and Erosion Control Blankets

Geotextiles assist in promoting vegetation and thus achieve control of erosion (Ranganathan, 1995). A geotextile is a flexible erosion control product designed to hold seeds and soil in place until vegetation is established.

Geotextile blankets may be woven from straw or fibres such as coir. They are environmentally sensitive bioengineering applications. Natural looking, high-strength polypropylene mesh can protect the soil surface from water and wind erosion while offering partial shading and heat storage to accelerate vegetative development. This material is used as a low cost alternative to bioengineering geotextiles constructed from exotic natural fibres. A geotextile can protect soil against erosion on swales or ditches and on steep slopes. It is degradable and provides surface soil stabilisation, which protects seeds and young woody plants in the critical early phases of growth. As the plant material develops, roots and shoots pass unrestricted through the degrading material.

Steep slopes, soil-lined ditches, channels and banks require ground cover to protect the soil, seeds and new vegetation from rainfall erosion, runoff and blowing wind. Straw blankets provide the traditional ground cover solution. They are low cost, easily laid by hand, provide short-term ground cover and rot away in a few weeks or months. However, steep slopes and high flow velocities often result in washout of the ground cover or the soil beneath. In some locations, temporary ground covers degrade and lose their effectiveness before vegetation can be established.

Blankets of artificial fibres can provide the long-term ground cover necessary to prevent erosion on these sites. Like organic, biodegradable blankets, they act as mulch to protect the ground surface from washout and accelerate the growth of seedlings. Unlike organic blankets, the artificial fibres do not absorb water, so all available moisture can be absorbed by the soil, where it is needed most. The blankets reduce the impact of rainfall, slow down the flow of surface water and reduce the erosive effects of runoff.

Cellular Containment System

When annual rainfall is under 300 mm, the use of vegetation to aid in erosion control of slopes and gullies is reduced significantly. In these sites, less conventional forms and materials are applied for erosion control.

Fig. 14: Cellular containment system ('Volta-Cell' system)

Cellular containment systems (Fig. 14) can be used in such environments. A typical cellular system is a plastic or concrete panel with cells of various sizes, from 10 to 69 cells per m². Cell height is 10 cm. The panels are anchored to the slope with metal posts. The cells can be filled with gravel or, if establishment of vegetation is desired, filled with soil. After that, grass is sown or trees and shrubs planted. In this way, the treated slope is protected from erosion, even if the vegetation has not established. Where the slope length is greater than 10 m (i.e., one panel length), the panels are continuously stapled to enable rapid installation by way of pulling the top panel up the berm face. Due to space constraints at the top of the berm, all soil filling operations are undertaken from the bottom. Upper reaches of the slope require the utilisation of crane and cable to facilitate the filling of cells.

Prefabricated Concrete Blocks

These seek to provide a protective armour as a matrix of individual concrete units. The main function of the blocks is to protect and stabilise under design conditions. The two types of blocks described in this text have been constructed and recommended by G. Stanojević. They are:

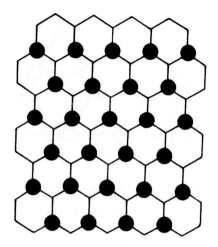

Fig. 15: Wire-concrete network.

1) wire-concrete network,
2) triangular concrete block.

Wire-concrete network

This product is an improved variety of a common wire network. The flexibility of a common wire network has been retained and the stabilising effect of the third dimension has been improved by moulded concrete blocks (Fig. 15).

The network is manufactured and installed by segments, which can be transported manually. The joining of segments is performed by metal hooks and the network is fixed to the ground by metal anchors or by posts. It is possible to produce another variety, i.e., the network made of composite materials, concrete block and the network as the base. This product is intended for erosion control on steep slopes covered with loose material. It can also be used as a stabiliser of coarse materials used for mulching.

Triangular concrete blocks

Industrially designed concrete blocks can be applied in two ways:
1) as prefabricated blocks for drainage,
2) as blocks for stabilisation of land on slopes of artificial origin.

For drainage purposes, concrete elements are joined in line series and fixed by clamps made of water-resistant material. For the stabilisation of soil of artificial slopes, concrete blocks are placed with their openings

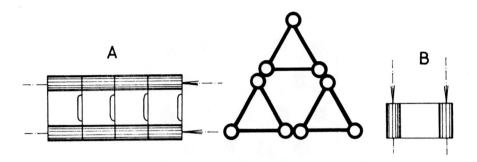

Fig. 16: Triangular concrete blocks.

turned upwards. The blocks are fastened by posts made of impregnated wood (Fig. 16).

CONCLUSIONS

The application of technical (mechanical) erosion control measures for reconstructed or degraded steep land dates back to ancient times. The techniques of contouring and terracing are still widely utilised nowadays (Lujić, 1973; Huanchen and Baopin, 1992; Stanev, 1979; Vučićević, 1995; Critchley and Bruijnzeel, 1995; Carver, 1995). Research in several countries has shown that the effect of these erosion control measures on steep slopes remains very important. Among the terrace systems, gradons are the most successful measure for the afforestation of eroded terrains (Djorović, 1973).

Terraces eliminate slope effects as well as land-use effects almost completely. Surface runoff and, accordingly, soil loss are observed only on the risers of bench terraces and often only during their first 2 to 3 years. After that, where risers are well protected by natural vegetation cover (grass), erosion is slight. Further, eroded soil from the risers is mainly deposited on the bench (flat part of the terrace) and not transported beyond, especially on back-sloping terraces under continuous cultivation. Research confirms the great significance of bench terraces as very successful water erosion control measures. Bench terraces may decrease erosion losses from very steep slopes from strong and medium erosion categories (strong erosion 20–50 $t \cdot ha^{-1} \cdot yr^{-1}$; medium erosion 10–20 $t \cdot ha^{-1} \cdot yr^{-1}$), to below the erosion tolerance of 2.0 $t \cdot ha^{-1} \cdot yr^{-1}$ (Djorović, 1990).

In Yugoslavia terraces and rock barriers with small walls have proven very effective in reducing surface runoff and soil loss. Studies showed that surface runoff reduced to 67% and 61% of control plot values during the first year and 60% and 57% in the second year respectively. Soil loss was reduced to 83% and 81% of control plot values in the first year and 80% and 78% in the second year (Dano and Siapno, 1992).

Along with many positive qualities there are some negative effects. The main problem is that these terraces must be permanently maintained and repaired. Where this is not done, as in the case of the highland terrains of Yemen, the results can be catastrophic. None of these measures should be applied to slopes liable to landslides as they can activate landslips. This refers first of all to contour ditches. Due to such problems, as well as the cost factor, contour ditch construction has all but ceased in Yugoslavia's hill country.

Another disadvantage arises from erosion of the land between the terrace benches. The 'terrace riser problem' is another constraint to the effectiveness of terracing and evidently one which has largely been overlooked, or at least underestimated (Critchley and Bruijnzeel, 1995).

Programmes, which promote terracing—or terrace rehabilitation—as blanket remedies for soil and water management in upland or reconstructed areas, need to reconsider their strategies in view of the limitations of these structures.

The application of more recent constructions of prefabricated erosion control elements made of concrete, nets, geotextiles etc. is still restricted to small areas—slopes adjacent to roads, slopes of cuts and fills of drain and natural watercourses etc. High cost is a major reason.

Technical erosion control measures for the reconstruction of steep lands play a significant role in the strategy of watershed management. The choice of measures and works must bear in mind the natural circumstances of the site as well as the characteristics of the particular measures. In this way, the best effects will be achieved in erosion control and in reconstruction of degraded steep lands.

Land reclamation programmes can improve their effectiveness by carefully monitoring the on-site and downstream consequences of intervention. Monitoring needs to be accompanied by adaptive research of an interdisciplinary nature and must accommodate the land-user perspective.

Finally, it should never be forgotten that the durability and continuous effectiveness of these structures usually depends on their permanent maintenance.

References

Carver, M. 1995. Indigenous land management: erosion control in the middle mountains of Nepal. *Int. Erosion Control Assoc., Proc.,* 28: 231–238.

Cohn, W. 1995. A review of landfill applications of articulating concrete block revetments. *Int. Erosion Control Assoc., Proc.* 28: 95–109.

Critchley, W.R.S. and Bruijnzeel, L.A. 1995. Terrace risers: erosion control or sediment, pp. 529–541. In: R.B. Singh and M.J. Haigh (eds.). *Sustainable Reconstruction of Highland and Headwater Regions.* Oxford & IBH, New Delhi, 681 pp.

Dano, A.M. and Siapno, F.E. 1992. The effectiveness of soil conservation structures in steep cultivated mountain regions of the Philippines. *Erosion, Debris Flows and Environment in Mountain Regions Proc. Chengdu Symp.* Int. Assoc. Hydro. Sci., Publ. 209: 399–405.

Djorović, M. 1973. Eksperimentalno *Uturdivanje Antierozionog Dejstva Gradona.* Institute za šumarstvo I drvnu industriju, Zbornik radova 11, Belgrade.

Djorović, M. 1990. Experimental study of erosion and crop production on bench terraces on sloping land, pp. 531–536. In: J. Boardman, I.D.L. Foster and J.A. Dearing (eds.). *Soil Erosion on Agricultural Land.* John Wiley & Sons, New York, 687 pp.

Gavrilović, S. 1972. *Inžinjering o Bujičnim Tokovima I Eroziji.* Izgradnja, spec. ed., Belgrade, 293 pp.

Goldman, S.J., Jackson, K. and Bursztynsky, T.A. 1986. *Erosion and Sediment Control Handbook.* McGraw-Hill Book Company, New York, 454 pp.

Huanchen, Z. and L. Baopin, 1992. A study on land degradation and restoration in mountain environments in Liaoning Province. *Erosion, Debris Flows and Environment in Mountain Regions Proc. Chengdu Symp.* Int. Asso. Hydro. Sci., Publ. 209: 477–485.

Kostadinov, S. 1996. *Bujični Tokovi I Erozija.* Šumarski Fakultet Univ. u Beogradu, Beograd, 510 pp.

Lujić, R. 1973. *Šumske Melioracije.* Naučna Knjiga, Beograd, 409 pp.

Morgan, R.P.C. 1986. *Soil Erosion and Conservation.* Longman Sci. Tech./John Wiley & Sons, New York, 298 pp.

Ranganathan, S.R. 1995. Jute geotextiles as efficient promoter of vegetation for erosion control. *Int. Erosion Control Assoc., Proc.,* 28: 239–248.

Rosić, S. 1960. *Bujice I Njihovo Uredivanje.* Naučna Knjiga, Beograd, 130 pp.

Saccardy, L. 1951. Notes sur le calcul de banquettes de restoration des sols. *Revue Terres et Eaux* 11. Alger.

Sheng, T.S. 1977. Protection of cultivated slopes terracing steep slopes in humid regions in *Guidelines to Watershed Management 1.* FAO, Rome.

Sheng, T.S. 1989. *Soil Conservation for Small Forests is the Humid Tropics.* FAO, Rome Soils Bulletin 60

Stanev, S.I. 1979. *Počvena Erozija i Borbota s Neja.* D.I. 'Tehnika', Sofija, 466 pp.

Vucićević, D. 1995. *Uredenje Bujičnih Tokova.* Društvo bujičara Jugoslavije, Beograd, 441 pp.

Ward, L.E. 1995. Kolb road project design and installation of a cellular containment system. *Int. Erosion Control Assoc., Proc.* 28: 51–62.

7

Hydrological Management of Reclaimed Opencast Coal-Mine Sites

Marianne P. Kilmartin

Abstract

The success or failure of land reclamation depends, to a large extent, on the management of surface runoff and drainage. The existence of compacted layers close to the surface on many reconstructed lands, especially those developed on shale and mudstone overburdens, limits infiltration and fosters increased surface runoff and through flow. This situation is responsible for many of the erosion and runoff control problems which afflict reclaimed lands. It causes the characteristic stream hydrograph, which has relatively high flood peaks, low base flows and short lag times, compared to most pre-mining landscapes. Several research studies, conducted on small plots, suggest that this is due to the low storage capacity of the upper layer of the mine spoils and/or artificial topsoils on such sites. Catchment scale studies on reconstructed surface coal-mine sites are very few. However, they show that the link between rainfall and stream runoff is much more complex. The hydrological response is affected by the development of soil pipes, ponding (often associated with soil settlement), variations of soil depth across reconstructed slopes, and possible flowlines linking waters held above and below the compacted surface layers of the reclaimed lands. The balance between these processes changes during the years following land reinstatement. The implications of these research findings for drainage design and erosion control are assessed, with special reference to a U.K. context.

Land Reconstruction and Management Vol. 1, 2000, pp 137–158
ISSN 1389-2541
ISBN 90 5410 793 6
A.A. Balkema, Rotterdam, The Netherlands

INTRODUCTION

Although opencast coal mining only began in Britain in 1942, production had risen to 15.1 million tons in 1987/88 (British Coal Corporation, 1988). This output was maintained through 1993, although the number of mines decreased (British Coal Corporation, 1993; British Coal Corporation, Opencast Executive, 1994). This represents a land requirement of approximately 2000 hectares a year. This area will need to be restored to a stable and aesthetically acceptable use afterwards.

Reclamation of surface-mined land is a complex process, which may be described as a rebuilding or reconstruction of the land profile followed by a series of agricultural activities to revegetate the land surfaces (Ward et al., 1983). In fact, reclamation planning is an integral part of surface-mining industries in many parts of the world, and the development of approaches such as 'concurrent reclamation' are being led by the industry itself. Many technological institutes are now engaged in devising improved methods of drainage, reseeding, soiling and reshaping, revegetating and neutralising the acid runoff. Much of this work has developed as a result of legislation designed to establish and enforce minimum standards of restoration (Kilmartin and Haigh, 1988). In the U.K., statutory controls began with the Town and Country Planning Act of 1947 which required 'any spoil deposit within the worked-out area to be placed at a level below that of the adjacent unworked land and the worked-out area to be left in a level and tidy condition' (Allen, 1983). This has been superceded by more detailed Acts. Once a mining project is initiated an environmental impact assessment (EIS) is prepared, which includes selected hydrological parameters for assessment of the ground and surface water (Clarke, 1995). Careful reclamation planning is now being considered at the outset as the local authorities, the mine owners and the mine operators want to minimise future liabilites (Ricks, 1995). Clarke (1995) stresses the importance of water management after the decommissioning of the mine, so that contamination and erosion do not occur. This ensures that, when the final handover to the new tenants or landowners takes place, usually after a statutory 5-year period, the land is returned in a condition that is reasonably fit for its after-use, often commercial farming. The operations that are 'normal' within a restoration programme are: installation of new watercourses, ditches, fences and hedges, establishing the first temporary grass 'ley', installing fairly intensive under-drainage in the fourth year and then reploughing and cultivating to reduce unevenness caused by drainage operations and resowing to a permanent ley (Bragg, 1981).

In common with many environmental matters, reclamation involves many disciplines but, as Aspinwall (1976) suggests, whether the land is to be used for agriculture, industrial development, dry recreation development, or afforestation, one of the most important considerations is the hydrology of the area. The exposed coalfields of Great Britain receive between 600–900 mm rainfall per annum, with parts of Wales sometimes receiving up to 1800 mm (Carolan, 1985). The control and drainage of water contributes significantly to the success or failure of a reclamation project.

Research in the U.K. had been sponsored mainly by the national coal company, British Coal. It concentrated on designing economically viable drainage systems to remove excess surface water and aid soil development (Bragg, 1982; Bragg et al., 1984: Scullion and Mohammed, 1986: Rands and Bragg, 1988). Most projects were based either on small plots (Trafford and Twocock, 1972; Scullion, 1984) or natural catchments, where only a part was affected by surface mining. (Addis, 1982; Curtis, 1974). Much of this work on reclaimed sites assumed that the short time-lag between rainfall and runoff, and the steeply rising hydrograph limb that resulted, were caused by surface runoff (Trafford and Twocock, 1972; Bragg, 1983; Hodgkinson, 1989; Addis, 1982). Explanations emphasise the presence of a shallow 'active' layer of topsoil which, being the same depth over the whole area, becomes uniformly saturated. However, although this presupposes a simple hydrological system, calculating runoff continued to confound land reclamation engineers, especially in areas of high precipitation such as South Wales. This is reflected in the large number of cases where flooding occurred due to inadequate drainage and the many instances where unanticipated volumes of surface runoff caused problems such as accelerated soil erosion, gullying and ponding (Haigh, 1992).

Research into the hydrology of reclaimed opencast mined-land falls into four main areas: surface runoff, infiltration, groundwater and drainage.

Surface Runoff

The amount of surface runoff from a catchment depends on the intensity and duration of the storm, antecedent soil moisture conditions, time of year (snow or rain), slope, surface storage capacity, infiltration and vegetation cover. All these quantities affect, in one way or another, the recorded stream hydrographs. They can be valuable sources of information about the behaviour of a surface mine reclamation before, during and after the operation. A variety of indirect methods have been used to estimate runoff per unit area or the stream flow characteristics of ungauged streams. Early efforts included the transfer of stream-flow records from gauged to nearby ungauged basins and the estimation of runoff from catchment area and average annual rainfall. Later, numerous basin characteristics and multiple-regression analysis were used to estimate parameters of discharge (Osterkamp and Hedman, 1979). Further improvement of these techniques led to the development of various models relating rainfall to runoff, the most popular one in the U.K. being that given in the Flood Studies Report

(Natural Environment Research Council, 1975). This model provides a method for estimating a design hydrograph with a peak of a specific return period. The approach has been developed further to give more detailed information relating to a particular catchment (Boorman and Reed, 1981; Boorman, 1985, 1986; Reed 1987). In the USA, the Soil Conservation Service runoff curve number method is widely used for estimating total runoff from total rainfall; the method incorporates precipitation values, land-use and antecedent moisture conditions.

Gregory and Walling (1973) looked at the effects of human actions on runoff quantity, showing the particular impact when land is disturbed, cleared of vegetation and compacted. Flow modifications that result include increased storm runoff, heightened flood peaks and changes in lag-time. Obviously where spoil material provides increased storage capacity and infiltration, increased base flow and reduced runoff may occur.

Changes in runoff response due to surface mining have been fairly well documented in the USA (Curtis, 1974; Schroeder, 1987; Gilley et al., 1977; Hamon et al., 1977, 1979). Collier et al. (1970) carried out a particularly long running experiment between 1955 and 1966. This was a paired-catchment study in which 10% of the site had been affected by mining. Even with only partial disturbance, the mine-affected catchment exhibited higher peak discharges, greater flow range, lower base flows and shorter lag-times.

Many other studies have been carried out on a plot scale, such as Schroeder (1987) who tried to develop reliable estimates of runoff curve numbers for reshaping fine-textured spoil areas in North Dakota. The aim was to determine whether spoil surface conditions, slope gradients and sodium adsorption values had a significant effect. Schroeder (1987) worked on plots of 1.9 × 4.9 metres at gradients of 3°, 6° and 9° on three spoil sites. He found no significant difference between the runoff from slopes of differing gradient. However, freshly reshaped slopes had a lower runoff than crusted slopes when dry, but much the same when wet.

Haigh and Sansom (1999) carried out rainfall simulation tests in Yorkshire to determine variations in runoff generation on different spoils. These tests involved simulating a very heavy rainstorm, which delivered the maximum 2-min intensity over a 5-min period across metre-square runoff plots. In this study both age and the presence of vegetation greatly affected the percentage runoff. The newly laid unvegetated spoil shed almost 45% of the rainfall and the older unvegetated site shed between 25–35%. By contrast an older 'compacted and degraded' vegetated site only shed 10% in the first test and none at all in a further test.

Gilley et al. (1977) studied 4.1 m × 22.1 m plots in North Dakota comparing the erosion and runoff from rangeland, spoil and sites reclaimed with 610 mm of topsoil. They reported only 12% of applied rainfall being measured as runoff from the rangeland control site, but 66–74% from spoil sites, declining to 43–50% when topsoil was added.

Osterkamp and Hedman (1979) measured channel geometry to evaluate stream flow characteristics on ungauged surface-mined areas in Kansas. They argue that although rainfall runoff models may be useful for estimating runoff from 'limited areas, generally in humid regions' hydraulic geometry offered a 'more reliable and easily applied' source of empirical evidence (Leopold and Maddock, 1953). This work assumes that channel size and shape are determined directly by water and sediment discharged. The authors see a direct relationship that is especially useful for mined areas when hydrologic information is sparse. They offer examples of power-function relations that can be applied directly to new reclaimed catchments. However, their method does not appear to take into account the bank erosion caused by trampling or the impacts of the unusually large percentage of surface runoff present on many reclaimed catchments.

Much of the research carried out in the U.K. has been undertaken by industry, so access to the results is limited. Most published studies have been directed to designing adequate drainage. The exception is a study in the Forest of Dean (Addis, 1982) where three small catchments (0.238, 0.246 and 0.87 km^2) were instrumented. The paired-catchment approach was adopted, while another much larger basin was studied in order to demonstrate contrasts in the magnitude and timing of runoff. Only 50% of this larger basin had been affected by mining. Despite problems caused by changes in land-use during the study, Addis found that in summer 90% of flow in the affected catchment was due to surface runoff and only 38% in the unaffected (compared with 71% and 58% in the winter period). The paired catchments showed greatest contrast after a dry spell. Addis also confirmed that the affected area produced extremely peaked hydrographs with fast rising runoff responses compared to the moderated flood peaks of the more slowly responding control catchment.

Infiltration

One of the main components of a reclaimed hydrological catchment is the infiltration of rain water. This varies according to two general factors. First, whether the final profile or the surface layer is porous or highly compacted. This will depend on the nature of the replacement operation and the type of subsoil. Second, the hydraulic character of any artificial topsoil layer added to the site. This depends on the depth of the applied layer, the topsoil material and the handling procedures. Research results to date concerning infiltration in reconstructed spoils in laboratory tests and tests on site show widely different results (Table 1).

So many factors influence these water infiltration rates that wide variations can occur within the same site area, as well as between sites.

Table 1: Infiltration rates

Source (Date)	Area of Site	Rate of infiltration $(mm \cdot h^{-1})$
Farmer and Richardson (1976)	South-eastern Montana	3–36.1
Grandt (1984)	Illinois	1.2–58.4
Jorgensen and Gardner (1987)	Pennsylvania	7.0–40.4
Massey (1970)	Northumberland	22.2–232.2
Sanchez and Wood (1989)	Western New Mexico	331–667
Rogowski (1980)	Pennsylvania (reconstructed soil)	144–765

However, in general, reclaimed mine soils do tend to exhibit lower infiltration rates than the natural soils they replace—except in the lignitic fields of Central and East Europe.

Potter et al. (1988) studied reconstructed mine spoils 4 and 11 years after reclamation. They found that the saturated hydraulic conductivity of the reclaimed topsoil was about 25% of that of the undisturbed A horizons and the reclaimed subsoil was less than 10% that of the undisturbed B horizons. No significant difference was established between the constructed soils 4 and 11 years after reclamation.

Jensen et al. (1978) suggest reshaping the soil surface to increase infiltration and to reduce the rate of surface runoff by intercepting and storing runoff. They refer to experiments by Hodder (1972) who concluded that this technique effectively channelled additional water into the soil and also aided in the re-establishment of vegetation. However, surface manipulation involving mechanical stress may not be beneficial on soils with a high expanding clay content. Such action may encourage soil aggregate breakdown and so result in even less infiltration. It must also be remembered that artificial slopes can change shape relatively quickly after recontouring (Haigh, 1978, 1988).

Prunty and Kirkham (1979, 1980) were also concerned with the topography of reclaimed land and how it affected the seepage of infiltrated water and the movement of possible pollution leachates. Their drainage models consisted of three surface topographies: A) a uniform slope, B) a four-terrace topography and C) a one-terrace topography. Saturation of the overburden was assumed so the results depicted maximum seepage. They found that the total seepage rate for the four-terrace topography was 10 times greater than for uniform-slope topography (see Kostadinov, this volume).

Vegetation

The amount and kind of vegetation cover protecting spoils are also important determinants of the infiltration, surface runoff and erosion behaviour of reclaimed land, as well as of the hydraulic resistance of the surface to overland flow (Fujita and Shutto, 1987). The rooting depth of vegetation has a significant effect on the surface hydrology, especially in arid and semi-arid climates (Wright, 1978; Soil Conservation Service, 1982). Overland flow will be greatly decreased if the watershed is covered with deep-rooted vegetation that can dry out the soil by evapotranspiration (Lyle 1987).

Chong et al. (1986) examined the different hydrological characteristics of mine spoil with and without topsoil. They found that it was inadequate to use only infiltration-related properties in the assessments, as a soil with a high clay content could give a fast infiltration rate because of water flow through macropores and desiccation fissures rather than through the soil matrix. They then measured the hydraulic conductivity, macroporosity and drainage rate and found these to be higher in the topsoiled mine spoil. Ward et al. (1983) found that the infiltration rate through dry applied topsoil was controlled by surface conditions.

Until recently, little research had been done on the stripping, storage and respreading of topsoil and few guidelines existed as to when, and under what conditions, the movement of soil is best undertaken. At present, the advice given to avoid serious compaction is to ensure that the soil is moved only when it is at a moisture content 5% drier than the lower plastic limit (Bragg, 1982). On most sites, approximately 250 mm of topsoil is stripped and stored in mounds on the perimeter of the site (Carolan, 1985). Where available, at least 600 mm of subsoil is saved, the objective being that a new 900 mm thick soil profile may be reinstated after mining by respreading (Carolan, 1985). The soil may remain in these mounds for several years. Their height may be 10–15 metres, the internal conditions may become anaerobic and the soil system degraded. Further problems may be caused by the machinery used both during and after resoiling. Both Bragg (1983b) and Downing (1972) mention the excessive wheel rutting that causes gullying on restored sites. Obviously, in areas such as South Wales, surface coal mines are located on sites of poorer land quality, where soils are thin and sometimes non-existent. In Britain, the policy of the former British Coal Corporation was to recover any suitable local materials including glacial sands and gravels, to make up for any deficiency in soil volume (Bish, 1986). Adams (1987) gives a detailed account of the progressive restoration taking place on the Ffyndaff site in South Wales and the problems encountered due to the difficult climate and soil conditions.

Groundwater

The impacts of mining on surface waters have been given serious attention, with more emphasis given to water quality than to quantity. However, the impacts of mining on groundwater have been studied relatively little.

During the process of surface mining dewatering operations take place, depressing the groundwater level (Ngah, Reed and Singh, 1984). After mining, pumping ceases and the water table begins to recover its natural level. The rate and degree of groundwater re-establishment in overburden are defined by several interrelated factors (Herring, 1977).

a) *Climate*: Normally, recharge to shallow aquifers is accomplished by local precipitation, which percolates downwards through the soil according to the permeability of these near-surface layers. Ringler (1984) looked at spoil aquifer resaturation in Montana, USA. He found that under conditions of low precipitation and high evapotranspiration, surface recharge would not play a significant role in spoil aquifer recharge. Anecdotal reports from speleologists working beneath erstwhile opencast coal mines in Wales suggest that the cave systems are unusually dry. However, site investigations show that the body of the fill in former open-pit mines is loosely packed with many voids but the surface layer is dense and relatively impermeable. In such circumstances, lateral recharge may be very important. Nevertheless, climates with great seasonal variations affect groundwater levels. Both Norton (1984) and Reed and Singh (1986) refer to groundwater lowering in the summer and rising in the winter.

b) *Overburden characteristics*: The process of surface mining creates spoils that have quite different properties from the natural material they replace. These differences include the chemical composition and physical properties. Due to the disruption of the rock strata, great variation in the fill permeability may occur, varying from being unable to hold water during a 'pump-in' test to being more impermeable than the original rock. This can be affected by the degree to which the spoil has been affected by weathering. A recovering water table will accelerate the processes of physical and chemical weathering by the processes of scouring, swelling, air breakage and freeze thaw (Charles et al., 1977). This will result in consolidation of the fill, thereby causing settlement. Both Reed and Singh (1986) and Norton (1984) refer to the Horsley experiment in Northumberland, U.K. in which changes in groundwater levels had a significant effect on backfill settlement.

c) *Aquifer coefficients*: The hydraulic conductivity and storage coefficients for aquifers vary considerably depending on the geological strata in the area being mined. Herring (1977) found that aquifer transmissivity in overburden ranges from 60.42 lpd $\cdot$ m^{-1} (litres per day) in shale and sandy shale, to 27.68 lpd $\cdot$ m^{-1} in sandstone. He confirms that, during the mining process at least, more water enters the pit from the cast overburden side than from the high-wall side—which could be due to greater transmissivity and/or greater recharge. Reed and Singh (1986) looked at groundwater recovery in ten sites in three different areas. They found that the rate varied from 144 days in a shallow site to several years in a much deeper and larger site. Recovery, in one site, was still continuing eight years after the sequence was initiated. Such information is of great importance because, until the water table reaches a static equilibrium level, settlement under subsidence is likely to continue (Norton, 1984).

d) *Topography*: The effects of surface mining on topography are well known. In the past, the typical outcome was steep-sided ridges of loosely compacted spoil with depressions in the form of inclines and final cut pits. Today, the restored land is graded and contoured to form long gentle slopes to emulate the surrounding countryside. As Riley (1976) and Sawatsky et al. (2000, this volume) note, there has emerged a philosophy that only natural systems are correct and suitable. However, this does not mean that the ecosystem must and should be returned to its original condition. Herring (1977) points to the disadvantages of this kind of thinking. Reclaimed opencast mined land is often compacted during and after reclamation, so an almost impervious surface layer is formed. The net result is that more storm water runoff develops than would have occurred in nature and this requires the design of a different drainage system.

Toy and Hadley (1987) refer to various studies that have shown that erosion is greater on the uniform hill slopes created by reclamation than on natural slopes. Haigh (1978) finds that it is the upper convexity of artificial slope profiles that suffers the most erosion. In Oklahoma, Goodman and Haigh (1981) showed that younger (30-year-old) sites suffer more erosion that older (60-year-old) sites and that the zone of maximal erosion migrates from the crest towards the midslope as a slope ages. They also detail a sequence of associated geomorphological changes. These factors should be considered when planning post-mining use. Herring (1977) suggests that, because of the impermeable surface of many reclaimed sites, the creation of a topography suitable for the development of lakes should be given careful consideration. Sawatsky et al. (2000, this volume) argue that the morphological design of hillslopes should reflect our understanding of final stable forms.

e) *Groundwater recharge*: Ringler (1984) states that groundwater recharge comes from three different sources: 1) laterally from unmined coal beds, 2) from a lower aquifer with a higher piezometric head than the spoil and 3) via surface infiltration. Recharge to a reclaimed area depends on a number of general factors, including climate, topography, soils, vegetation and geology. In areas where little consumptive use of groundwater takes place, the amount of recharge can be fairly easily estimated by hydrograph separation techniques or base-flow measurements of streams draining the area. Herring (1977) refers to Walton's experiments in 1965 which showed that groundwater runoff may vary from 5.1308×10^{-3} to 18.324×10^{-3} metres^3 per second per square kilometer. He found that groundwater runoff was significantly greater in basins where the bedrock was permeable than where it was relatively impermeable.

Herring then relates this to surface-mined areas by suggesting that since mining leaves broken, unconsolidated and permeable overburden, it is reasonable to assume that considerably greater groundwater recharge and runoff will occur. Obviously, this will only happen when the spoil is not compacted.

Norton (1984) mentions the problem of groundwater 'rebound' where the surface mine has been excavated in an area of high natural groundwater table and where surface springs may issue from, or adjacent to, the backfilled void. Successful prosecutions have been instigated by the Regional Water Authorities against British Coal (previously NCB) in respect of pollution downstream.

Aldous et al. (1986) worked in the Forest of Dean, U.K. where small-scale mining activities continued after deep mining was abandoned. The possibility of unrecorded workings, the threat of random collapse and associated ponding, together with uncertainty over the hydrological behaviour of the substrata, increase the hazards of pollution of previously clean groundwater.

Schwartz and Crowe (1987) demonstrate how groundwater flow models can be used to study the long-term response of groundwater flow in and around strip mines. Their two-dimensional, finite-element model can be used on two different scales—either with the mine occupying the major part of the system or at a smaller scale, where individual hills formed during reclamation are the main point of interest.

f) *Water quality*: In 1972, 13% of the total groundwater abstraction in England and Wales was from coal measure aquifers, of which 43% was from abandoned coal mines (Aldous et al., 1986). Of this only 3.5% was used, the remainder being disposed to surface watercourses. Since the shales and mud-stones of British opencast mine backfill weather quickly when exposed to air and water, a polluted discharge can be produced, especially when pyritic material oxidises to form acidic and ferruginous drainages (see Haigh 2000 this book, chapter 9). The resultant waters can be low in organic matter and high in dissolved metal salts causing great concern over the environmental and economic effects. Further information can be found in Addis et al. (1984), Bird (1987), Evangelou and Karathansis (1984) and Trouart and Knight (1984).

Drainage

In the U.K., despite the development of improved Codes of Practice covering the principles and objectives of reclamation in the late 1970s, it became apparent that there were serious inadequacies in restoration techniques and practices' (Carolan, 1987). To improve this situation, the British Coal Opencast Executive initiated and financed a programme of research, employing universities, research institutions and MAFF to undertake the work. A programme of experiments looking at early underdrainage of reclaimed sites was initiated by the Field Experimental Unit under this scheme. The broad hydrological were to monitor the effects of different drainage systems and to in the relationship between surface water flow and drain flow.

The longest running experimental site was the 'Radar' site in Northumberland, U.K., which was set up in 1968. At first, this was designed to monitor the performance of alternative plastic drain systems and to compare them with traditional clay tile systems (Trafford and Twocock, 1972). Some of the drains included gravel as a permeable backfill and some had an impermeable backfill. The site was subsoiled in 1981, when compaction of the soil was identified, so the results show a comparison of before and after subsoiling. The methods and results are documented by Bragg (1982), Bragg, Arrowsmith and Rands (1984), Bragg, Griffiths, Jones and Bell (1984) and Hodgkinson et al. (1987). The conclusions from the 'Radar' experiment and a similar study carried out at Hirwaun in Wales show that the flow in the drains with permeable backfill (e.g. 'French' drains) was considerably greater because they provided a positive hydraulic connection through all zones of compaction. The study also showed that moling (creating unlined or mole drains by means of a subsoil plough) was generally worthwhile. Further research was needed to determine the spacing and size of lateral drains with respect to future use of the land and to establish the balance between cost and benefit. Such a study had to take into account the proposed land management practices in order to avoid silting up of gravel backfill, scouring of the soil around the pipe, slaking of material into subsoiling fissures and increased surface runoff, among other difficulties encountered (Scullion, 1984; Scullion and Mohammed, 1986). Experiments in Belgium on the drainage of agricultural land confirm that silting up of drains will occur in soils that are not well structured (Dierickx, 1987). Drainage trials frequently lack any statistical evaluation, because of the practical difficulties of setting up random replicated plots and the work involved in collecting detailed moisture measurements. However, Scullion et al. (1986) devised a technique for statistical evaluation that aimed to 'enhance the value of field drainage experimentation without using any marked increase in workload'. They employed a system of scoring to establish surface wetness categories. These were then calibrated against soil moisture, as measured by an estimation of waterlogged pore spaces.

Kemp (1988) set up a catchment size drainage experiment in Cumbria. This 48–hectare catchment had a comprehensive underdrainage system that employed 60 mm diameter lateral drains at 10 metre spacing. He reports that very short lag-times are observed and that the peak flow was 'frequently substantially higher than expected'.

However, unless field drainage is really effective in achieving better soil-water control, the creation of reclaimed land is likely to lead to an increase in flood frequency downstream. Open ditching is generally adverse, the response to rainfall is accelerated and the catchment becomes sensitive to shorter, more intense storms (Reed, 1987). Moling or subsoiling can provide a storage buffer but this is strongly influenced by the rainfall regime and the characteristics of the parent soils (Robinson, 1987). Mine spoils, which contain a high shale content, can cause the drains to silt up in a very short time (Bragg, 1984; Hodgkinson et al., 1987). It is well known that opencast mining can have a detrimental effect on flood frequency. Hence, it is inevitable that allegations will arise when flooding in a particular locality is perceived to increase, even when that flooding is actually due to entirely natural processes such as an extreme rainfall creates. It is therefore advisable to overdesign runoff control for such sites.

Most runoff and drainage studies have used small experimental plots, which may not be representative of the whole catchment response. Roels and Jonker (1985) carried out a statistical study to calculate the accuracy of the erosion data from plot experiments. They found a large scatter in the data. Soil loss estimates that deviated by 50–100% from the subpopulation values were the rule rather than the exception. Given differences in infiltration rates, bulk density, topsoil depth, slope angle and vegetation cover in a catchment, it would be reasonable to assume that the total area may not show the same hydrological response as a small part of it. Most of the few catchment projects that have been undertaken to date have examined sites that have only been partially disturbed by opencast coal mining. Drainage experiments, which might be successful on a plot scale, may prove either inefficient or economically impossible over the whole catchment area. Perhaps the size and spacing of the drains should vary according to the position on the site.

Another problem of looking at any runoff data is the natural variability of the rainfall. Reed (1987) suggests that a screening test should always be applied to check whether any apparent change in flood frequency is, in fact, due to a shift in rainfall frequency.

Although there are few catchment scale studies, their results suggest that these sites have a more complex response than previous research has indicated (Kilmartin, 1994). The preferred route for the runoff is not surface runoff, as often assumed, but interflow at the boundary between the applied artificial topsoil and the mine spoils beneath. Due to topsoil erosion, this boundary is near the surface at the crest of the slope and deeper near the base of the slope (Haigh, 1979). In Wales, Kilmartin (1994) found that most runoff occurs either as 'throughflow' through the macropores or as 'rapid response runoff' through a system of 'soil pipes' or 'fissures', which appear to be in the upper layers of the mine spoils. The process is further complicated by the presence of surface ponding—due to settlement—and artificial troughs, which overflow in severe events, adding a sudden extra input to the hydrograph.

whole of the runoff on the upper slopes to the main stream channel as surface wash. This could be *one* of the reasons why the flood peaks tend to be high. It could also explain why, in Wales and in rainfall events with >4 mm·hr^{-1} intensity, some peaks show a much greater increase with rainfall intensity than others.

There are two reasons why this effect is likely to be more dramatic than in a natural catchment. First, almost none of the rainfall is lost to groundwater aquifers. Second, the flow due to the 'pipe' system and the thin vegetation has already reached a high velocity. This was shown to be exceptionally high when there was a period of high intensity rainfall late in the storm duration.

Examination of the hydrographs also gives evidence of a low 'shoulder' on the recession limb in some storm events. This could be the result of seepage from the increased storage area.

Third Process (Fig. 1C)

Some storm events generate particularly high hydrograph peaks. The nature of a reclaimed surface is that it tends to be uneven and to undergo irregular subsidence. The result is ponding midslope. During heavy rains, these ponds can overflow causing 'sheetflow' down the slope. If this should coincide with saturation of the base zone, not only will it increase the contributing area by the size of the pond, but also by all the slope above that has been draining into it.

SOIL PIPES OR FISSURES

Although there does not appear to have been any work done specifically on 'fissures' in reinstated land, Armstrong and McGowan (1984) have looked at the swell/shrink processes that could lead to the establishment of fissures. Nicolau (1996) describes how this process can lead to soil-pipe generation in late spring causing discontinuous infiltration and the possibility of reinfiltration. Trafford and Twocock (1972) note a considerable variation in infiltration rates and the presence of fissures on the Radar site but attribute this to the 'subsoiling' that had taken place eight years earlier. Bragg Arrowsmith and Rands (1984), however, caution that fissures and pores in reinstated land may be random packing voids rather than continuous channels. However, Ritchie (1963) confirms that 'tunnelling' often affects artificial landforms, where the soil materials are dispersible and inadequate vegetation has led to uneven infiltration.

However, the above phenomena are not new in reports on groundwater flow in spoil materials. Caruccio (1984) identified two types of flow—slower porous media flow, which occurs under steady state or low flow conditions, and rapid turbulent flow, which occurs through conduits or large voids during periods of hydrostatic stress ('pseudokarstic flow'). Further work by Hawkins and Aljoe (1992), and Aljoe and Hawkins (1992), confirms the existence of these conduits and stresses their significance when modelling groundwater flow in reclaimed surface-mined land. This work has not been linked to the studies of surface runoff on the soil/water processes in the upper spoil layers.

Kilmartin (1994) observed fissures of 'soil pipe' outlets both at the midslope, slope base and at the channel bank, suggesting that original fissures created by the reinstatement process in the upper shale layer have evolved slowly, thereby increasing the infiltration rate in that area. She suggests that, *in the early years of restoration*, this world lead to extremely 'peaky' hydrographs on the site and could possibly contribute to the peak drain flow rates that regularly exceed the design rate, which were reported in drainage experiments of the Field Drainage Experimental Unit (Hodgkinson et al., 1988).

Further scrutiny of the previous research has revealed a number of inconsistencies that could be explained by the presence of flow lines. Trafford and Twocock (1972) noted that on one of their plots the drain flow exceeded the rainfall. This they attributed to seepage from outside the metered zone. Although they had thought that a hedge and tall coarse grass would have prevented surface runoff onto the plot, in the absence of obvious subsidence upslope, they decided that this must have been the case. They also observed wetter areas on the plot for which they could not offer a definite explanation. Hodgkinson et al. (1988) working in South Wales on drainage plots, also noted that 'foreign' water had entered the monitoring area. This, they concluded, was 'probably' during periods of high-intensity rainfall. Hodgkinson (1989), working in Shropshire, commented that the 'magnitude of the surcharge' from one of the plots suggested that it was collecting water that was being shed from another plot. Addis (1982), authoring the only other catchment project in the U.K., also found output exceeded input in *all* his catchments but could draw no conclusion as this may have been due to pumping by the mining contractors. Kemp (1988), who set up a catchment size drainage experiment in 1984 in Cumbria, reports that the peak flow was 'frequently substantially higher than expected'.

HYDROLOGICAL CHANGES

This already complex picture is further complicated by the emerging view that, in the years following land reinstatement, the soil/water processes change. As mentioned earlier, previous studies have mainly been carried out on plot-size sites; thus detection of any change in the total slope area was not possible. Again, Hawkins and Aljoe (1992) identified similar processes in spoil affecting groundwater. They suggest that the age of the spoil, particularly one with a high shale content, has a strong influence on its hydraulic conductivity. 'Large conduits and voids' had formed within 30 months of reinstatement.

A number of studies have noted that the runoff coefficient tends to be lower than expected (Kilmartin, 1994; Gilley et al., 1977; Hawkins and Aljoe, 1992). There appears to be a closer correlation with the longer term climatic conditions rather than the 5-day rainfall preceding the event. Kilmartin (1994) also notes other changes within the catchment that support the possibility that flow *through* the mine spoil contributes to the stream flow. These included: clay 'hummocks' appearing at the base of the slope, soil-pipe outlets and midslope 'ponding' of a clay/coal fines suspension.

In the years after reclamation, changes in soil development, slope evolution and drainage pattern are already taking place, posing different drainage problems from those recognised initially. It is possible that the topsoil storage on the upper slopes will decrease but this may be balanced by an increase in the ponding caused by local settlement and also an increase in the base of slope storage. A more effective pipe system may develop but may eventually be blocked by sediment. Uncurbed poor drainage will encourage the growth of excessive moss on the upper slopes and *Juncus* near the channel and in the hollows on the slope. Possible fissuring in the subsoil after restoration could substantially increase the hydrologically 'active' area.

These factors have important implications for the design of underdrainage. Any open pipes installed under the spoil will quickly become silted up, as has been found, with the ones at the crest of the slope eventually becoming exposed. Pipes installed downslope could just be adding to the natural soil-pipe complex so *increasing* the flood peaks.

CONCLUSION

Detailed examination of the hillslope hydrology on reclaimed opencast sites suggests an alternative explanation for the extreme peakiness of their hydrographs and the unpredictability of reclaimed catchment response. This involves a complex relationship between different slope sections, the hydrological performance of the wedge of permeable topsoil and the slope foot depositional area, a 'soil pipe' fissure system, surface and subsurface ponding on the mainly impermeable mine spoil subsoils, and rainfall intensity.

The 1991 UNECE Policies and Systems of Environmental Impact Assessments emphasised the importance of identifying environmental problems in advance. There is a need to reappraise the drainage of reclaimed land both in the short term, to avoid flooding downstream, and in the long term, to allow for natural changes in the catchment. In view of the instability of the spoil, it is rather optimistic to assume that the grass ley will continue to improve beyond the five-year management period. Kendle (1994) suggests that either bonds or the establishment of management trusts by developers are essential to ensure long-term management support. Ricks (1995) offers a programme for mine closure that extends the statutory 5-year 'active care' period to include a 2-year 'passive care' period. He suggests that, during this time, sampling and monitoring should be carried out on the surface runoff, mine-water issues, spoil runoff and plant growth, to show that the active care has been successful.

References

Addis, M.C. 1982. The environmental impact of opencast coal mining at the Woodgreens site in the Forest of Dean on surface runoff and its quality. M.Sc. thesis, Univ. Bristol, Bristol (unpubl.).

Addis, M.C. Simmons IG and Smart P.L. 1984. The environmental Impact of an opencast operation in the Forest of Dear, England *Journal of Environmental Management.* 19: 79–95

Adams, J.N. 1987. A case study of contractual restoration: Ffyndaff site, *Min. Eng. (London).* 146(307): 638–641, 643.

Aldous, P.J., Smart P.L., and Black J.A. 1986. Groundwater management problems in abandoned coal-mined aquifers: a case study of the Forest of Dean, England. *Quarterly Journal of Engineering Geology.* 19: 375–388.

Aljoe, W.W. and Hawkins, J.W. 1992. Application of aquifer testing in surface and underground coal mines. *Proc. Focus Conf. Eastern Regional Groundwater Issues.* Nat. Groundwater Assoc., Mass., pp. 541–555.

Allen, D.F. 1983. *Quarry Site Rehabilitation and After-use in Surface Mining and Quarrying.* Inst. Mining and Metallurgy, London.

Armstrong, M.J. and McGowan, M. 1984. Soil shrinkage values from opencast sites. ADAS Field Drainage Experimental Unit: Trumpington (unpubl. ms., cited in Bragg, N.C., 1982).

Aspinwall, R. 1976. Hydrogeology in land reclamation. *Proc. Land Reclamation Conf. In Grays, Essex,* Thurrock Borough Council, pp. 355–367.

Bird, S.C. 1987. The effects of hydrological factors on river suspended solids: contamination from a colliery in South Wales. *Hydrological Processes.* 1: 321–338.

Bish, G.M. 1986. Opencast coal mining in South Wales. *International J. Mine Water.* 5(3): 1–12.

Boorman, D.B. 1986. A micro-computer package to aid design flood estimation in the United Kingdom, *International Assoc. of Hydraulic Research. Conference on Computer Aided Design in Hydraulic and Water Resource Engineering, Budapest. July 1986, Proceedings:* 349–357.

Boorman, D.B. 1985. A review of the flood studies report rainfall-runoff model parameter estimation equations. *Institute of Hydrology (Wallingford, Oxon) Report.* 94: 1–58.

Boorman D.B., Acreman M.C.. and Clayton M.C. 1989. *A Study of Percentage Runoff on Scottish Catchments: A Report to the North of Scotland Hydro-Electric Board.* Institute of Hydrology, Wallingford: 25 pp.

Boorman, D.B. and Reed D.W. 1981. Derivation of a catchment average unit hydrograph. *Institute of Hydrology (Wallingford, Oxon) Report.* 71: 1–59.

Bragg, N.C. 1981. *Opencast Mining. A Review with Specific Interest in Research and Development Work on Restoration Techniques.* ADAS Field Drainage Experimental Unit, Trumpington: 6 pp.

Bragg, N.C. 1982. Land drainage and restoration of land after N.C.B. opencast mining, *Soil and Water* 10(2): 13–15.

Bragg, N.C. 1983. The study of soil development on restored opencast coal sites. *MAFF, ADAS, Land and Water Service, Research and Development, Field Engineering Report* RD/FE/09 TFS 773: 1–12.

Bragg, N.C. 1983b. Restoring land to agriculture. *Soil and Water*. 11. (4): 21–23.

Bragg, N.C. 1984. Reseeding effect on mole drains. *Soil and Water* 12(2): 10–11.

Bragg, N.C., Arrowsmith R. and Rands J.G. 1984. The re-examination of 'Radar' opencast coal drainage experiment. *U.K. Ministry of Agriculture, Food and Fisheries, Land and Water Service, Research and Development, Report* RD/FE/19: 6–10.

Bragg, N.C., Griffiths C., Jones, A. and Bell S. 1984. A study of the problems and implications of land drainage on reinstated opencast coal sites. *North of England Soils Discussion Group, Proc.*, 19: 37–59.

British Coal Corporation, 1988. *Report and Accounts 1987/88*. British Coal, Doncaster: 71pp.

British Coal Corporation, 1993. *Report and Accounts 1992/93*. British Coal, Doncaster: 71 pp.

British Coal Corporation, Opencast Executive. 1994. Personal Communication, Public Relations Department, Mansfield, U.K.

Carolan, I.G. 1985. Ransomed, healed, restored, forgiven—the restoration of opencast sites. *Soil and Water* 13 (1): 8–11.

Carolan, I.G. 1987. A drain on our assets: the underdrainage of restored opencast coal mining sites—review. *Min. Eng. London* 146: 644–646.

Caruccio, F.T., Geidel, G. and Williams R. 1984. Induced alkaline recharge zones mitigate acidic seeps. *1984 Symp. Surface Mining, Sedimentology and Reclamation*. University of Kentucky, pp. 27–36.

Charles, J.A., Hughes D.B. & Burford D. 1977. The effect of a rise of water table on the settlement of backfill at Horsley restored opencast coal mining site. pp. 229–251. In: *Large Ground Movements and Structures, Third Conference*. Univ. Wales Institute of Science and Technology. Cardiff.

Chong, S.K., Becker M.A., Moore S.M. and Weaver G.T. 1986. Characterization of reclaimed mined land with and without topsoil. *J. Environ. Quality* 15.(2): 157–160.

Clarke, L.B. 1995. *Coal mining and water quality*. IEA Coal Research London, pp. 81–88.

Collier, C.R., Pickering, R.J. and Musser, J.J. 1970. Influence of strip mining on the hydrologic environment of parts of Beaver Creek Basin, Kentucky, 1955–66. *U.S. Geol. Survey, Prof. Paper* 427-C: 1–80.

Curtis, W.R. 1974. Terraces reduce runoff and erosion on surface mine benches. *J. Soil and Water Conservation* 26 (5): 198–199.

Dierickx, W. 1987. Aspects physico-pedologiques due drainage. *Revue de l'Agriculture* 40 (4): 995–1001.

Downing, M.F. 1972. Drainage and erosion control. pp.36–43. In: *Landscape Reclamation I*. University of Newcastle upon Tyne, U.K.

Evangelou, V.P. and Karathanasis A.D. 1984. Reactions and mechanisms controlling water quality in surface-mined spoils, pp. 213–247. In: *Symposium on the Reclamation of Lands Disturbed by Surface Mining, Proceedings*. University of Kentucky and American Society for Surface Mining and Reclamation, Lexington, Kentucky.

Farmer, E.E. and Richardson B.S. 1976. Hydrologic and soil properties of coal-mine overburden piles in southern Montana, pp. 121–131. *4th Symposium on Coal Mining and Reclamation*. *Proc. National* Coal Association, Loiusville, Kentucky.

Fujita, K. and Shutto H. 1987. Experiment on the role of grass for infiltration and runoff processes. *J. Hydrology* 90: 303–325.

Gilley, J.E., Gee, G.W., Bauer, A., Willis, W.O. and Young, R.A. 1977. Runoff and erosion characteristics of surface-mined sites in Western North Dakota. *Ameri. Soc. Agric. Eng., Trans.*, 5: 697–704.

Goodman, J. and Haigh M.J. 1981. Evolution of slopes on abandoned spoil banks in Eastern Oklahoma. *Physical Geography* 2(2), 160–173.

Grandt, A.F. 1984. History of Reclamation Research. *Amer. Soc. Surface Mining and Reclamation: Symp. Reclamation of Lands Disturbed by Surface Mining, Proc.*, pp. 164–187.

Grandt A.F. 1951 cited in : Grandt, A.F. 1984. History of reclamation research, pp. 164–187. In: *Symposium on the Reclamation of Lands Disturbed by Surface Mining, Proceedings.* American Society for Surface Mining and Reclamation.

Gregory, K.J. and Walling D.E. 1973. *Drainage Basin: Form and Process.* Edward Arnold, London.

Haigh, M.J. 1978. Evolution of slopes on artificial landforms, Blaenavon, U.K. *University of Chicago. Department of Geography. Research Paper* 183: 1–293.

Haigh, M.J. 1979. Ground retreat and slope development on plateau-type colliery spoil mounds at Blaenavon, Gwent. *Inst. British Georgraphers, Trans.* (N.S.) 4(3): 321–328.

Haigh, M.J. 1988. Slope evolution on coal-mine disturbed land, pp. 3–13. In: *Environmental Geotechnics and Problematic Soils and Rocks*, Balasubramaniam A.S. Chandra, S., Bergado, D.T., and Prinya Nutalaya (eds). A.A. Balkema, Rotterdam.

Haigh, M.J. 1992. Degradation of 'reclaimed' lands previously disturbed by coal-mining in Wales: causes and remedies. *Land Degradation and Rehabilitation* 3(3): 169–180.

Haigh, M.J. 2000. Soil stewardship on reclaimed coal lands. *Land Reconstruction and Management* 1: 165–274.

Haigh, M.J. and Sansom, B. 1999. Rainfall simulation on reclaimed coal-lands in Doncaster, England *J. Balkan Ecology* 2(1): 62–70.

Haigh, M.J., Gentcheva-Kostadinova, Sv. and Zheleva, E. 1994. Evaluation of forestation for the control of accelerated runoff and erosion on reclaimed coal-spoils in South Wales and Bulgaria, pp. 127–149. In: *Environmental Restoration Opportunities Conf. (Munich, Germany), Proc.* Deutsche Aerospace/ADPA, Arlington, Va., 476 pp.

Hamon, W., Haghiri, F. and Knochenmus D. 1977. Research on the hydrology and water quality of watersheds subjected to surface mining. pp. 37–39. *5th Symposium on Surface Mining and Reclamation. Proc.* Louisville, Kentucky.

Hamon, W., Bonta, J.V., Haghiri, F. and Helgesen J. 1979. Research on the hydrology and water quality of watersheds subjected to surface mining: premining hydrologic and water quality conditions, pp. 70–98. *Surface Coal Mining and Reclamation Symposium.* McGraw Hill, New York.

Hawkins, J.W. and Aljoe, W.W. 1992. Pseudokarst groundwater hydrologic characteristics of a mine spoil aquifer. *Mine Water and Environment* 11(2): 37–52.

Herring, W.C. 1977. Groundwater re-establishment in cast overburden, pp. 71–87. *7th. Symposium on Coal Mine Drainage Research, Proceedings,* University of Kentucky. Lexington, Kentucky.

Hodder (1972) cited in Jensen et al. (1978) *op.cit.*

Hodgkinson R.A. 1989. The drainage of restored opencast coal sites. *Soil Use and Management* 5(4): 149.

Hodgkinson, R.A., Bragg, N.C.. and Arrowsmith, R. 1987. *Surface and Sub-surface Drainage Treatments for Restored Opencast Coal Sites.* Trumpington. Field Drainage Experimental Unit. Research and Development Service, 1.

Hodgkinson, R.A., Bragg, N.C. and Arrowsmith, R. 1988. Surface and Subsurface drainage treatments for restored opencast coal sites. *Proc. Seminar Land Restoration Investigation and Techniques.* Univ. of Newcastle-on-Tyne. British Coal Opencast Executive, Mansfield, Notts, pp. 3–15.

Jensen, I.B., Hodder, R.L. and Dollhopf D.J. 1978. Effects of surface manipulation on the hydrologic balance of surface-mined land pp. 754–761. In: M.K. Wali (ed.), *Ecology and Coal Resource Development*, p. 754–761. Pergamon, New York.

Jorgensen, D.W. and Gardner, T.W. 1987. Infiltration capacity of disturbed soils: temporal change and lithologic control. *Water Resources Bull.* 23(6): 1161–1172.

Kemp, J. 1988. Observations on the effect of modern restoration techniques on stream flow at Outgang opencast coal site, Cumbria. *Proc. Seminar Land Restoration, Investigation and Techniques.* Univ. of Newcastle-on-Tyne. British Coal Opencast Executive, Mansfield, Notts, 26–36.

Kendle, A.D. 1994. The restoration of species-rich grassland on reclaimed land. *Proc. Int. Land Reclamation and Mine Drainage Conf. and Third Int. Conf. Abatement of Acidic Drainage, Vol. 3:* Reclamation and Revegetation, pp. 57–66. US Dept. of the Interior.

Kilmartin, M.P. 1994. *Runoff generation and soils on reclaimed land, Blaenant, South Wales.* Ph.D thesis, Oxford Brookes University, Oxford, pp. 69–75 (unpubl.)

Kilmartin, M.P. and Haigh, M.J. 1988. Land reclamation policies and practices, pp. 441–467. In: S.C. Joshi and G. Bhattacharya (eds.). *Mining and Environment in India.* Himalayan Research Publ., Nainital, U.P.

Leopold LB and Maddox T. Jnr. 1953. The hydraulic geometry of stream channels and some physiographic indications. *United States Geological Survey Professional Papers.* 152: 1–56.

Lyle, E.S. (Jnr). 1987. Surface Mine Reclamation Manual. Elsevier, New York.

Massey, W. 1970. *A Note on the Validity of the Infiltration Experiment on Clay Soils.* Agricultural Development Advisory Service, Field Drainage Experimental Unit, Trumpington, U.K pp. 1–5..

Natural Environment Research Council 1975. *Flood Studies Report.* HMSO, London: 5 Volumes.

Ngah, S.A. Reed S.M. and Singh R.N. 1984. Groundwater problems associated with surface mining in the United Kingdom. *International J. Mine Water.* 3 (1): 1–12.

Nicolau, J.M. (1996). Effects of topsoiling on erosion rates and processes in coal-mine spoil banks in Utrillas, Teruel (Spain). *Int. J. Surface Mining, Reclamation and Environment* 10(2): 73–78.

Norton, P.J. 1984. Groundwater rebound effects on surface coal mining in the U.K. Bristol. *International. J. Mining and Metallurgy* X: 255–263.

Osterkamp, W.R. and Hedman E.R. 1979. Discharge estimates in surface-mine areas using channel-geometry techniques, pp. 43–49. *Symposium on Surface Mining Hydrology, Sedimentology and Reclamation, Proceedings.* University of Kentucky, Lexington, Kentucky.

Potter, K.N., Carter, F.S. and Doll E.C. 1988. Physical properties of constructed and undisturbed soils. *Soil Science Soc. America, J.* 52: 1435–1438.

Pringle, J. 1959. Opencast coalmining: the restoration problem. *Int. Symp. Soil Structure, Proc.,* pp. 64–72.

Prunty, L. and Kirkham D. 1979. A drainage model for a reclaimed surface mine. *Soil Science Soc. America, J.* 43: 28–34.

Prunty, L. and Kirkham D. 1980. Seepage vs. terrace density in reclaimed mineland soil. *J. Environ. Qual,* 9. (2): 273–278.

Rands, J.G. and Bragg, N.C. 1988. *The design of subsurface drainage systems in England and Wales of fine textured soils on which opencast mining operations have created extremely impermeable barriers at a shallow depth.* ADAS, Field Drainage Experimental Unit Internal Report. Trumpington, Cambridge, pp. 2–37.

Reed, D.W. 1987. Engaged on the ungauged; applications of the FSR rainfall-runoff method. *BHS National Hydrology Symp.,* 2: 2.1–1.19.

Reed, S.M. and Singh R.N. 1986. Groundwater recovery problems associated with opencast mine backfills. *International. J. Mine Waters* 5. (3): 47–74.

Ricks, G. 1995. Closure considerations in environmental impact statements. *Mineral Industry International* 1022: 5–10.

Riley, C.V. 1976. Surface mined land reclamation research and legislation—a paradox, pp. 106–115. *4th Symposium on Surface Mining and Reclamation, Proceedings.* National Coal Association, Louisville, Kentucky.

Ringler, R.W. 1984. Spoil aquifer resaturation following coal strip-mining. *American Society for Surface Mining and Reclamation, National Meeting (Kentucky,) Proceedings:* 274–291.

Ritchie, J.A. 1963. Earthwork tunnelling and the application of soil-testing procedure. *J. Soil Conservation, New South Wales* 19: 111–129.

Robinson, M. 1987. Agriculture drainage can alter catchment flood frequency. *BHS National Hydrology Symp.* 6 xx–yy.

Roels, J.M. and Jonker, P.J. 1985. Representativity and accuracy of measurements of soil loss from runoff plots. *Amer. Soc. Agric. Eng., Trans.,* 28 (5): 1458–1464.

Rogowski, A.S. 1980. Hydrologic parameter distribution on a mine spoil, pp. 764–780. *Symposium on Watershed Management, Boise, Idaho, Proceedings,* American Society of Agricultural Engineers, St. Joseph, Michigan.

Sanchez, C.E. and Wood M.K. 1989. Infiltration rates and erosion associated with the reclaimed coal-mine spoils in western New Mexico. *Landscape and Urban Planning* 17: 151–168.

Sawatsky, L., McKenna, G. Keys, M.J. and Long, D. 2000. Towards minimising the long term liability of reclaimed mine sites. *Land Reconstruction and Management* 1: 37–59.

Schroeder, S.A. 1987. Runoff curve number curve estimations for reshaped fine-textured spoils. *Reclamation and Reveg. Res.* 6: 129–136.

Schwartz, F.W. and Crowe A.S. 1987. Model study of some factors influencing the resaturation of spoil following mining and reclamation. *Journal of Hydrology.* 92: 121–147.

Scullion, J. 1984. The Assessment of Experimental Techniques Developed to Assist the Rehabilitation of Restored Opencast Coal-mined Land. Ph.D. thesis, Univ. College of Wales, Aberystwyth (unpubl.).

Scullion, J. and Mohammed, A.R.A. 1986. Field drainage experiments and design on former opencast coal mining land. *J. Agric. Sci., Cambridge* 107(3): 521–528.

Scullor, J. Mohammed, ARA and Ramsey, E.A. 1986. Statistical evaluation of drainage treatments in simple field trials with special reference to former opencast coal mining land. *Journal of Agricultural Science (Camb)* 107: 515–520.

Soil Conservation Service 1982. *Plant Materials for Use on Surface-mined Lands in Arid and Semiarid Regions.* United States Department of Agriculture.

Trafford, B.D. and Twocock, J.G. 1972. The drainage experiment on 'Radar' restored opencast coal site. *ADAS Field Drainage Experimental Unit Bull.* 72 (6): 1–22.

Toy, T.J. and Hadley R.F. 1987. *Geomorphology and Reclamation of Disturbed Lands.* Academic Press, Orlando Florida.

Trouart, J.E. and Knight R.W. 1984. Water quality runoff from revegetated mine spoil, pp. 16–35. *Symposium on the Reclamation of Lands Disturbed by Surface Mining, Proceedings,* University of Kentucky, Lexington, Kentucky.

Ward, A.D. Wells, A.G. and Phillips, R.E. 1983. Infiltration through reconstructed surface mined spoils and soils. *Amer. Soc. Agric. Eng., Trans.,* 5: 821–829.

Wright, D.L., Perry, H.D. and Blaser R.E. 1978. Persistent low maintenance vegetation for erosion control and aesthetics in highway corridors, pp. 553–583. In: F.S. Schaller and P. Sutton (eds.), *Reclamation of Drastically Disturbed Lands.* American Society of Agronomy/Crop Science Society of America/Soil Science Society of America, Madison, Wisconsin.

8

Case Study: Role of Grass Colonisation in Restricting Overland Flow and Erosion of Clayey Surface Coal-mine Spoils: Laboratory Results

Martin J. Haigh and Ben Sansom

Abstract

Soil loss increments from grass surfaces, on which little surface runoff developed, were two to three hundred times smaller than on unvegetated surfaces. Even after 6 months in the laboratory, infiltration rates into grass-covered spoil remained much higher than into unvegetated spoil. Infiltrated waters discharged along the junction between the topsoil and the relatively impermeable floor of the soil tray, much as they do at the junction between the applied topsoils and compacted clayey subsoils on reclaimed sites.

INTRODUCTION

Field tests suggest that there are major differences in the response of vegetated and unvegetated mine-spoil surfaces to overland flow (Nicolau, 1996; Kilmartin 2000, this Volume). This case study tries to demonstrate these differences by means of simple laboratory modelling.

SITES

This study examines spoil collected on land reclaimed after surface coal mining in the Heads of the Valley region of Wales (Haigh, 1978, 1992). These lands suffer, or have suffered, from serious erosion problems and their condition is one reason why surface coal mining remains controversial in Wales (Walley, 1994).

Land Reconstruction and Management Vol. 1, 2000; pp 159–164
ISSN 1389-2541
ISBN 90 5410 793 6
A.A. Balkema, Rotterdam, The Netherlands

Pwll Du Opencast was the first opencast coal-mine in Wales. It was opened by MacAlpines in 1942 and reclaimed in 1947/48. The site lies near Blaenavon, 325–500 m asl, on the north-eastern outcrop of the South. Wales Coalfield (51°31'N 03°07'W). The climate is mild, maritime and monthly mean air temperature ranges from 2.5°C to 15°C. Rainfall is 1420–1680 mm · yr^{-1}. The heaviest daily rainfall to be expected in one year is around 40 mm and 60 mm in ten years (Smith and Trafford, 1975). Soils are at field capacity 285–325 days every year. The soil moisture deficit (potatoes) ranges from 10 to minus 31 mm. The land is capability class 5 due to wetness (ADAS, 1990).

Premine geology consisted of Carboniferous/Pennsylvanian age Lower Coal Series strata: friable mudstones, coal shales and hard sandstones. Field surveys of the spoils on the disturbed coal lands and of artificial topsoils derived from these materials, found medium to heavy clay loams and occasionally silty clays. The spoils tend to be very stony with many cobbles of shale, sandstone and ironstone (Kilmartin and Haigh, 1997).

METHOD

Two soil trays (dimensions: 0.5*0.7*2.0*0.15 m) were taken to Pwll Du. The intention was to retrieve undisturbed slices of soil. The sample site selected lay on the margins of a dormant gully, where relatively well-grassed spoils (40% cover) lie adjacent to entirely unvegetated (0% cover) spoils. Slices of spoil, 15 cm thick were cut from two adjacent sections of a 28° slope on the site margins and transferred directly into the test trays. A physical distance of less than 5 metres separated the two samples. The grassed sample came from a (concave planform) midslope and the unvegetated from an adjacent (convex planform) part of the same midslope.

The soil trays were trucked to the laboratory and installed side by side in the sand-bed of a Plint TE91 Bank Erosion Channel (Fig. 1; also see Haigh and Kilmartin, 1987). This TE91 channel is designed to permit three-dimensional dynamic modelling of soil erosion and surface runoff processes. It consists of a 2.4 m long, 1.5 m wide tray capable of being tilted through 40° by hydraulic jack. The system generates surface runoff by means of a recirculating pump. Water can be pumped at rates between 5 and 30 l · min^{-1} across a hydraulic entry faring and baffle system, which allows the generation of thin uniform flows.

Despite care taken during transfer, apart from some grass tufts, the soil could not be transferred and relaid without some disruption. Consequently, it was necessary to allow the soil to resettle and restructure in the laboratory. The test-bed was levelled to the horizontal, the trays flooded to a surface depth of 1 cm, and the bed left for a week to drain.

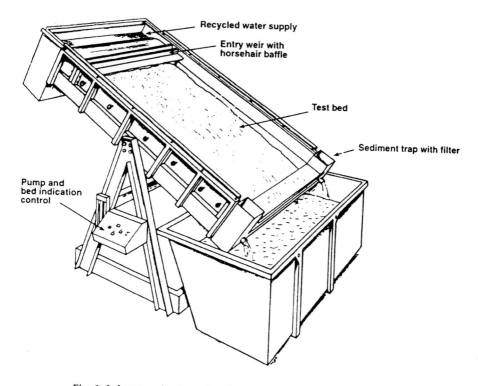

Recycled water supply

Entry weir with
horsehair baffle

Test bed

Sediment trap with filter

Pump and
bed indication
control

Fig. 1: Laboratory bank erosion flume (after Haigh and Kilmartin, 1987).

An initial series of runoff tests was conducted 2 weeks after collection and are recorded in a British Government information film (COI Films, 1993). The tests involved adjusting the inclination of each tray by 5° units, pumping 25 $1 \cdot min^{-1}$ across each tray in turn and recording the time taken for overflow and the amount of sediment eroded.

These trials found that, while the tray of unvegetated spoils supported surface runoff and quickly developed erosion rills, the tray with the grassed cover would not support surface runoff. Instead, the water ran through the spoils and flowed along the impermeable floor of the soil tray. Thus a major difference was evident between the vegetated and unvegetated spoils. However, the tests leave open the possibility that the result was an artifact of disturbance during sample collection.

To eliminate this possibility, it was decided to allow the two samples to rest, bed-in and settle for an extended period. Regular saturation of the soil bed continued for 3 months, after which the soil trays were allowed to drain and rest for a further 3 months. The final series of tests was undertaken 6 months after sample collection.

Prior to testing, both trays were saturated and allowed to drain to field capacity—a condition wherein there is no seepage from the base of the tray. Tests involved an input of 25 l · mm^{-1}. The trays were closed when overflow from the base had continued for 1 minute. The inclination of the soil tray was increased between each test: from 2° to 5° then 10° and 15°. Tests were conducted sequentially from the point at which drainage ceased.

Bulk density testing was conducted for both trays before and after the final test run, using the small ring method (Baize, 1992). Infiltration rates were recorded for both trays in advance of the test runs, using a simple single-ring constant-head infiltrometer.

RESULTS OF LABORATORY TESTS

Bulk density testing found that, on completion of the tests, the density of the spoil in the unvegetated tray was 1.3 g · cm^3 while that in the veg-etated tray was 1.2 g · cm^3. At field capacity, each sample contained 0.29 and 0.36 g · cm^3 of water respectively. The rate of water infiltration into the unvegetated spoils remained low at < 1mm· h^{-1} while that into the vegetated tray was around 80 mm · h^{-1}. Results from the runoff tests are presented in Table 1.

Table 1: Results of laboratory runoff tests

Laboratory	Unvegetated Tray	Unvegetated Tray	Grassed Tray	Grassed Tray
Slope: (Degrees)	Weight of eroded sediment (g)	Time to throughflow in seconds (depth of flow at weir/and midtray)	Weight of eroded sediment (g)	Time to throughflow in seconds (depth of flow at weir/and mid-tray)
2	60.4	30 (4/3 mm)	0	127 (9/19 mm)
5	1459.6	13 (3/3 mm)	0	116 (7/16 mm)
10	5755.7	7 (>1/>1 mm)	19.6	30 (3/4 mm)
15	3327.1	6 (>1/10 mm in rill*) [*Rill width: 65 Rill Depth: 9 mm]	10.2	24 (2/3 mm)

DISCUSSION

The tests demonstrate that both runoff and erosion develop more rapidly on unvegetated spoils than on spoils vegetated with grass. Soil loss records, which must be regarded as cumulative rather than sequential since the tests were run on the same soil-bed, suggest the characteristic exponential increase of soil loss with slope angle. They also demonstrate that soil loss increments from the vegetated tray, in which little surface runoff developed, were two to three hundred times smaller than on the non-vegetated tray.

Since the runoff tests are sequential, each test is conducted on a soil-bed already modified by erosion. Many of the more easily eroded particles have already been removed. So, the total volume of sediments recorded is less than would be the case if each test were the first in the series. Logistic difficulties meant that it was impossible to avoid this problem. Further, it should be noted that these tests are defined by runoff from the beds rather than input to the beds. Much greater volumes of water were required to generate the small sediment yield from the high infiltration capacity of the vegetated bed (see Haigh and Blake, 1989).

The test programme was discontinued at 15°. During this test, rills developed on the unvegetated soil bed. Of course, when runoff becomes concentrated in rills, running water is removed from large parts of the tray. Once the main process of rill incision is completed, the rate of sediment removal declines. Research into badlands geomorphology includes many reports of stabilised gully systems as witness to this process (see Haigh, 1978).

Finally, it should be noted that these tests employ surface wash in the absence of rainsplash. It is suggested that a great proportion of the sediment transported by sheet-flood is mobilised by rainsplash detachment (Luk et al., 1988). It is also recognised that sediment transport in overland flow is conditioned to a large measure by rainsplash induced turbulence (RIFT). During rainfall events, raindrops impact on the flow. In flows less than 4 mm deep, medium and large size droplets may punch right through the water to impact on the soil beneath. In deeper flows, the rainsplash impacts disturb the flow and create pockets of turbulence. In a heavy storm, there may be so many impacts that this turbulence may be the dominant characteristic of the flow (Kinnell, 1991; Kinnell and Wood, 1992). Such raindrop perturbed quasi-turbulent flow may be typical of natural sheet-wash. However, it is not a feature of these laboratory tests, which therefore tell only part of the story.

On vegetated soils surface waters infiltrate more rapidly. They run through the soil and flow at the junction between this and the impermeable floor of the soil tray. The pattern replicates that observed on reclaimed sites, where artificial, low-density topsoils overlie dense clayey subsoils (see Haigh et al., 1994). Nicolau (1996) confirms that infiltration rates into mine spoils increase dramatically with the development of vegetation. In his Teruel, Spain, case study, infiltration rates on a 3-year-old bank were just 20–25% those on a 7-year-old bank where grass was well established.

CONCLUSION

In sum, the final analyses confirm those of the initial tests. Even at 15°, it took 24 rather than 6 seconds to generate runoff from the vegetated tray. Runoff ran into and through the vegetated spoils. On the unvegetated spoils, much more ran across the surface and mobilised much more soil. Conditions in the vegetated tray mimic the conjunction of impermeable subsoil and permeable topsoil found on fully reclaimed sites. Infiltrated waters discharge along the junction between the topsoil and the relatively impermeable floor of the soil tray.

References

ADAS (Agricultural Development and Advisory Service). 1990. *Agricultural Land Classification: Physical Characteristics and Soil Forming Materials Survey. Report to the Public Inquiry into the Proposed Pwll Du Opencast Coal Site, Blaenavon, Gwent.* Gwent Country Hall, Cumbran, pp. 1–13.

Baize, D, (ed). 1992. *Referentiel Pedologique.* AFES/INRA, Paris.

COI Films (1993). *Perspectives: Repairing the Damage.* London: Central Office of Information Films: 28 min.

Haigh, M.J. 1978. *Evolution of Slopes on Artificial Landforms, Blaenavon, UK.* Univ. Chicago Geography Research Paper 183), Chicago, ILL., 374 pp.

Haigh, M.J. 1992. Problems in the reclamation of coal-mine disturbed lands in Wales. *Int. J. Surface Mining and Reclamation* 6:31–37.

Haigh, M.J. and Kilmartin, M.P. 1987: Teaching soil conservation in the laboratory using the "Bank Erosion Channel" flume. *J. Geog. Higher Education* 12 (2): 161–167.

Haigh, M.J. and Blake, L. 1989. Soil degradation on reclaimed surface-mined coal-land in South Wales: an hypothesis, pp. 12–19. In: M. Penkov (ed.). *Melioratsi Pochv: Int. Scientific Conf. Soil Melioration, Proc.* Higher Institute of Architecture and Civil Engineering, Sofia, Bulgaria, 310 pp.

Haigh, M.J., Gentcheva-Kostadinova, Sv. and Zheleva, E. 1994. Evaluation of forestation for the control of accelerated runoff and erosion on reclaimed coal-spoils in South Wales and Bulgaria, pp. 127–149 In: *Environmental Restoration Opportunities Conf.* (Munich), *Proc.* ADPA, Arlington, VA. 476 pp.

Kilmartin, M.P. 2000. Hydrological management of reclaimed opencast coal-mine sites. *Land Reconstruction and Management* 1: 137–158.

Kilmartin, M.P. and Haigh, M.J. 1997. Initial statistical comparisons between natural and reclaimed soils in the South Wales coalfield. *Lesobudska Misul* (in press).

Kinnell, P.I.A. 1991. The effect of flow depth on sediment transport induced by raindrops impacting shallow flows. *Amer. Soc. Agric. Eng., Trans.,* 34 (1): 161–168.

Kinnell, P.I.A. and Wood, J.T. 1992. Isolating erosivity and erodibility components in erosion by rain-impacted flow. *Amer. Soc. Agric. Eng., Trans.,* 35 (1): 201–205.

Luk, S., Abrahams, A. and Parsons, A. 1988. *Field Experiments on Overland Flow and Sediment Transport.* Univ. Toronto (Video:20 min)

Nicolau, J.M. 1996. Effects of topsoiling on erosion rates and processes in coalmine spoil banks in Utrillas, Teruel (Spain). *Int. J. Surface Mining, Reclamation and Environment* 10 (2): 73–78.

Smith, L.P. and Trafford, B.D. 1975. Climate and Drainage. *Ministry of Agriculture, Fisheries and Food, London, Technical Bulletin* 34: 62–63.

Walley, C. 1994. Carving out a future? *Rural Wales,* Summer/1994: 22–24.

9

Soil Stewardship On Reclaimed Coal Lands

Martin J. Haigh

Abstract

*Natural soils are tightly coupled integrations of physical and organic
processes. They are biocybernetic systems. Their character and quality
may be constrained by their source geology, climate, topography and time,
but they are regulated and controlled by their biota. By contrast, the soils
on reclaimed lands are none of these things. They are artificial or
immature constructions of soil forming components that have yet to
achieve the self-regulating properties of natural soils. The art of soil
management on reclaimed lands is to help these soils develop the
characteristics of natural soils, while preventing their deterioration due
to the negative aspects of mine spoils and restored topsoils. These negative
properties result from the physical and chemical characteristics of the
spoils and from damage sustained during land reclamation and by
subsequent land management. They include: poor physical characteristics,
notably soil compaction and structural instability; poor chemical
characteristics, including infertility, toxicity and acidity; poor
hydrological characteristics, including poor water-holding properties and
often accelerated runoff and erosion; and poor biological characteristics,
including a weakened and depleted soil biological system. This chapter
describes the main features of the soil system on reclaimed land, dealing
with its eight components in turn. It evaluates the problems that result
from human interventions including trafficking, forestation, soil
reconstruction etc. It discusses soil compaction, a common cause of land
degradation on sites reclaimed after hard-coal surface-mining, advising
that successfully reclaimed land should have soil bulk densities that do
not exceed $1.6 \ g \cdot cm^3$ within 50 cm of the surface and $1.8 \ g \cdot cm^3$ within
100 cm. The value of evaluating soil strength by means of the*

Land Reconstruction and Management Vol. 1, 2000; pp 165–274.
ISSN 1389-2541
ISBN 90 5410 793 6
A.A. Balkema, Rotterdam, The Netherlands

e-penetrometer index is considered. Soil acidity, in the surface metre reclaimed lands, should be constrained between pH 3.5 and 8.5. The causes of acid mine drainage, both physical and microbiological, are described. The biotic processes are accomplished by autotrophic acidophilic bacteria that create leachates that can have pH 2.0. This leachate mobilises metals contained by the waste rock, especially iron, that are precipitated in streams as a toxic orange sludge. Techniques of AMD prediction and control are evaluated. ARD/AMD control has four foci: chemical inhibition of acid generation; isolation of the acid generating rock; suppression of the bacterial catalysts of ARD reactions; and chemical or biological treatment of runoff. Many reclaimed lands suffer additional contamination by metals. The sources of contamination include; acidification; air pollution and fertiliser application—especially the use of sewage sludge. The threats posed by different metals at different concentrations are described and some of the critical thresholds for recognising legal contamination are listed.

I. THE SOIL SYSTEM

'... soil not only contains biological activity, but actually is a living resource... soil [is] the most important resource... the basis and capital resource for agricultural, forestry, industrial, recreational and infrastructural use. Never in historical times has the soil been more threatened by degradation and irreversible destruction on a global scale than in this century...the destructive processes are accelerating more rapidly than the restorative efforts carried out by numerous farmers, experts, planners, researchers—in short people like us. Hans Hurni (1991): President's Report—The Soil as a Living Resource. *World Association of Soil and Water Conservation, Newsletter* 7(2): 1.

Soils are a Living Resource

Treated properly, most soils renew, even improve, their quality. Treated improperly, they suffer progressive degradation. They lose their fertility, their structure and ultimately, even their skeleton of soil particles is washed or blown away. In general, the objective of soil stewardship is to preserve a soil resource that with judicious use of additional available resources of water, favourable climate, plants and technology can meet the present and future needs of the land-user.

On reclaimed land, the focus of soil stewardship is the reinstatement of soils as a self-sustaining, or at least sustainable, living system. Frequently, this task must be accomplished through revitalising topsoils that have been collected and stored from the pre-mine environment and that have suffered serious degradation during years of storage. Sometimes, this task must be accomplished by building a new soil profile in spoil materials discarded during the mining process. Soil plays a vital role in reclamation. It provides the substrate for life and it mitigates hydrological processes. As Volk, (1998: 109) remarks: 'Without soil, every sunny day would be a drought, every rain a flashflood'. As Kilmartin (2000, this book) demonstrates, this describes the hydrological character of much reclaimed land.

The central problem on reclaimed land arises from the fact that the soil organic system is either absent or seriously depleted. Natural soils are complex dynamic, evolving, biologically controlled open systems. They develop in response to the vertical and horizontal movements of organisms, organic materials, water, chemicals and soil particles. The self-creation and evolution of natural soils is conditioned by five factors:

- Biological processes, the activities of organisms that grow on and in the soil, together with the chemical impacts of their secretions and waste products.
- Geological conditions, which through weathering, determine the character, availability and amount of material from which the soil skeleton is created.
- Geomorphological and hydrological processes that are determined by the position of the soil in the landscape system, which affect the amount of erosion, drainage or deposition it has experienced.
- Climatic and microclimatic processes that control the activities of the soil biological system and that encourage physical processes through wetting/drying, freeze-thaw and the potential for erosion by wind, rainsplash and runoff.
- Time, which determines how long all these factors have had to act, and which may allow the soil to retain relict features created by environmental conditions in the past.

Finally, soils may have been altered by land-use (Bridges and de Bakker, 1997). Most soils have been disturbed by human activities and, in many cases, these activities have been sustained for many decades through cultivation, forest farming, or grazing.

By contrast, the soils that exist on reclaimed land, if they exist at all, are the mixed remnants of natural soils that evolved in a completely different context.

- The vegetation that covers reclaimed land is rarely the same as that in the landscape it replaces and the microecosystem (microcoenosis) in the soil is severely depleted after its period in storage. This is why stockpiled topsoils in storage berms, like those at Turija 'D' in Serbia's Kolubara, have the greatest amount of organic matter and nutrients in their surface layer and much less at depth (Institute of Forestry Belgrade, 1997: 41).
- The geological foundation of the soils is different. The natural strata of their lower layers have been replaced by fractured mine spoils with different physical and probably different chemical characteristics.

- The topographic context is different. On a natural hillslope, even in an agricultural field, there are differences between the soils found at the top of a slope and those found at the foot. These are conditioned by differences in drainage, sediment supply, microclimate, vegetation etc. (cf. Gerrard, 1981). By contrast, no-one segregates stockpiled soils by topographic site and no-one knows if the soil that is relaid upon a site evolved at the hill-crest or hill-foot. Most agencies do not even segregate the layers of the original soil profile, Usually, the best that can be done is to spread the stored soil mixture evenly, like a carpet.
- The climate and moisture conditions differ since the soils are relaid in a new geomorphological, hydrological and microclimatic context.
- The soil is immature. Soils on natural sites have structures and properties that result from many decades, sometimes many centuries, of evolution on site. The soil that exists on reclaimed land is freshly deposited and has had little time to adapt to its circumstances. Research in East Europe suggests that it may take a century or more for key soil processes such as the recycling of organic matter to become fully functional on new lands. Most reclaimed coal lands are rather younger.
- Finally, the new or reinstated soil may be subjected to an entirely different land-use, one that is determined by negotiation between the mining contractor and local community, rather than by any consideration of the intrinsic qualities of the new land.

The end-result is a soil that is a mismatch between its character and circumstances and which has a depleted or disturbed soil organic control system. Inevitably, this soil system is in a weakened state and actively engaged in the processes of adjustment to its new context. In many cases, this involves the construction of a new soil profile for the new environment. These processes can be observed on sites with reinstated topsoils as well as those where there is no topsoil. It involves the gradual evolution of a new soil ecosystem, new organic horizons, new eluvial and indurated layers, new structures that appear within and sometimes contradict the original horizons of the reclaimed land. These, of course, tend to be rather simple. There may be the restored or manufactured layer of artificial topsoil. There may be some kind of subsoil, often created by ripping and mixing of soil-forming materials and mine spoils. There is the weathered upper surface of the mine spoils that grades downwards into unaltered mine spoils beneath (Fig. 1: p 180).

Classification of Soils on Reclaimed Lands

The nature of the morphological differences between the soils on reclaimed land and those in nature can be controversial. It is hard to distinguish between true pedogenic features and artifacts of the land reclamation process. This is one reason why soil surveyors find it difficult to include the soils on reclaimed sites in conventional soil taxonomic systems.

The case study in Box 1 describes a field survey in which professional soil surveyors, using standard British soil reporting techniques (Hodgson, 1976), were sent out to tackle a landscape that included large tracts of coal-mine-disturbed land. This work was undertaken as part of a project that sought resources of soil-forming materials for the future reclamation of a proposed new mine at Pwll Du Wales. This had the distinction of being the largest opencast coal-mine plan ever submitted in Wales and resulted in Wales longest-ever Government Public Enquiry, which is less important than the fact that these results provide an official viewpoint on the differences between natural soil characteristics and those of disturbed lands.

Box 1. Case Study: Differences between Natural Soil Characteristics and those of Coal-mine Disturbed Lands, Pwll Du, Wales (Kilmartin and Haigh, 1997)

In 1989, the British Government's Agricultural Development and Advisory Service (ADAS) were contracted to survey the 635.5 ha site of a proposed new opencast coal-mine at Pwll Du, S. Wales (Haigh, 1995). This site included many areas that had already been disturbed by upwards of 40 years surface coal-mining activity and more than 200 years of underground mining and patch-working (Haigh, 1978).

Professional soil surveyors, from the British Government's Agricultural Development and Advisory Service (ADAS), mapped this area and recording 10 descriptive variables from 106 soil profile pits. The variables identified at each soil profile were: the number of horizons in each profile; depth; texture (assessed by hand texturing); clay content of each horizon; the degree of stoniness; size of stones in the uppermost soil horizon; soil colour at the soil surface, at 25 cm and at 100 cm depth; and finally, topographic slope angle. Soil structure was recorded with reference to the *'Soil Survey Field Handbook'* (Hodgson, 1976) and the soil colour according to *'Munsell Soil Colour Charts'*.

Statistical comparison of the data for the artificial and natural soils allows the rejection of the null hypotheses—that there is no significant difference between the two data sets. Table 2 records the results of t-testing. The *t*-tests' confirm that, if a threshold probability of $p = 0.005$ is accepted, then most of the properties recorded for the two populations differ significantly and if $p = 0.02$ is accepted, then all differ significantly.

Discriminant analysis (Rao's Method) is a standard multivariate statistical technique that is often used to identify those variables which best differentiate between classes (Klecka, 1975). In this case, the key discriminating variables were the colour and clay content of the surface layer. Subordinate variables include depth and stoniness, slope angle and number of soil horizons (cf. Kilmartin, 1994).

Contd.

Contd. Box

Table 1: Comparison of undisturbed and disturbed soils at Pwll Du: Surveyors Records (ADAS, 1990)

Variable	Undisturbed soil	Disturbed soil
Number of horizons	2.85	1.15
Depth of top horizon (cm)	37.6	79.7
Texture of top horizon	Peat	Clay loam with occasional silty clay
Clay content of top horizon	Usually 0 and rarely over 27%	27%–30%
Stone content in top horizon	< 6% with little variation	55% mean with a range of 45–65%
Stone size in top horizon (cm)	< 2	6–20
Colour at the surface	Very dark brown/black	Brown/Dark yellowish-brown
Colour at 25 cm	Dark brown/Very dark grey/Greyish-brown	Dark grey/Dark yellowish-brown/ Dark greyish-brown
Colour at 100 cm	Grey/Dark grey/Yellowish-brown	Dark grey/Dark yellowish-brown/ Dark greyish-brown
Slope angle (°)	0–3	0–3 (rarely 3–7)

Table 2: T-test of undisturbed and disturbed soil records

Variable	Undisturbed-Mean [s.d.]	Disturbed-Mean [s.d.]	t [$p =$]
Number of horizons	2.85 [1.04]	1.15 [0.36]	9.22 [0.000]
Depth of top horizon (cm)	37.63 [29.37]	79.71 [25.67]	– 7.16 [0.000]
Texture of top horizon (index)	10.94 [2.03]	7.50 [0.79]	9.55 [0.000]
Clay content in top horizon (index)	0.11 [0.55]	2.29 [1.14]	– 13.36 [0.000]
Stone content in top horizon	0.60 [1.42]	4.59 [1.88]	– 12.14 [0.000]
Stone size in top horizon (cm)	0.56 [1.13]	3.12 [1.15]	– 10.79 [0.000]
Surface colour (value/chroma)	2.29 [0.64]	4.33 [0.85]	– 13.70 [0.000]
Colour at 25 cm	3.32 [1.63]	4.16 [1.10]	– 2.71 [0.008]
Colour at 100 cm	4.96 [1.83]	4.11 [1.10]	2.49 [0.014]
Slope Angle (°)	1.38 [0.62]	1.77 [1.02]	– 2.45 [0.016]

Additional bivariate analysis confirms the greater variability of the natural soils. A Pearson product-moment correlation matrix generated for each population displayed far greater numbers of significant correlations between variables in the subpopulation of natural, undisturbed soils than in the disturbed soils of the coal-mined lands. This suggests that these natural soils are better-integrated systems (cf. Chorley and Kennedy, 1971).

Taxonomy of Soils on Reclaimed Lands

Soil taxonomists find it difficult to incorporate the soils on reclaimed lands, and former coal-mine-disturbed lands in general, into their standard systems. The Russian/FAO system escapes the problem by simple description. The soils on reclaimed lands are Anthrosols, man-made soils, and those on reclaimed lands are 'Technogenic recultisols' (FAO-UNESCO, 1988). Anthrosols are defined as soils in which human activities have resulted in profound modification or burial of the original soil horizons through removal disturbance of surface horizons, cuts and fills, secular additions of organic and mineral materials, ploughing, irrigation etc. (FAO-UNESCO, 1988 and Table 3).

Agreement on the subdivision and classification of Anthrosols has yet to be reached. In Bulgaria, Class Anthrosol is included in the base soil taxonomy at the highest rank and three types are recognised: Agrogenics, Urbogenics and Technogenics (Gencheva 1994, 1995). At the third level, these are defined according to their specific diagnostic horizons, such as the distinctive layer of applied topsoil on reclaimed lands, and the litter and humus layers that often rest abruptly on weathered spoils on sites reclaimed without topsoiling. The FAO-UNESCO (1988) recognises several additions and variants to this scheme as well as the Agric diagnostic horizon. This is an anthropogenic layer created by cultivation, which contains significant amount of illuvial silt, clay and humus. It is found

Table 3: Properties of anthrosols (after Bridges and de Bakker, 1997)

Site Characteristics:

- Sites have been altered technogenetically by levelling, terracing, embanking etc. (cf. Haigh, 1978: 1–10).
- Soil water relationships have been modified by drainage, irrigation adjustments of the water table etc. (cf. Haigh, this book).

Profile Characteristics:

- Where present, surface horizons are relatively deep (300–500 mm in depth) and enriched with organic materials.
- Disturbed and mixed soil layers may lie abruptly over, or adjacent to, material of a completely different nature.
- Horizons of the original soil profile may be buried under other sediments.
- Chemical conditions often include artificial enrichments with plant nutrients, especially phosphorus and potassium, and other chemicals and metals, often as contaminants.
- Often includes dense layers, pans due to trafficking or ploughing. These may also be associated with gleying or mottling, often with iron.
- Subsoils are compacted, often through the use of heavy machinery. Some formerly compacted horizons may show signs of shattering by deep cultivation.

immediately below the plough layer, where these materials form thick lamellae, lining pores and structure faces (Soil Survey Staff, 1994).

Another diagnostic feature of evolving soils on sites that are not suffering continuous disturbance is an organic epipedon, a surface layer of undecomposed or partially humified litter. The incorporation of organic matter within the soil has been considered a first indication of soil formation. In young soils, organic matter accumulates rapidly at the soil surface. Later a more balanced condition obtains and the organic matter is worked deeper into the soil profile. In the USA, studies of forest soils have indicated that a topsoil horizon (A1) with a steady organic content develops in 20–40 years, while on floodplain grasslands, mature-looking soil develops in 100–120 years (Nielsen et al., 1975). It took 50 years for North Dakota mine spoils to develop a 150 mm deep topsoil layer (cf. Haigh, 1978).

An abrupt lower boundary to the organic layer has been considered a diagnostic feature of many Anthrosols, especially those on coal-mine-disturbed lands. Abrupt boundaries have been described between 4–6 cm (Taranov et al., 1979); at 2 cm and from 3–5 cm depending on the period of management (Varela et al., 1993); at 5 cm under self-established grass and forest and at 20 cm arable land (Banov et al., 1995). Even on 70-year-old surface-coal-mine spoil banks, the main part of the organic matter is concentrated in the litter (Nakariakov and Trofimov, 1979). The boundary often indicates that the litter has poor contact with the soil and is prone to drying out, which retards microbiological assimilation (Schafer et al., 1980). However, it is also one reason that the Americans tend to classify these soils with the Entisols.

Calling mine spoils 'Entisols' recognises that mine spoils are more than just loose debris and that they are immature true soils. This is still a debated point. However, 20 years ago, Merriwether and Sobek (1978) sought to classify mine spoils with the Orthents—the Entisols of recently eroded surfaces. These soils are characterised by truncation, such that all original horizons are removed. They have textures that tend to the bimodal with peaks in the (fine) sand fraction and in the coarse gravel range. Their organic content declines rapidly with depth. Mine-spoil-derived orthents, Spolic Orthents, or Spolents were supposed to share at least three of nine other distinctive properties (Table 4).

American soil taxonomists have rejected this system, so it is mainly of historical interest. However, it provides a useful practical morphological approach to the kinds of materials in the new soils forming on coal-mine-disturbed lands, in the absence of topsoils. In fact, it is easy to find soils that conform to one or the other of the subgroups. Indeed, different subgroups may be identified at different points on almost any hard-coal-derived spoil in Britain. The main problem is that most new soils in coal spoils are not Entisols. They may begin as Entisols but the instability of many of their primary components is such that many

Table 4: A heterodox taxonomy of Entisols-derived-from-mine-spoils (after Merriwether and Sobek, 1978)

Taxonomic Group	Diagnostic Characteristics
Order: **Entisol**	'Recent soils' that are azonal and cover around 12.5% of the land surface. These mineral soils have no distinct *pedogenic* horizons within 1015 mm (40″) of the soil surface. They do have horizons, includng an organically enriched layer at the soil surface, but not eluvial or illuvial horizons (Steila, 1976: 83).
Suborder: **Spolent**	Most of the properties of spolents are a direct consequence of the technical aspects of their creation. Spolents share three of the following 9 properties: 1. Disordered coarse fragments. [Gravel size fragments are more than 10% by volume and more than 50% of these have long axes between 20 and 250 mm in length that are oriented at least 20 degrees form any recognisable plane in the section]. 2. Colour mottling occurs and has no relationship with depth or texture of the material. 3. Coarse fragments have splintery, rather than smooth edges. 4. Voids may be bridged by coarse fragments. 5. A coarse surface stone layer is immediately superimposed above a layer (soil crust) with an unusually high percentage of fines (> 2mm). The depth of this layer ranges from 25–100 mm. 6. Pockets of dissimilar materials are mixed irregularly in the profile. 7. The profile includes artifacts: worked metal, iron, glass etc. 8. The profile contains fragments of the appropriate worked mineral, such as coal. 9. The distribution of oxidisable organic carbon in the profile is irregular.
Great Groups of the **Spolents**	**Cryospolents** are found in cold areas where frost action is the dominant process. **Tropospolents** are found in tropical environments. **Udispolents** are found in udic environments. These are environments wherein the soil is neither dry for 90 cumulative days nor for 60 days when the temperature at 508 mm depth exceeds 5 degrees. **Usticspolents** are found in drier ustic environments. **Xerospolents** are found in dry environments, where the soil contains no plant-available moisture for half the year and the soil temperature at 508 mm is higher than 5 degrees.
Subgroups of the **Udispolents**	These are described according to the composition and morphology of the materials thus: **Pyrolithic udispolents** have at least 50% of the coarse fragments altered by burning of the carbolithic materials. These fragments are pink, red, cream or yellow.

Table 4: *Contd.*

Taxonomic Group	Diagnostic Characteristics
	Matric udispolents have less than 10% coarse fragments < 2 mm.
	Plattic udispolents have more than 65% of the coarse fragments (< 2 mm) as sandstones.
	Fissile udispolents have more than 65% of the coarse fragments (< 2 mm) as finely laminated (> 2 mm) shales that are breaking into smaller fragments.
	Shlickig udispolents have more than 65% of the coarse fragments as shales or mudstones with lamina (< 2 mm) and show no reaction in 10% HCl.
	Kalkig udispolents have at least 65% of the coarse fragments (< 2 mm) calcareous enough to react with 10% HCl.
	Carbolithic udispolents have more than 50% of the coarse fragments with a streak of 3 or less in the Munsell charts. The reason is that they have many coals or carbolithic shales and mudstones.
	Typic udispolents are not dominated by any one rock type, have more than 10% coarse fragments (< 2 mm) and have not undergone combustion.
	Lithic udispolents are thin veneers of less than 500 mm above unaltered bedrock. etc.

undergo accelerated weathering. Eluviation of the products of this weathering, a true pedogenetic process, leads to the creation an illuvial horizon, which is known to land reclamation workers as the 'clay cap'. Down (1975), working on deep mine spoils in Britain, found that in spoils 178 years old, there was little sign of rock breakdown below 20 cm. However, this clay layer, which can form within weeks of spoil being dumped, extends through depths 300–500 mm into the soil. In fact, the existence of this illuvial clay layer is one of the most abiding features of soils on reclaimed lands. Its recognition, however, remains controversial, even though its formation can easily be demonstrated in the laboratory (Haigh and Blake, 1989; Kilmartin, 1994). Certainly, in the field, its presence is often obscured or conflated with technogenic artifacts such as traffic and plough pans in the profile (Bragg 1983, Bragg et al., 1984).

Soil Systems

Soil systems include three phases: solid, liquid and gaseous and four fractions: mineral, organic, aqueous and gaseous (Koorevaar et al., 1983: 1). Natural soil systems arise as the integration of eight functional components (Table 5). Almost all of these have new and unusual characteristics on reclaimed sites.

Primary Grains and Rock Fragments

Soils are biocybernetic systems. Their development is controlled and steered by their biota (cf. Van Breemen, 1983). So, functionally, the least

Table 5: Eight components of the soil system

Components of the Soil System

Control System:
- soil biota: the sum of its living organisms

Soil Architecture

- soil structural spaces, cracks and pores—the factory floor where the activity of the soil takes place
- soil aggregates of organic and mineral particle—the soil structure that preserves the soil pores

Transportation Systems

- soil waters
- soil atmosphere

Building Blocks, Cement and Structural Stability Regulators

- primary grains and rock fragments—the soil skeleton of larger particles (< 2 mm)
- clay minerals and other secondary weathering products
- organic and mineral-organic (chelate) inclusions

interesting and important components of a natural soil system are the primary particles that make up the soil skeleton. These fragments provide the bulk of the soil but they are the least chemically and biologically active. Despite this, because they are the most overt and easily measured parts of the soil, and because of the way thinking in soil science and soil mechanics has developed, they are given undue prominence in the literature (Feda, 1992). This prominence continues to affect thinking in soil conservation; to this day, soil conservationists worry more about the loss of the soil skeleton than the more active components of the soil system (cf. Shaxson, 1995). Of course, this situation is very different in many coal-mine spoils. Here, the organic system is weak while the primary particles are very active. In mine spoils, many primary particles, especially shales and mudstones are very unstable. This instability lends them a very important role in soil formation and soil compaction.

In natural soils, primary particles are produced by physical breakdown. Mechanical breakdown, often caused by frost action or wetting (hydration), reduces rocks first into smaller and smaller fragments and finally into individual mineral grains, most often quartz. Since natural soils evolve through long periods of time, the primary particles that survive in the coarse fractions do so because they resist breakdown, either physical or chemical. Quartz grains and their aggregates, which have little reaction with the surrounding soil water in most environments outside the tropics, tend to dominate the larger size fractions. Mica, which is much less stable, tends to disappear into the finer fractions. Inside these finer

Table 6: Increase in specific surface of soil particles with decrease in size (after Verigo and Razumova, 1963)

Particle Diameter (mm)	Specific Surface (cm^2.g)
2.0–0.2	45
0.2–0.02	446
0.02–0.002	4458

fractions, the proportions of quartz decrease and those of weathering products tend to increase. The finer silts and clays are dominated by weathering products such as amorphous hydrated/anhydrated silicic acid, and clay mineral crystals (Kachinskii, 1958; Plyusnin, 1964).

As weathering proceeds the surface area of the grains increases (Table 6). This increased surface area allows greater opportunity for chemical attack and it is chemical weathering that produces the smallest size particles. Clay minerals are, of course, radically different structures to their larger cousins. In addition, electrochemical rather than mechanical processes dominate their behaviour. This textbook tale, of course, may be altered by characteristics of the source bedrock. This may, for example, supply clay-size fragments (cf. Dang Xhi et al., 1994).

The degree of size reduction also remains a key determinant of the name given to a soil by soil scientists. Most soils are given texture-based names that are determined by the size of their constituents. They are sands, silts or clays. A very few are named for their organic component, which is usually relatively small by volume. More are named for the relative balance between the sand and clay particles. If their behaviour is dominated by sand, they are light loams. They are heavy loams if their behaviour is dominated by clay. The huge surface-to-volume ratio of clays makes them electrochemically active. Their charged surfaces tend to stick together, aided by surface water, and as a result they are cohesive. The traditional field test to divide a heavy loam from a light loam involves trying to roll a millimetre-thick ribbon of soil and curl it into a ring. If an unbroken ring can be formed, clay dominates and the soil is a heavy loam. If the ribbon is hard to form and tends to disintegrate, then sand dominates and the soil is a light loam (Kachinskii, 1958).

The engineering definition of a soil extends to all non-cohesive deposits, including mine spoils. However, mining engineers distinguish between mine spoils that act as if the entire mass were loose rock and those that act as if they were entirely soil (Dawson and Morgenstern, 1995 and Fig. 3). The division separates rock-like deposits, which have less than 30% passing the 25 mm sieve, from soil-like spoils, which have more than 30% passing the 25 mm sieve. More traditionally, engineers have characterised a soil by the size of the largest primary particle in the smallest 10% of the soil. This is D_{10}, the 'effective particle size'. This index has been linked to the geotechnical strength of a soil and also to its capacity to transmit water. The effective conduit diameter of a soil is defined as $D_{10}/5$. These measures ignore the fact that most soil primary particles are integrated within complexes called soil aggregates and do not function independently. However, soils are named and classified according to their effective particle diameter. A variety of textural classifications of soils have been used (Table 7).

The particle size distribution of many mine spoils (including the reinstated topsoils on some newly reclaimed sites) is bimodal. The first mode is dominated by particles of relatively large size. On many newly reclaimed sites, these may be large gravel and cobbles. However, they may be large grains, visible to the naked eye, sands, and rock fragments, or those smaller grains, visible only under a hand lens, called silts. Many soil scientists distinguish between soil particles larger then 2 mm, often called the soil skeleton, and those smaller, which are considered the chemically and biologically active body of the soil (Gentcheva, 1995). The second mode dominates heavily trafficked reclaimed sites, sites where a thick topsoil has been established and sites constructed from initially fine textured materials. This is dominated by fine silts and clay minerals. These provide a matrix between the larger cobbles.

In natural soils, the condition of the larger soil particles and the final soil texture is stable. The process of reduction by weathering tends to be long and slow, measured in centuries. This is not the case for the soils and spoils that contain mine stones, coarse cobbles brought to the surface by mining (Haigh and Blake, 1989). These cobbles, broken from deeply buried strata, weakened by excavation and decompression, often include rocks that have little or no capacity to resist weathering. Many of the poorly consolidated shales and mudstones that are associated with hard-coal-deposits are easily ruptured by frost action and some are unstable when immersed in water (Hawkins and Pinches, 1992). Exposed to wetting and freeze-thaw processes for the first time, they undergo accelerated physical weathering (cf. Dang Xhi et al., 1994; Taylor, 1988; Taylor and Spears, 1970). In extreme cases, newly exposed mudstone cobbles break down to their constituent clays within weeks by slaking and dispersion (Taylor, 1974; Haigh and Sansom, 1999). Harder shales and ironstones may survive for years. The net result is that initially loose, bouldery, mine spoils become flooded by fines released through accelerated weathering. Thus, the particle-size distributions of many spoils, and also artificial topsoils that include admixtures of mine stones, change quickly with time.

This situation makes that fundamental type of physical soils analysis, particle size analysis, something of a nonsense (British Standards Institution, 1975). The fragility of freshly exposed minestones is such that they are peculiarly vulnerable to the laboratory procedures used for particle size fractionation. These involve mechanical shaking, washing and sometimes the settlement of particles in a water-filled sedimentation tube.

Table 7: Some soil textural classifications (after Kachinskii 1958; Brady, 1984)

Effective diameter (mm)	United States Department of Agriculture	United States Public Roads Admin.	Russia: Kachinskii (1957)	Effective diameter (mm)	British Standards Institute	International Society of Soil Science	Atterberg
> 3	Gravel	Gravel	Stones	> 6	Gravel	Gravel	Coarse Gravel
3–2			Gravel	6–2			Fine Gravel
2–1	Very coarse sand	Coarse sand		2–0.6	Coarse sand	Coarse sand	Coarse sand
1–0.5	Coarse sand		Coarse sand	0.6–0.2	Medium sand		Medium sand
0.5–0.25	Medium sand	Fine sand	Medium sand	0.2–0.06	Fine sand	Silt	Fine sand
0.25–0.10	Fine sand		Fine sand [Physical sand > 0.1]	0.06–0.02	Coarse silt		Sandy silt
0.10–0.05	Very fine sand		Fine Sand [Physical clay < 0.1]				
0.05–0.005	Silt	Silt	Coarse [> 0.01] silt / Medium [< 0.01] silt	0.02–0.006	Medium silt	Clay	Loamy silt [<0.1]
0.005–0.002		Clay	Fine silt	0.006–0.002	Fine silt		Colloidal silt and clay
0.002–0.0002	Clay		[>0.001] / Clay [<0.001]	0.002–0.0002	Clay		
< 0.0001			Colloids	< 0.0002		Colloids	Colloids

These procedures are perfectly capable of accelerating the breakdown of a gravelly minestone-based deposit into silt and clay.

Soils and weathered spoils on reclaimed land include both clay, which makes them prone to aggregation, and primary particles, mudstones and shales, which are water unstable (Berkovitch et al., 1959). Consider the two basic approaches to particle size analysis (e.g. Smith, 1971). The material can be mechanically sieved while dry. However, dry sieve data from weathered mine spoils tends to be polluted by microaggregates of clay and the mechanical disturbance disrupts some of the more fissile shale particles. The alternative is to wet the sieve and to use a dispersant to break down the aggregates. However, here, data is corrupted because the breakdown of the soil aggregates is accompanied by the breakdown of water-unstable primary particles. Thus, it is possible to attain standardised results by rigorous adherence to standard procedures but hard to gain results that reflect the situation in the field.

A third approach to particle size analysis, routinely used for fine sediments and also employed by many automated systems, is based on the rate of particle sedimentation in water. However, in addition to their water-unstable primary particles, coal spoils include very high-density ironstones and very low-density coals—which means that normal sedimentation procedures cannot be applied. Some coals and their associated colloids have a density that is less than one. Fines and even much larger particles never settle from suspension.

Despite this, several workers define mine spoils in terms of their particle size. The American Office of Surface Mining, Reclamation and Enforcement (OSMRE, 1992), following Strohm et al. (1978), suggests that spoils with more than 20% of their materials passing the 4.75 mm sieve, or with slakeable rock, should be considered soil-like rather than rock-like. Canada's Dawson and Morgenstern (1995) refine this by suggesting that spoil piles that have more than 20% (10–30%) sand (> 2 mm), behave geotechnically more like soil than rock materials. CANMET (1977) reports that Canadian mine spoils have sand contents that range from 10–78%.

The problems of conducting laboratory tests on these spoils, however, do lead directly to some insights into the role of primary particle instability in pedogenesis on reclaimed lands. Box 2 describes a study of initial soil forming on unvegetated, pyritic, deep-mine spoils at Cadeby Washings, Doncaster, England (Haigh and Sansom 1999. Dyckhoff et al., 1993). The study explains the differences between the soil profile cartoons in Fig. 1. Naturally, the process of accelerated breakdown of unstable mine stones is that which liberates the fines that undergo eluviation or accumulate *in situ* to create the dense illuvial layer found in many reclaimed lands at circa 300–500 mm depth (Haigh and Blake, 1989). At Pernik, Bulgaria, similar results emerge. Heavily weathered spoils, in the 0.-20 cm layer,

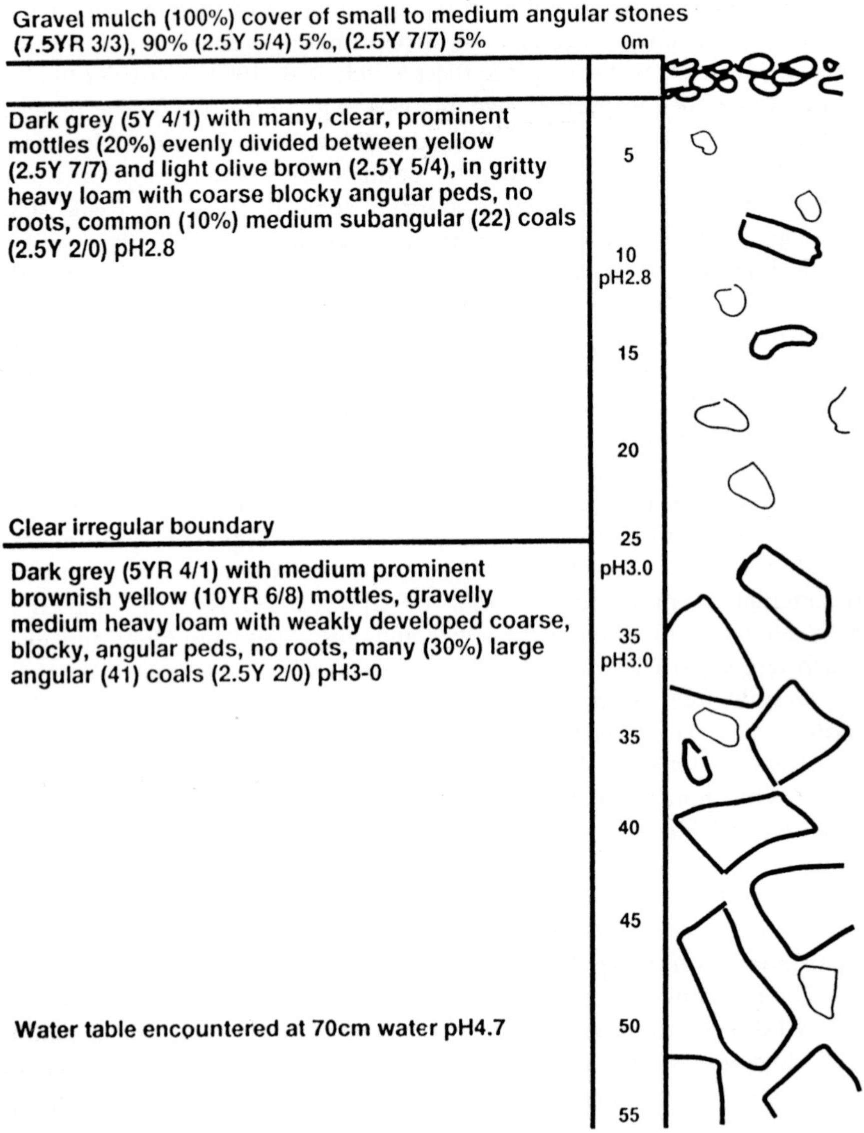

Gravel mulch (100%) cover of small to medium angular stones
(7.5YR 3/3), 90% (2.5Y 5/4) 5%, (2.5Y 7/7) 5%   0m

Dark grey (5Y 4/1) with many, clear, prominent
mottles (20%) evenly divided between yellow   5
(2.5Y 7/7) and light olive brown (2.5Y 5/4), in gritty
heavy loam with coarse blocky angular peds, no
roots, common (10%) medium subangular (22) coals
(2.5Y 2/0) pH2.8   10
                                                               pH2.8

                                                                  15

                                                                  20

Clear irregular boundary   25
                                                               pH3.0
Dark grey (5YR 4/1) with medium prominent
brownish yellow (10YR 6/8) mottles, gravelly
medium heavy loam with weakly developed coarse,   35
blocky, angular peds, no roots, many (30%) large   pH3.0
angular (41) coals (2.5Y 2/0) pH3-0

                                                                  35

                                                                  40

                                                                  45

Water table encountered at 70cm water pH4.7   50

                                                                  55

Fig. 1: Field sketches of two freshly laid mine spoils at Cadeby Washings, Doncaster, England (after Dycroft et al., 1993).
a) Cadeby Colliery, older regraded deep-mine coal spoils on 4° slope with failing plantations of stunted *Betula pendula* (2-3 m high).

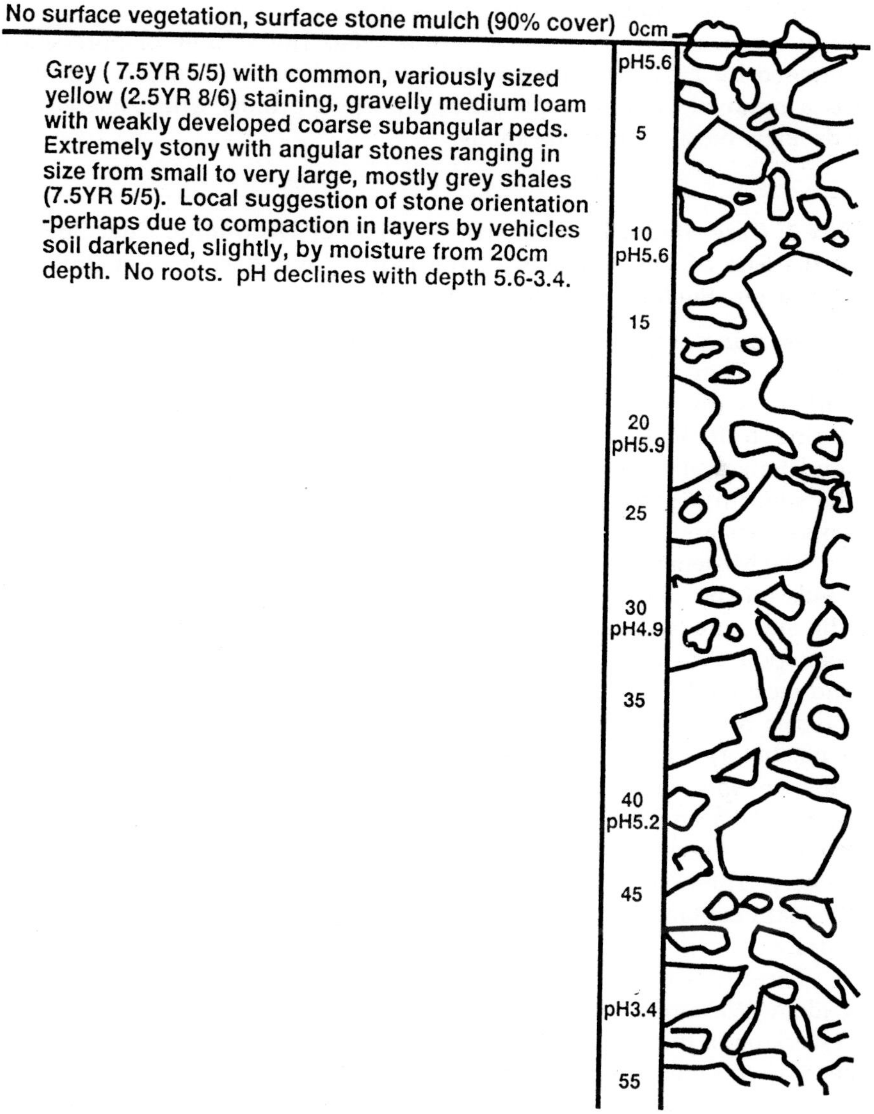

No surface vegetation, surface stone mulch (90% cover) 0cm

Grey (7.5YR 5/5) with common, variously sized yellow (2.5YR 8/6) staining, gravelly medium loam with weakly developed coarse subangular peds. Extremely stony with angular stones ranging in size from small to very large, mostly grey shales (7.5YR 5/5). Local suggestion of stone orientation -perhaps due to compaction in layers by vehicles soil darkened, slightly, by moisture from 20cm depth. No roots. pH declines with depth 5.6-3.4.

pH5.6

5

10
pH5.6

15

20
pH5.9

25

30
pH4.9

35

40
pH5.2

45

pH3.4

55

b) newly regraded (1993) reworked deep-mine spoils on upper convexity (5°) Cadeby Washings.

contain 66% clay, while unweathered spoils, those buried deeper than 100 cm, contain 52% clay (Kandev, 1987; cf. Gentcheva and Hubenov, 1988).

Pionke and Rogowski (1979: 93) offer similar findings to those in Box 2 from their examination of the effects of the first application of water to newly reconstructed spoils at Kylertown, Pa. The volume of pore space decreased by 25%. The change in density declined with depth but there was a removal of some topsoil into the spoil profile, which altered the hydrologic behaviour of the spoils.

Box 2: Case Study: Particle Size Analysis of Deep-mine Coal Spoils at the Site of the 'Museum of the Earth' Land Reclamation Project, Doncaster, England (Haigh and Sansom, 1999).

British mines spoils undergo a fairly predictable pattern of textural change. They begin as ungraded mixtures dominated by cobbles, perhaps with included layers of fines created by particle disruption during trafficking. After a few months, less stable cobbles, mainly mudstones, siltstones and shales, have broken down and transformed the spoil to a scatter of more resistant rock fragments: harder shales, sandstones and ironstone concretions, in a matrix of silt and/or clay.

The Museum of the Earth Project is developing an area of deep-mine spoils on the site of the former Cadeby Colliery, Doncaster, England. This site's mudstone geology is buried beneath up to 5.5 m of deep-mine coal spoil. At Cadeby Tip, these spoils have been reshaped, topsoiled and reclaimed with grass. Where this land has been carefully managed, a dense turf and soil mat survive. Elsewhere, the site has degraded to a patchwork of grass and bare spoil. Locally, acid drainage and soil pipe erosion have destroyed the vegetation exposing gullied mine spoils. Adjacent, the Cadeby Washings are mantled with fresh spoils from active reworking of some never-reclaimed coal tips from the same colliery. These spoils were being washed to remove some of the coal they still contain and the residue was being used to recontour the land surface.

Three sets of samples were subjected to particle size analysis. At Cadeby Washings, samples of unweathered spoil exposed during the reworking of the coal spoils but prior to treatment were taken. A second set was collected from the top 10 cm of shale after the shales had been treated by washing and immediately after they had been deposited from dump trucks. A third was collected from the rainfall-simulation test site, where the soils had been in place for around 6 weeks. Prior to washing and reclamation treatment, newly exposed spoils were non-cohesive and dominated by cobbles. Dry-sieving found just 2% passing the 2-mm sieve. However, high percentages of these cobbles were water unstable.

Contd.

Contd. Box

When the same samples were wet sieved, the percentage passing the 2-mm sieve soared from 2 to 99%. The second set of samples, collected immediately after washing, were less vulnerable to breakdown. Moving from dry to wet-sieving raised the per cent passing a 2-mm sieve from 48 to 59%. The third set of samples showed how, after several weeks, the situation is changed when surface weathering and pressure release begin to affect the spoils. Wet sieving the third sample, which had been exposed for some weeks, found the per cent passing the 2-mm sieve rising from 13 to 28%. By contrast, analyses of old spoils of the 1960s-reclaimed Cadeby Tip, found the per cent passing the 2-mm sieve rising from 10% to 90%, indicating the further progress of accelerated weathering.

Clay Minerals and Other Secondary Weathering Products

The clay minerals are some of the more exotic and interesting components of the soil. Clay minerals may be liberated by the breakdown of geological mudstones and clays. However, in natural soils, most are the products of chemical weathering. They dominate the smallest fractions of the soil that lie beyond the circa 0.002 mm minimum of materials produced by mechanical weathering. Clay mineral crystals differ dramatically from the macromolecular strands that characterise organic materials and the compact spheroids of mechanically rounded quartz sand grains. Many are presented as thin, mica-like flakes with major to minor axis ratios of between 10:1 and 50:1. Other clay minerals may be polygonal, e.g. kaolinite, or they may resemble needles, e.g, atapulgite (Table 7).

Clay minerals are composed of hydrated aluminium, iron or magnesium silicates that are combined in complex crystalline structures. The most common clay minerals are those of the kaolinite group (kaolinite, halloysite), montmorillonite group (montmorillonite, beidelite and nontonite) and the hydromica group (illite). The main building blocks of the clay minerals have been portrayed as layers of silica tetrahedra, involving silicon cations and 4 oxygen anions and aluminium octahedra involving 6 hydroxyl or oxygen anions and cations of aluminium, or sometimes magnesium, calcium etc. (Table 6). The two layers have similar dimensions and are held together by covalent bonds.

Clay minerals include a large proportion of the surface area in a soil. Their specific surface is huge compared to the sand and gravel fraction. It is 700–800 $m^2 \cdot g^{-1}$ for montmorillonite, 300–500 $m^2 \cdot g^{-1}$ for mica and 5–100 $m^2 \cdot g^{-1}$ for kaolinite (Koorevaar et al., 1983: 15). The clays, along with the other fine fractions of the soil (<2mm) cover the surfaces and contact points between the particles of the soil skeleton. They are the main

Table 8: Properties of some common clay minerals

Clay mineral	Structure	Properties
Kaolinite	One layer of silica tetrahedra and one of aluminium octahedra very tightly bonded together.	A very common clay mineral that is stable and not prone to shrink-swell. Particles tend to clump into thick domains. The geological source rock tends to be crystalline, such as granite.
Illite	One layer of aluminium octahedra between two of silica tetrahedra with a sheet of water molecules between the layers. Potassium provides a strong binding link between the layers.	Irregular flakes that deform more easily than kaolinite. Illite is not prone to shrink-swell except in the absence of potassium. Marine clays and micaceous rocks are common geological sources.
Montmorillonite	One layer of aluminium octahedra between two of silica tetrahedra and sometimes with a weakly bound layer of water temporarily. Iron or magnesium may replace aluminium in the octahedra and aluminium the silica in the tetrahedra.	Presented as irregular flakes or fibres, the weak bond between the layers and negative charge resulting from the molecular substitutions allows the mineral to gain and lose a layer of water. The mineral has a high shrink-swell capacity. Geological sources include volcanic ash and the weathering of ferromagnesium rocks in humid tropical environments.
Chlorite	One layer of aluminium octahedra between two of silica tetrahedra with a layer of aluminium providing a strong bond.	Presented as irregular plates, the mineral is non-expansive and forms from the weathering of well-drained soils and micaceous rocks.

link between the solid and liquid phase of the soil. This link is dominated by electrical forces that govern processes such as shrink-swell, aggregation, flocculation and dispersion in the solids, as well as the movements of soil-water and air.

The growth of the crystal lattice of clay minerals is never entirely finished. There is always a free electrical charge at the edge of the lattice. If, during the growth of a clay crystal, particular ions are in short supply in the soil solutions, they may be replaced by another of similar size—a process called isomorphous substitution—which becomes a permanent mutation of the lattice. In general terms, for clay at pH 7 or less, crystal faces develop with negatively charged faces while the edge of the lattice may have a positive charge at pH < 7 (Koorevaar et al., 1983:16).

Thus, normally, in the absence of molecular bonding, clay minerals tend to repel each other except in edge-to-face contact. Normally, clays

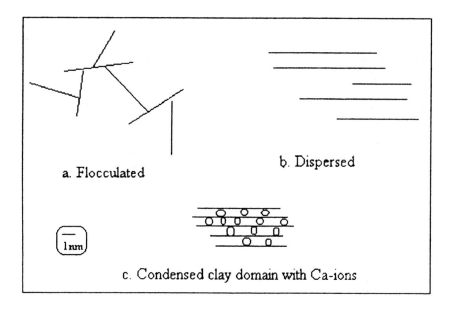

a. Flocculated

b. Dispersed

1 nm

c. Condensed clay domain with Ca-ions

Fig. 2: Clay mineral structures in the soil.

settling out of suspension will do so with edge-to-face crystal contacts. This is called a flocculated structure (Fig. 2). Sometimes the flocculation of clays is inhibited by the presence of dissolved salts or other clay minerals; for example the presence of small amount of montmorillonite and illite can inhibit the flocculation of kaolinite.

Clays that have been reworked, remoulded or disturbed, as by rainsplash, tend to adopt a more parallel arrangement. This face-to-face structure, which may result from pressure or disruption of the clay bonds by wetting, is called dispersion.

Nature tends to preserve electrical neutrality, so the negative charge of clay minerals tends to be compensated by positively charged cations. These may be adsorbed onto the lattice or swarm in the adjacent soil solution. Water is also an electrically charged dipole. The surface tension meniscus is a manifestation of its free electrical charge. Water molecules also tend to bond with the clay surfaces.

If the distance between clay crystals is strongly reduced, as by desiccation, cations in the water layer that surrounds them may be forced into the central plane. These positively charged cations then bond with the negative charged ions on the crystal face and a very large mutual bond is constructed. If the cation is polyvalent, such as calcium, then the attraction may lead to a stable condition called plate-condensation. In this circumstance, the clays form a dispersed structure called a clay domain. A clay domain is a group of clay crystals that are so close together that they behave as a single unit in water and will not break up unless mechanically disturbed. When water enters between the crystal layers, they may swell as a whole.

Some of the clay minerals may be useful in reclamation. In Bulgaria, Dimitrov (1992) found that natural zeolite (clinoptilolite), applied in conjunction with $N_{200}P_{200}$, improved the structure and fertility of spoil dumps at the Maritsa East lignite mine. The effect seemed to last for 6 years but during this period crop yields of wheat, sunflower and fodder were similar to those from undisturbed lands. The technique is widely used. Zeolites play an important role in cation exchange. They preserve their structure and enhance this function in the face of attack by humic acids (Filcheva et al., 1998).

Organic and Mineral-organic Inclusions

A particular feature of coal-mine spoils is the presence of mineral carbon that is associated with the coals themselves. The amount of unextractable mineral carbon in mine spoils can be very high. In the Middle Urals, Nakariakov and Trifomov (1979) suggest that the amount of unextractable carbon normally ranges between 50% to 67% and reaches 75% of the total carbon in newly formed soils on coal wastes. These high scores are echoed in studies from reclaimed lignites in the Maritsa-Iztok region (Banov et al., 1989). In alkaline conditions, lignite coal facilitates humus formation and accumulation (Banov et al., 1989). The humic acids in oxidised coals are thought to be similar to those found in natural chernozems and those from lignites are sometimes used to create synthetic spoil conditioning agents (Kommissaroff, 1979; Haigh, 1993).

Bulgarian researchers hope to reduce the depth of topsoil required for agricultural recultivation by using a biostimulator 'Biomine', an organic compound extracted from lignite, in combination with other additives (Nicolova et al., 1992). A Russian equivalent is 'Biogum' produced by the microbiological oxidation of lignites. This creates a material which has a relatively high level of plant-available minerals and a vigorous microbiological flora. After an initial application of 120 t $\cdot ha^{-1}$, annual applications of 13.5 t $\cdot ha^{-1}$ proved sufficient to increase maize/corn yield 23% over that achieved by normal PK fertiliser on a low-humus natural chestnut soil. Similar results were achieved in other crop experiments (Serebriakov and Sharshovets, 1992).

Sohnitzer et al. (1969) argue that pH is a significant determinant of the interactions between humic acids and clay minerals. Komissaroff (1979) suggests that this may come about through the immobilisation of a mobile 'easily peptised complex' by a stable adsorption on mineral surfaces. This is probably due to hydrogen bonding between humic acids and oxygen atoms on the clay mineral surface. It is thought very unlikely that humic acid macromolecules penetrate the internal layers of clay minerals. However, detailed investigations have yet to be carried out.

Soil Organic System

By definition, soils are not inanimate collections of minerals,. They are living ecosystems. The term for a soil that contains no life is *regolith*, if it forms in place through physical and chemical weathering, and *sediment*, if it is moved to another location. In academia, Civil Engineers may use the term to describe any deposit that is not a rock but the other soil sciences include life in their definition. Volk writes that 'without life there is no soil' (Volk, 1998: 109).

Natural soils are living systems. Van Breemen (1993) defines the soil as 'a biotic construct favouring net primary productivity'. Every true soil is a tightly linked integration of physical, chemical and most importantly, biological processes. 'Soil is where the crucial links churn in the cycles of biochemicals' (Volk, 1998: 109). In a real sense, soil is alive and it is the life in a living soil that controls the quality of that soil and, in general, which makes the soil a better place for living things. Biological processes tightly coupled with physical and chemical ones, organise nutrient supply, chemical buffering, soil density, porosity, aeration and water-holding capacity. They effect soil aggregate stabilisation, detoxification and soil self-creation. In nature, soil organisms are so thoroughly integrated within the soil physical system that some soil scientists conceive the soil as a kind of 'living organism' or superorganism (Van Breemen, 1993).

On reclaimed land, soil life plays an ambiguous role. It is the process by which land reclamation specialists hope to achieve a self-sustaining final result. It is also the process that is most liable to destroy that reclaimed land.

Natural soils are full of living organisms. Each cubic centimetre of healthy soil may contain a billion bacteria, perhaps 750 million actinomycetes or ray fungi, a million protozoa, 10,000–100,000 algae and 100,000–1 million fungi with anything from 10–100 metres of fungal hyphae. In addition, each hectare may contain up to 1 million earthworms, 7 million arthropods—mites, springtails, insects—as well as larger organisms such as snails, slugs, mammals etc. and, in addition to the aforesaid, a healthy soil is densely packed with root hairs and their associated life forms (Steila, 1976; Brady, 1984). Each of these organisms has a special place in the soil system, an ecological niche, a role in an efficient recycling of organic materials and waste products, and often in the transmutation of minerals weathered from the rocks or chemicals exchanged with the atmosphere. Like a forest, an ocean and the whole biosphere, natural soils are capable of a sophisticated self-regulation.

The key area in a living soil is the root zone or rhizosphere. This may have 100 times more biological activity than the rest of the soil (Brady, 1984). It is one reason why healthy soils are associated with active plant growth. So, it is not simply that plants grow well on healthy soils. To a large degree, healthy soils are created and preserved by the organisms that live in them; hence, by and large, the more organisms, the better the soil.

However, micro-organisms are also responsible for hazards in mine wastes, not least the production of acidity, acid mine (rock) drainage [AMD/ARD] and the consequent mobilisation of metals—topics which are discussed later. The implication is that the micro-organisms in mine wastes need to be examined and also managed (Ledlin and Pederson, 1996). To date, microbiological studies of soils have dealt with determination of species, measurement of soil biological indicators such as respiration or enzymatic acitivity, and evaluation of the system's capacity of break down and mineralise organic materials; (Visser, 1985). In almost all cases, the productive microcoenoses on reclaimed lands have been shown to be impoverished, inactive and ineffective compared to natural soil systems.

In the spoil banks of Serbia's Kolubara complex, soil pH ranges from 8.1 down to 2.7 in the presence of coals. The soil microbiological community is greatest in the zones of moderate to neutral pH and this also has the highest number of actinomycetes, suggesting rapid mineralisation. Where pH is extreme, the microcoenosis is restricted to a low number of ammonifying bacteria and both oligotrophic micro-organisms and actinomycetes are absent. Biological activity is low in clayey spoils, except for some grey clays and gleys, which have high numbers of ammonifying bacteria (Institute of Forestry Belgrade, 1997). When vegetation colonises these spoils, the population of micro-organisms increases rapidly, especially in the rooting layer (<300-mm).

Organic matter accumulation is often touted as one indicator of increasing soil quality on reclaimed lands. Certainly, litter accumulation affects soil hydrology. The litter from *Quercus* spp. can absorb water up to nine times, *Fagus* spp. eight times and *Pinus* spp. five times its weight. It is also suggested that mild humus absorbs more water than acid humus because more readily water absorbent (Verigo and Razumova, 1963).

Box 3 provides another illustration. This summarises results from a project involving research into the workings of the organic system and initial soil-forming processes on surface coal-mine spoil banks created by brown-coal extraction in western Bulgaria.

Box 3: Case Study: Impact of Forestation on Organic Accumulation and Microbial Action in Surface Coal-mine Spoils, Pernik, Bulgaria (Filfcheva et al., 1998; Noustorova et al., 1998).

In Bulgaria, around 80% of lands disturbed by surface coal mining await reclamation. Topsoils from the pre-mine environment are reserved for the recultivation of gently sloping agricultural benches. Steeper slopes are reclaimed through forest fallowing.

The impacts of two tree species on initial pedogenesis are examined on opencast coal-mine spoil banks near Pernik. Thirty-year-old test plots have been planted for 25 years with: 1) Black Pine (*Pinus nigra Arni*) and 2) Black Locust (*Robinia pseudoacacia* L.); a third control plot remains largely unvegetated. Litter accumulation under pine is 7.68 t·ha^{-1} (pH 4.86) compared to 5.53 t·ha^{-1} (pH 7.07) under black locust. Beneath, an organomineralic zone at the surface of the mine spoils is 16.18 t·ha^{-1} (pH 6.76) under black pine and 8.21 t·ha^{-1} (pH 7.25) under black locust. Litter originating from the deciduous (broad-leaved) species undergoes more rapid degradation and mineralisation than that from the conifers (Vos and Stortelder, 1992:328). This explains why there is a lower volume of leaf waste beneath black locust than black pine despite black locust litter having a higher ash dry weight (7–15% vs 2.5–3.5%) (Vos and Stortelder, 1992). The total amount of litter is tiny compared to that of a mature forest (Remezov and Pogrebynak, 1965: 114).

The more rapid transformation of black locust litter creates more (and more mobile) organic substances that migrate further into the mineral profile. However, all the major components of soil organic matter are present: humic acids, fulvic acids and unextractable organic carbon. The three sites share high levels of common carbon, which is attributable to the presence of coal as much as to pedogenesis. However, total organic carbon and humic acid is greatly enhanced under the trees.

Black pine supports a greater microflora, both in the litter and the organomineralic layer beneath the litter; a greater microflora is found in the 0–10 cm layer of the spoils under black locust. Non-sporous bacteria dominate. The spore-forming bacilli are typical of natural forest soils but their variety and numbers are much reduced. *Bacillus cereus* dominates the control plot and that planted to black pine. *Bacillus mycoides* dominates under the black locust. These results confirm the positive impacts of forestation on initial soil-forming processes in surface coal-mine spoils and also that the creation of mature soils by forest fallowing is a long slow process. The most prevalent components of the

Contd.

Contd. Box

bacillic microflora are typical of natural forest soils, but their variety and numbers are much reduced (Mishustin and Mirsoeva, 1968). However, forest biological recultivation establishes the preconditions for self-sustaining natural soil development.

Research in Germany indicates that the annual increase of common organic carbon on reclaimed coal lands should average about 0.0055% over the first 30 years (Insam and Domsch, 1988). However, results from other contexts report rates as low as 0.018%. In North America, Anderson (1977) reports accumulation rates of 28.2 $g \cdot m^2 \, yr^{-1}$, while Schafer et al. (1980) give 135 $g \cdot m^2 \cdot yr^{-1}$ for the 6-year-old mine spoils and 45.4 $g \cdot m^2 \cdot yr^{-1}$ for 50-year-old spoils under grass. It is difficult to compare units. However, accumulation is clearly affected by the climate, the productivity of the vegetation cover and the age of the site, with organic accumulation at the surface being greater on more immature sites.

One British study suggests that the amount of organic carbon in topsoils (stored from the premining environment then restored after mining) may return to normal within 5 years (Harris et al., 1989; cf. Banov et al., 1989). Felton and Taraba (1994) confirm that the rate of organic matter degradation in a reconstructed B horizon on reclaimed land was far lower when stockpiled soils were used than when that same soil was hauled directly from its natural location. Work elsewhere indicates that it may take many decades to restore the equilibrium of the soil organic system. Bagautdinov (1996) considers 100 years the minimum period for the complete recycling of humus in either a Grey Forest Soil or Chernozem (also Schafer et al., 1980). This may explain why levels of carbon in restored soils can drop rapidly when such soils are brought into cultivation (Tate, 1985).

Table 9 compares the variation of total organic carbon (%) with depth in some from surface coal-mine sites in the USA (Illinois, Montana) and Bulgaria (Pernik, Maritsa-Iztok—lignite) (Thomas and Jansen, 1985; Schafer et al., 1980; Filtcheva et al., 1998, Banov et al., 1989, 1995). In general, the sites show a general pattern of organic enrichment of the surface horizons, which declines with depth. This overall pattern is most clearly displayed under the sites reclaimed to grass and deciduous trees. It is less clear on the Pernik site reclaimed to pine, perhaps because of relative inactivity of the microbial system (Schafer et al., 1980). Compared to the natural soils, roots in the mine spoils with both natural and artificial topsoil covers tended to be found nearer to the surface and to be more obviously associated with coarse fragments (Flege, 2000, this book). Roots were

Table 9: Total organic carbon in soil profiles on reclaimed coal lands

Site	Age (years)	Vegetation	Depth:				
			0–5 cm	5–10 cm	10–20 cm	40–60 cm	90–120 cm
Bulgaria: Pernik	30	None	2.36	2.32	1.89	2.26	4.40
Bulgaria: Pernik	30	Pine	4.31	4.46	4.49	3.56	4.98
Bulgaria: Pernik	30	Deciduous	3.12	2.67	2.49	3.01	3.86
Bulgaria: Maritsa-Iztok	20	Deciduous	3.70		0.94 (5–20 cm)	0.22 (20–40 cm)	
Bulgaria Maritsa-Iztok	10	Deciduous	1.56		0.16 (5–20 cm)	0.23 (20–40 cm)	
USA: Illinois	50	Deciduous	1.80	0.77	0.36	0.23	0.18
USA: Illinois	55	Deciduous	1.34	1.12	0.58	0.83	0.25
USA: Illinois	30	Grass	0.88	0.60	0.42	0.20	0.12
USA: Illinois	64	Grass	1.24	0.39	0.18	0.14	0.10
USA: Montana	50	Grass	3.5	1.1	0.45	0.30	0.18
USA: Montana	6	Grass	2.0	0.9	0.21	0.12	0.11
USA: Montana	Natural	Grass	6.0	1.2	0.95	0.60	0.21

infrequent in the undisturbed and unweathered spoils. However, Pedersen et al. (1978) attribute the lack of roots to high density and lack of moisture as well as lack of soil structure and nutrients.

Chemical and Biological Stimulants

Coal spoils and the soils on most reclaimed lands are often seriously deficient in plant nutrients. Part of the problem is the inactivity of the soil organic system, which does not achieve sufficient fixation and mineralisation of nutrients. However, while some mine spoils contain minerals and carbon compounds that encourage biological production, many others include compounds that neutralise those available plant nutrients that may exist. As a consequence, most reclaimed lands demand sustained chemical fertilisation (cf. Box 4).

Box 4: Case Study: Reclamation of the Blaenant Opencast Coal-Mine, near Pwll Du, Brynmawr, Wales (Fairfax, 1985; Kilmartin, 1994)

Blaenant Opencast Coal-Mine operated from 1973/4 until final closure before 1978. The Welsh Office Agriculture Department's Land and Water Service (WOAD) took the site for reclamation between September 1975 and August 1977. It was reshaped and resoiled with around 100 mm of artifical topsoil, manufactured from weathered shales and glacial drift discovered in the workings. This was supplemented

Contd.

Contd. Box

with 5.021 t·ha^{-1} ground limestone, 1.255 t·h^{-1} basic slag and 0.735 t·ha^{-1} 8N:20P:16K compound fertiliser. The site was sown with 31 kg·ha^{-1} of hillland seed mix. Management included stone picking and the application of 0.625 t·ha^{-1} 20N:10P:10K fertiliser annually (Fairfax, 1985).

The site suffered some erosion during establishment that resulted in rilling, the trenching of drains and the removal of much topsoil from the slope-crests. Eventually, between 1982–1983 the problems of accelerated runoff were countered by a major investment in soil conservation engineering works. In 1984–1985, sections of the site were damaged by an outbreak of insects and had to be treated with pesticide (Dursban 4) and reseeded (Fairfax, 1985). WOAD's regime of intensive managment was gradually withdrawn between June 1983 and November 1985. Much of the land remained in the control of British Coal until the later 1980s when it was returned to a mixture of private and common management (Kilmartin, 1994). The Coal Authority began selling the land from January 1998.

Today, the land supports a dense mossy turf mat on a thin 50 to 150-mm layer of topsoil. The soil suffers severe compaction, accelerated runoff, water-logging and poor drainage. However, erosion problems are restricted to heavily grazed and trampled steep slopes, drainage ditch margins, wheelings and rilling from the site margins. Mass movement has become a problem on some steep slopes.

Several chemical elements are required in large amounts for healthy plant growth. These include nitrogen (N), phosphorus (P), potassium (K), sulphur (S), calcium (Ca) and magnesium (Mg). Many others are required in smaller amounts. On reclaimed sites, most or all of these may be available in very small amounts. The main deficiencies are usually nitrogen and phosphorus (Wilson, 1985: 10). In most soils, the main source of nitrogen is the breakdown of organic matter by microbial action. However, phosphorus must usually be out-sourced and provided as fertiliser. In Wales, Broad (1979) suggested that nitrogen fertilisers produced better results than adding either potassium or phosphorus in the cultivation of conifers on a range of man-made sites. However, on acidic and pyritic sites, phosphorus can be rendered unavailable to plants and the deficit requires the addition of fertilisers (Jobling and Stevens, 1980).

Wilson (1985: 10) repeats the maxim that deficiencies of micronutrients are unlikely to be encountered on reclaimed land since the amounts required are small and there is a lot of freshly shattered source rock around. However, the author's tests plots on the Varteg opencast site, Blaenavon, Wales, contain dramatic demonstrations of the effectiveness of a 'remineralisation agent', in this case ground basic igneous rock. This rock

Table 10: Some common organic amendments and mulches for reclaimed lands (Wilson, 1985; Slick and Curtis, 1985)

Organic fertiliser	Problems
General animal manure	Farmyard manure (FYM) contains variable NPK content, depending on the source, and sometimes high C:N ratio (nb. cattle: C:N 25:1, pH 6.6). If the C:N ratio is high it may depress plant growth unless supplemented with N fertiliser. It often contains weed seeds but may add fibre as straw. May wash into watercourses and cause off-site pollution.
Pig manure	As above but contains high levels of Cu. May wash or leach into water courses.
Chicken manure	As above but especially prone to high ammonia. This may suppress seed germination as well as plant growth. Nitrate nitrogen accumulations can be toxic and cause problems with access to off-site waters. Poultry manure contains, on average, 10 times as much N and 15 times as much P as cattle manure, while K is much the same.
Spent mushroom compost	Good quality with high lime content. Supplies are expensive and hard to find. High nitrogen content.
Straw	Long fibre cellulose from farms or grain stores, pH 5.6–7.1, C:N ratio: wheat up to 150: 1, oats 50:1. Durability, one year. Application rate: mulch: 10 $t \cdot ha^{-1}$ erosion control: 7 $t.ha^{-1}$. problems include weed seeds that are fellow travellers, and its tendency to be blown off site. May be anchored by use of bitumen emulsion.
Alfalfa hay	Hay may be used as a mulch in the same way as straw and can be blown off site. High applications (< 75 mm depth) are a fire hazard but the hay of leguminous alfalfa (C:N 18:1) aids nitrification in soils.
Wood residues	These have a high C:N ratio (130–930:1) but the bark adds texture to heavy soils and provides a moist microhabitat for soil micro-organisms. The pH ranges from 4.2–6.4 and is higher if composted. Tends to cause a temporary N deficiency in soils. Improved by composting prior to application. Application rates of 10–15 mm suffice but 50 mm is better for erosion control and 100 mm as mulch. Grass will grow directly on top of bark mulch. Hardwood chips cause less denitrification than bark but softwood rather more and have a lower pH. Sawdust is less effective than wood chips and more prone to cause N problems. Leaves can also be used, as can commercial wood fibre mulches that can be applied hydraulically. A wood-chip or coarse bark mulch could last 5–10 years.
Municipal solid waste	Composted, shredded municipal solid waste benefits plant growth and is often useful as a soil conditioner. It promotes soil aggregations and better waterholding properties in amended soils. Its pH is nearly neutral. However, C:N ratios can be high and municipal wastes may be contaminated with non-biodegradable fragments of glass, plastic and metal as well as contain considerable heavy metals and other contaminants. Shredded paper mixed into the soil is supposed to improve water-holding properties.

Table 10: Contd.

Organic fertiliser	Problems
Sewage sludge	This is a good source of nitrogen and potassium but not phosphorus. Its pH ranges from 5.8–6.8 and C:N is usually around 25:1. Uncomposted sludge is not very useful. However, composted sludge produces dramatic improvements in plant growth, its colloids reduce nutrient leaching, improve cation exchange, and reduce salt and may help leaching toxins to groundwater. Unfortunately, it also often contains potentially high levels of metals that may supplement the heavy leadings often inherent in mine spoils. Most European nations set strict limits of how much sewage sludge can be applied to agricultural land (Blum, 1987). Its use on reclaimed land is widely advocated but, in general, it should be avoided. However, if sludge is to be used, application in winter should be avoided because of the possibility of nitrogen being transformed to nitrate and leached. Direct runoff and seepage should be avoided and sludge should not be added to steep, frozen, dry or cracked soils. Less digested sludges may contain weeds and pathogens.

contained no NPK and just a little magnesium. However, trees treated with the substance outperformed those treated with conventional fertilisers alone. This suggests that there may, in fact, be key micro-nutrients that play, if not a direct role, then a catalytic role in promoting bioproduction.

Organic fertilisers have the advantage that they release nitrogen slowly as the organic matter decomposes. However, the decomposers also consume nitrogen and, in the short term, their development may deplete the existing nitrogen reserves in the fertilised soils. In sum, the rotting of low organic nitrogen may deplete soil nitrogen and require that additional nitrogen be added as fertiliser (Wilson, 1985:12). Nevertheless, some workers suggest that organic wastes, ranging from flocculated sewage slurry to urea-treated chicken manure (Table 10), work both to improve soil fertility and eliminate potential pollutants from the environment (Dobrzanski et al., 1982).

An alternative approach to using organic fertilisers is to cultivate nitrogen-fixing species, such as legumes (Table 11). Field tests have shown that although species fix nitrogen for their own benefit, they also increase nitrogen availability in the soil. In field trials in South Wales, Hood and Moffat (1995) found that a stand of *Alnus glutinosa*, planted in coal spoils at 1.6 m spacing, raised the nitrogen status of the spoil. After 6 years, it could be detected in interplanted *Larix decidua* (see Flege 2000, this book). The rate of acceleration was between 8 and 20 times that from natural inputs. In Hungary (Fadygas, 1992), leguminous *Robinia pseudoacacia* proved most economical on sandy substrates because of its excellent regrowth (*sim.* Bulgaria—Haigh and Gentcheva-Kostadinova 2000, this book).

Table 11: Some nitrogen-fixing plant species (Wilson, 1985; Vogel, 1987)

Nitrogen-Fixing species	Problems
Robinia pseudoacacia	Used successfully in many nations across Eurasia and N. America. In Britain, it is prone to wind throw and animal damage. Tolerates pH 4.0.
Alnus spp.	Used successfully in many nations across Eurasia and North America. Waterloving, short-lived but tolerates pH 3.5.
Lupinus arboreus	Reputedly can fix 150 kg·ha·yr^{-1} of nitrogen (Williamson et al., 1982). It is not hardy in northern climates and has a tendency to suppress tree growth.
Ulex europaeus	Grows well in exposed conditions and resists grazing. It is very inflammable and may suppress the growth of young trees.
Sarothomnus scoparius	Broom has the same problems as gorse (*U. europaeus*)
Trifolium spp.	Traditional nitrogen fixer for pasture. Clover species have high recoverability after damage by trampling. Some species tolerate pH as low as pH 5.0 (*T. pratense*)
Medicago sativa	Alfalfa is a traditional, perennial fodder nitrogen fixer that is deep rooting and produces hay that is a valuable mulch. Its limiting pH is 5.5. Survives in low rainfall regimes down to 300 mm·yr^{-1}
Other forbs-legumes	*Lespedeza* spp. and *Lotus corniculatus* withstand pH down to 4.5, slow-growing *Lathyrus* spp. to 4.0 and *Polygonum cuspidatum* to 3.5. *Meliotus* spp. and *Coronilla varia* are also used in America's eastern coal lands. In the West, *Vicia americana* Mulhl ex Willd. resists rainfalls down to 350 mm. *Onobrychic viciaefolia* Scop. resists salts but is short-lived Vogel, (1987) lists several other species. Vogel (1987) lists several other species.
Local Species: *Astragalus* spp. *Oxytropis* spp. *Hedysarum* sp.	Recommended where possible native legumes are used rather than those found in standard seed mixes. These native high-altitude legumes were used in British Columbia. They proved not to be effective competitors of rhizomatous grasses but could coexist with bunch grasses (Smyth, 1997).

Sometimes it suffices that the vegetation be resistant to continuing pollution. Open-field sulphate-sulphur deposition rates in Europe may be as great as 30.7 kg · ha^{-1} · yr^{-1} in eastern Germany (declining to 18.9 Switzerland, 17.2 in Hungary, 16.2 in western Germany, 12.6 in France, 7.3 in the U.K. and 6.6 in Finland). Rates of ammonium-nitrogen deposition range between 7–15 kg · ha^{-1} · yr^{-1} and those of nitrate-nitrogen between 4.3–7.6 kg · ha^{-1} · yr^{-1} in central and industrial western Europe (Brechtel, 1992). The situation is made worse by trees that harvest air pollution nuclei, German trials on 51 spruce and 21 pine stands found spruce capable of accelerating annual deposition rates of NH_4-N, NO_3-N, SO_4-S and C1 by factors of 1.8, 2.0, 2.9 and 1.9 (spruce) and 1.7, 2.0, 2.8 and 1.4 (pine) respectively (Brechtel, 1992). In the west of the Czech Republic, Hruska and Kram (1992) found spruce forests capturing three times as much sulphur as open ground deposition (33 vs 11 kg · ha^{-1} · yr^{-1}). They relate the dangerously high release of inorganic monomeric aluminium from the Lysina Catchment to acid deposition on poorly buffered podsolic soils overlying granitic rocks.

In the Russian Federation, reclamation work often has to be undertaken on contaminated land and adjacent to major sources of industrial pollution. Pollution-resistant trees including the common birch (*Betula verrucosa* Ehrh.), box elder (*Acer negundo* L.), laurel poplar (*Populus laurifolia* Ledeb.) and willows (*Salix* spp.) are recommended along with the grasses *Festuca rubra* L., *Poa pratensis* L., *Bromus inermis* L., *Agrostis alba* L. and *Agropyron repens* (L.) Beauv. Kapelkina (1992) recommends that seeds of these species be mixed with topsoiling materials during spreading by a transverse milling earth spreader.

Another approach to the bioremediation of surface-mine spoils is to inoculate with (micro-) organisms that help plants grow (Tate, 1985; Scullion, 1992). The practice of restoring topsoils from the pre-mining environment, of course, is intended to serve this function. In addition to the fertility that remains in these soils after storage, chemical and humic, they are also reserves of the basic components of the soil microcoenosis. Topsoils inoculate the reclaimed land with bacteria, mycorrhiza, earthworms and many other organisms (Harris, et al., 1989). In the case of the bacteria, Harris et al. (1989) show that aerobic bacterial numbers, after initially increasing in the surface layers of stored topsoils, subsequently decline. Similar processses affect reinstated topsoils, where bacterial populations increase initially but after a few weeks or months begin to decline (Johnson, 1995). This can be countered by providing the bacteria with additional organic substrates mixed into the soil, such as straw, wood bark etc., and by improving the density, drainage and chemistry of the soil (Felton and Taraba, 1994).

Szegi and Voros (1992) studied the distribution of obligate vesicular arbuscular endomycorrhizas in reclaimed lignite spoils at Visonta, Hungary and in Poland. These fungi, which are important for plant uptake of phosphorus etc., appeared spontaneously by the sixth year in the non-toxic Visonta spoils but were inhibited by pyrite/zinc/lead toxicity in the Polish dumps. It is suggested that earthworms may be one agency in the spread of mycorrhizal spores (Scullion, 1992). Harris et al. (1987) have shown that mycorrhiza are heavily depleted in stored topsoils with the depletion being related to the time in storage. It is suggested that these (endophytic) symbionts increase the surface area available for the absorption of nutrients and also permit the exploitation of a greater soil area. In Britain and the USA, trees on reclaimed lands that were inoculated with mycorrhiza outperformed those that were not (Wilson, 1985).

Earthworms promote the mineralisation of organic material and concentrate nutrients in their feeding. Earthworm casts include higher levels of nitrogen and phosphorus than surrounding soils, their aggregates are much more water-stable and they contain much higher levels of biological activity. They are very effective at mixing organic matter from the surface into the soil and reducing soil densities, and their burrows are large continuous macropores that improve soil drainage (Scullion, 1992; Stewart et al., 1989).

In Britain, it may take more than 20 years for populations of deep-burrowing earthworms such as *Lumbricus terrestris* and *Aporrectodea longa* to reach natural levels in reclaimed lands. They may be inhibited by soil compaction (Rushton, 1986). Even in stored topsoils, their numbers are depleted. They tend to become restricted to the outer skin of such storage mounds. They may be deterred by the anaerobic conditions that often develop in the interior of such mounds, which can lead to the mineralisation of oxygen and the production of ammonium-N (Harris and Birch, 1989). Thus, in the field, underdrainage, deep-ripping/subsoiling and grazing, especially by sheep, and forestation, which also adds organic matter to the soil surface, encourage the growth of earthworm populations. Scullion (1992) suggests that the distribution of earthworms on such sites is proportional to the distance from undisturbed land. His technique, which involves sowing earthworms in slit trenches, is intended to overcome this problem.

Soil Aggregates

Soil aggregates are the basic units of soil structure. They are natural clusters of soil particles in which the forces that hold the cluster together are greater than those linking it to adjacent soil aggregates. Soil aggregates have greater strength, lower bulk density and lower rates of water transmissivity than do bulk soils (Horn, 1989:13). Their strength arises from the higher number of inter-particle contact points within the aggregate and the strength of those chemical, biochemical and electrochemical bonds. Their importance arises from the fact that while under some circumstances a soil aggregate will act as a single soil particle, under others it will break downs into a collection of smaller soil aggregates, and under still others it will break down into its primary particles, clays etc. (Plyusnin, 1964: 94).

Soil aggregates are created by an array of biological, physical and chemical processes (Sequi, 1978; Lynch and Bragg, 1985). They are complex systems of primary particles, quartz grains, clay, platelets, often bonded together as clay domains and organic matter. They are bound together by electrochemical bonds and by cementing agents including organic matter, clays, lime and sequioxides. Sometimes they are also enmeshed in a web of organic fibres, especially fungal mycelia.

The most important binding agents are organic. Some forms of organic binding are mainly physical. However, the main ones are electrochemical.

Humus, which can be colloidal like clay, adsorbs cations. If it contains a high proportion of Ca^{2+} or other polyvalent cations, its long organic polymer macromolecules can form bonds with each other and with the mineral components of the solid phase of the soil. It also binds clay domains to quartz sand grains. This generates a stable humus-clay complex and hence soil aggregate. In sodic and alkaline soils, dominated by monovalent Na+, or strongly acid soils, dominated by H+, the bonds are unstable and the humus dissolves (Koorevaar et al., 1992: 23). The clays may also bind aggregates together but their effect is much weaker. The degree of cementing depends on the electrical charge of the crystal faces and on the type of clay.

Electrochemical bonds are called long-range bonds because they re-establish and reform after mechanical disturbance. There are also chemical bonds. These are short-range bonds which, because they are cements, are slow to reform once broken. They include lime, sesquioxides of iron and aluminium, especially aluminium. Coal spoils are notoriously prone to brittle fracture and are characterised by a high proportion of weak short-range rather than long-range bonds between their aggregates (Dawson and Morgenstern, 1995). There are several reasons: they have weakly developed soil organic systems, their primary particles include many unstable mine stones that have been weakened by excavation and accelerated weathering, their aggregates are mainly the consequence of a mechanical process and they have not been in place long enough for really strong chemical bonds to precipitate. This makes the aggregates on reclaimed lands very vulnerable to damage by trampling or tillage and also prone to liquefaction under high pore-water pressures (Bishop, 1973). The latter was a contributing factor to the disastrous flow-slide failure of the Aberfan spoil tip in South Wales, 1966 (Dawson and Morgenstern, 1995).

The variety of soil aggregate responses to mechanical disturbance, wetting and soil chemical properties has been summarised for Australia by the eight-unit classification of Emerson (1967). The first division separates dry aggregates that slake when dispersed in water (Classes 1–6) from those that don't. Class 8 aggregates, which are found in clay pans, remain unchanged. Class 7, which are characterised by a high organic content, swell but do not slake. The second division distinguishes aggregates that slake but do not disperse on immersion (Classes 3–6). From the others, Class 1, which contain 17–55%, mean 32%, exchangeable sodium, disperse completely. Class 2, which contain more than 7% exchangeable sodium, undergo partial dispersion. Classes 1 and 2 aggregates are responsible for many soil piping-induced failures in embankments and impoundments. High salinity is a particular problem of land reclamation sites in the rapidly developing coalfields of Australia.

Aggregates of Emerson (1967) Classes 3–6 are distinguished on the basis of their stability under mechanical disturbance, with respect to the initial moisture content. Class 3 aggregates disperse when immersed in water after remoulding at a water content equivalent to field capacity. These aggregates are more than 25% illite, with admixtures of kaolinite and rarely montmorillonite. They have neutral pH and an exchangeable sodium content of around 6%. Class 4 aggregates do not disperse on immersion after remoulding at field capacity. They often have a high carbonate or gypsum content (20–24% by weight) and contain minerals that dissolve quickly enough to maintain the quartz-clay bonds in the aggregate. Class 3 aggregates are easily disrupted by tillage in wet conditions. However, the effect can be countered by adding lime to the soil.

Class 5 aggregates are those that disperse at a water content intermediate between field capacity and a suspension (1:5 soil/water ratio) after 10 minutes shaking. Class 6 aggregates also disperse but reflocculate after 5 minutes. Class 5 and 6 aggregates are both characterised by little exchangeable sodium or other soluble salt and mild acidity (pH 5.5 to 6). Class 6 aggregates are dominated by kaolinite while Class 5 always include some illite.

Repeated wetting and drying can enhance the strength of a soil aggregate; repeated shrink-swell changes the arrangement of soil particles creating a clay-silt outer coat over a coarse interior. The effect is greater, the more saturated the aggregate becomes in the wetting process (Horn, 1989).

Soil compaction is such a serious problem on reclaimed coal lands that this topic is treated separately here (see section II). Box 5 deals with another series of chemical and biochemical additives, which aim to improve functioning of the soil system, especially by improving soil aggregate stability.

Box 5: Chemical Soil Conditioning Agents

Soil conditioners are chemical agents that aim to improve soil structure, soil aggregate stability, soil erodibility and the hydrological properties of soils. They are used to upgrade soil structure to ensure a more friable tilth, which drains more freely and can be cultivated sooner after rain or irrigation and more easily (De Boodt, 1979). Many are chemicals that mimic organic polymers. Much like these natural soil organic compounds, they have a high resistance to decomposition, many electrically charged sites and hence bind soil particles together. Many conditioners help modify soil crusting and sealing. Hence, they can be used to prevent erosion, enhance water infiltration, soil drainage, aeration, soil microbial activity and soil fertility. The outcome is earlier seedling emergence, earlier maturation, fewer outbreaks of diseases associated with poor soil aeration and stronger root systems. Coloured conditioners adjust the soil's temperature regime, hence regulate

---

Contd.

evaporation and seedling emergence (De Boodt, 1979). In sum, soil conditioners help increase bioproductivity and reduce runoff and erosion (Brandsma, 1996).

Unfortunately, soil conditioners do not solve all soil problems. They supplement but do not replace the role of soil organic matter, provide little of plant nutritional value and, because of the relatively small dosage, do not affect the gross chemical characteristics of the soil. [The exception is the conditioner phosphogypsum, which raises soil calcium and phosphorous levels]. They are also expensive. 'Krilium', the first commercial soil conditioner, was introduced in 1951. After initial enthusiasm, its use declined to nothing. Soil conditioners did not prove cost-effective. Indeed, costs still remain prohibitive for general agricultural use (Wallace and Wallace, 1990; De Boodt, 1992). However, since the 1960s there has been a revival of interest, especially in soil-stabilising chemicals, tackifiers and crusting agents that help control erosion by wind and water. Nevertheless, the use of soil conditioners remains restricted to high-value projects. Popular applications include: in agriculture—the control of furrow erosion in irrigated agricultural systems; in construction and reclamation—the stabilisation of erodible surfaces until vegetation is established; and in active mining—the suppression of dust and wind erosion of spoil piles and roadways (IECA, 1994). Brandsma (1996:27) suggests that low-cost conditioners could also have value in horticulture and the prevention of erosion during forestation.

Many chemicals have been deployed as soil conditioners, viz. inorganic salts (especially calcium and silicate compounds), foams and polymers. Most popular are the synthetic polymers PAM (Polyacryla-mide), UF (urea formaldehyde), (Agri-SC) ammonium-laureth-sulphate, phosphogypsum, methacrylates and bitumen (Azzam, 1980; Sojka and Lentz, 1996; Mostaghimi et al., 1994; Agassi and Ben-hur 1992, Brandsma et al., 1996; Brandsma, 1996; Neururer and Genead, 1991). VAMA (vinylacetate maleic acid) and HPAN (hydrolysed polyacrylonitrile) and PAV (polyvinylalchohol), among others, have an extensive research literature (De Boodt, 1979). In review, Levy (1996) has offered that, of the polymers currently on offer, anionic PAM is the most effective and long lasting.

To be effective, a soil conditioner must be correctly targeted and appropriately applied to a soil susceptible to modification. The effectiveness of a soil conditioner is a function of the chemical employed, its electrochemical properties, its molecular weight, the mode of application, degree of incorporation and the properties of the soil

Contd.

Contd. Box

treated. These include soil moisture status, porosity, clay content, organic matter content, pH and other chemical properties (Brandsma, 1996: 18). If it is not a surface-coating agent, then the conditioner must be mixed into the soil. Once in the soil, conditioners function like natural organic polysaccharides and sesquioxides. They are more persistent than most organic chemicals and they help flocculate soil particles. Some uncharged polymers work by displacing water in the interlayers of expansive clay minerals (Theng, 1982). Most charged polymers link with the negatively charged clays or with cations adsorbed into the clay particles (De Boodt, 1979). They encourage adhesion and aggregation so creating larger, more stable soil aggregates. Consequently, they can permit medium to long-term improvements in soil architecture.

Hydrophilic polymer conditioners, such as PAM, are used on soils with low infiltration capacities and a high potential for crusting. These hydrophilic polymers draw water into the soil and so enhance drainage, reducing runoff, and improving soil water-holding capacity with its reserves of plant-availabe moisture. Hydrophobic conditioners, which include bitumen and polyvinyl acetate, are used as barriers to water movement. They create an impermeable layer that holds water in the soil, preventing evaporation and repelling surface waters. These chemicals are used to defend soil-water reserves in semi-arid and irrigated agriculture and to isolate acid-generating mine spoils.

On application, soluble polymers must be stabilised in the soil by chemical or electrostatic adsorption. The conditioners are applied in water to a moist soil. The applied water droplets must merge with the water tension meniscus that links soil particles and the polymers must be drawn into this meniscus. Subsequent evaporation or drainage from the soil concentrates the polymer chains, which coagulate, thereby binding adjacent soil particles. To be effective, binding must occur within as well as upon soil aggregates. The process is affected by the suction pressures within soil aggregate micropores but it demands soil aggregates with sufficiently large intra-aggregate pores. The mobility of polymers is very limited, especially when the chains exceed a molecular weight of 50,000 and become cross-linked with clays etc. (De Boodt, 1990). Once a polymer develops surface contacts, the number of these tends to increase, especially in the case of small particles held as colloids. This increase in linkages causes contraction, which draws soil particles closer together physically. In sum, the conditioner encourages flocculation and subsequently strengthens the aggregate so created

Contd.

Contd. Box

(De Boodt, 1990). Theng (1982) considered that anionic conditioners might be more effective in flocculation while non-ionic conditioners, which spread over the aggregate like paint, might be more effective stabilisers. This view remains controversial.

Inevitably, given the mechanisms described, the effectiveness of soil-conditioning polymers is greatly affected by the clay content of the soil. Different clay minerals have different capacities for absorbing polymers—illites may absorb but a tenth of sodic montmorillonite. This affects the number of binding sites. Once a soil's capacity to provide sites for the conditioner to bind is saturated, further additions have no effect.

Soil Structural Spaces, Cracks and Pores

The amount of unoccupied space in a soil normally ranges from about 25–60%. In some organic soils it can run as high as 90%. Frequently, the total pore space of a soil is 40–50% by volume. Shaxson (1992) has argued that the spaces in soils are their most important features because it is here that most of the important activities within soils take place: the movements of water, air, chemical reactions and biological processes. His advice was for soil scientists to attend to the soil architecture, which in this chapter is reflected by the attention paid to an inverse of soil space, namely soil density and compaction.

Rode (1952) described the total porosity of a soil as:

$$P = (1 - bd/sg) \cdot 100 \qquad \qquad ... (1)$$

where: bd is bulk density in grams per cubic centimetre, sg the soil particle specific gravity, often close to 2.65 or 2.7 g $\cdot$ cm^3 (2,650–2,700 kg $\cdot$ m^3), and P the total porosity as a percentage of soil volume.

In natural soils, the particle specific gravity ranges from 2.6–2.85 g $\cdot$ cm^3. Some minerals, including iron compounds such as magnetite can be more. In mine spoils, carboliths can be less, sometimes 0.9–1.2 g $\cdot$ cm^3. In general, the specific gravity of the solid, liquid and gaseous phase of a soil is taken to average 2.65, 1.00 and 0.0013 g $\cdot$ cm^3 (Koorevaar et al., 1983).

The voids in a soil are classified into two types. Inside soil aggregates are micropores. Between soil aggregates are the more dynamic macropores. These voids within a natural soil are created by four main agencies: clay shrinkage, biological activity, ice segregation and cultivation—human action. In reclaimed lands, there is an important fifth agency, which relates to the vagaries of emplacement. Many coal spoils are heterogeneous mixture that are dumped, levelled and sometimes compacted. These processes do not return the material to its original pre-mining density. The material contains many new voids, due to the packing of the mine-stone cobbles and their role in bridging spaces (Merriwether and Sobek, 1978).

Soil pores are maintained by the presence of clays and organic compounds, in fact anything that contributes to the stability of the main structural components of the soils, namely, the soil aggregates. Pore collapse is often caused by the slaking and dispersion of soil aggregates. Pore space can also be lost by changes in the packing density that follow mechanical soil compaction or natural vibrations consequent upon cyclic frost action or hydration.

A special type of pore formation is the fissuring that creates desiccation cracks. Soil shrinkage is due to the dehydration of the electrically charged surfaces of solid particles and especially the extraction of the water layer held by clay minerals such as montmorillonite. Shrinkage is the inverse of the process of hydration. It occurs when the forces drawing ions to the charged surfaces are greater than the bonding forces that hold these surfaces together. The force of expansion (P_e) can be expressed as the osmotic potential of the charged surface with its liquid contact. This potential energy is regulated by the amount of surface bonded water in the soil and the quantity of water needed to build a complete wetted film.

When the soil is entirely saturated, the volume of water and pore space is the same. At this point, the forces of interparticle attraction (P_{ipa}) balance those of osmotic attraction due to the solute coatings of the charged soil particles. The equilibrium of the system, including the charge of the particle surface (P_s), may be described by the equation:

$$P_e = P_s + P_{ipa} \qquad \qquad ... (2)$$

Any change in soil moisture creates a change in all three quantities. Soil swelling and shrinkage is required to re-establish the balance. The specific pore volume is a function, together with a component of the change in soil surface elevation, of the amount of shrinkage and swelling. However, there is a maximum to the loss of volume, the shrinkage limit, beyond which no further reduction occurs.

The shrinkage of an individual soil aggregate is regular. However, this is not the case for aggregates bound in soil masses. In these circumstances, contraction can only proceed by the fracture of the soil's weakest bonds, opening up interaggregate fractures and fissures. These weakest points are those wherein shrinkage pressure first exceeds the interparticle attraction.

Berezin et al. (1984) demonstrated that this process occurs as a series of structural jumps that occur at successive threshold energy levels and moisture contents. As each threshold is crossed, the soil mass ruptures and a new system of interaggregate cracks emerges, defining progressively smaller and more tightly bonded soil aggregates.

Table 12: Three objectives of tillage operations (Constantinesco, 1977)

Tillage operations have three objectives:

- **They reduce the density of the soil.** Usually, the soil is loosened to encourage infiltration and re-compacted to create a seedbed or root-bed of appropriate density.
- **They change soil composition and structure.** Surface soil crusts and growths of unwanted vegetation are obliterated. The soils may be mixed and turned to incorporate additions such as fertiliser or organic debris at the soil surface.
- **They adjust soil surface microtopography.** Unwanted features such as rills are obliterated. Surface roughness is increased to encourage infiltration or drainage and to create a surface form that encourages plant growth.

Plant roots create soil pores in two ways. The first is that during their growth they create root channels. When the root dies, these are vacated. The second is by compression. When root tips are forced into small pores, they create local soil compression. This closes up neighbouring macropores and causes a reorientation of soil particles, sometimes a remoulding of clays, so that their axes parallel the growth of the root. These structural features help ensure the persistence of vacated root channels.

Worms achieve a similar effect by eating their way through the soil. They contribute to the bioturbation of the soil and the mixing down of humic materials. Soil that has been digested by worms is highly charged with soil-stabilising organic compounds and is much more water-stable than undigested soil. Worm channels persist as macropores in the soil while their 'bioturbation' of the soil is associated with a significant reduction in soil packing density and the creation of pore space (Scullion, 1992).

Ice segregation creates soil pores. Frost heave is witness to the force of attraction of the crystal faces of a freezing ice crystal. Water is drawn to such a freezing plane, rising upwards as freezing progresses down into the soil and downwards if there is a permafrost layer. The expanding ice crystals create new interaggregate boundaries by drawing water from the soils and create new voids by expanding within soil macropores. In a susceptible soil, the changes to soil surface elevation, a measure of soil compaction, that follow winter frost action can amount to tens of millimetres.

Tillage is the traditional way of loosening soils. Tillage means altering the character of the soil to promote biological production and farmers expend large amounts of energy to break up the soil surface (Hillel, 1982 and Table 12). Recently, the value of agricultural tillage has been questioned. The 1980s saw the rise of conservation tillage and no-tillage agricultural techniques and systems. Certainly, the application of tillage to the cultivation of the weakly structured soils of reclaimed lands is problematic.

Tillage is often used on reclaimed lands, not least because many are returned to agriculture. However, deep ripping is also employed to loosen the soils prior to forestation. In Illinois, deep ripping prior to tree planting improved the survival of *Quercus rubra* L. from 9 to 61% and *Juglans nigra* L., from 42 to 74% over non-ripped plots in 12-year-old plantations. Tree growth was also increased from 2.2 to 4.5 m in *Q. rubra* and 2.6 to 5.5 m in *J. nigra* (Ashby, 1996). For gently sloping lands in Britain, Wilson (1985) recommends ripping to 0.5 m on clays and 0.75 m on sandy sites, with winged tines on a parallelogram mounting pulled by a 300–400 hp tractor. On steeper slopes, Wilson recommends creating a series of ridges 300 mm by 1.5 m and then cross-ripping. Slopes of 10–20 degrees are the steepest recommended for forestation and these should be benched at 20-m intervals to reduce the risk of erosion. In Britain, deep cultivation, involving the complete shattering of the profile down to 0.5 m or so, has been found to encourage the infiltration of rain water or reclaimed lands, where conventional cultivation tended to encourage the concentration of water and more erosion (Coppin and Bradshaw, 1982). Vehicle tracks can also lead to gullying (Wilson, 1985: 10).

Unfortunately, the effects of tillage are not always positive and, when the soil recompacts, the end-result may be reduced soil porosity. Powers and Skidmore (1984) examined the impact of soil tillage on samples of a fine silt loam (Typic Argiudoll) soil, both cultivated and non-cultivated. The study employed a variety of techniques including scanning electron microscopy (SEM). They found that tillage significantly reduced the structural stability of their soil and that the degradation was proportional to the degree of disturbance. Virgin soils suffered proportionately more degradation than those which had previously been subjected to cultivation. The SEM micrographs confirmed that the virgin soils included more insoluble, and probably organic, structural bonds between soil particles than did the disturbed soils. This gave the virgin soils a better aggregate structure with a larger proportion of water-stable aggregates. These soils were also less compressible.

Tillage increases the dry aggregate stability of soils but decreases the level of aggregate water stability. The soils on reclaimed lands, which have weak primary particles and a relatively ineffective soil organic system, tend to have structurally weak soil aggregates. They can suffer massive damage through tillage.

Tillage consumes a great deal of energy. The weight of the top 300 mm of soil may be 4000–4500 t · ha^{-1}. Tillage aims to loosen, pulverise, invert and recompact the soil. A traditional cultivator, working with a primitive plough, may walk 40 km across each hectare during the course of each tillage operation (Hillel, 1982). Tillage usually involves several operations, primary cultivation to break the soil, then a series of secondary operations to prepare the seedbed. The energy required for each operation increases rapidly as the depth of tillage and range of operations increase. Tillage aims to achieve appropriate soil disturbance with minimum force, if only for the reason that force requires energy, which costs money, time and labour. Hillel (1982) estimates that the cost of deep ploughing (to 50 cm) is four times that of normal ploughing (25 cm) and ten times that of shallow ploughing.

The classic formula of Goryachkin expresses the tractive force of a plough thus:

$$P = fG + kab + Eab \; V^2 \qquad \qquad \ldots (3)$$

where: a and b are the width and depth of the plough furrow, V the plough velocity, G plough weight, and *f, k* and *E* are scaling coefficients (Khachatryan, 1963: 196).

Later work includes the critical lateral component and emphasises the need for side restraint against slippage. Even on a level slope, tillage implements do not move in a straight path. Armenian studies found different assemblages of ploughs and harrows deviating from the true direction more than a hundred times during a 1000-m run on horizontal ground (Khatch-atryan, 1963: 13). The result is additional soil disruption and smearing.

The problem is that a great deal of the energy used in tillage is transmitted destructively into the soil. Sohne (1966) estimates that the efficiency of transmission between tractor wheels and the soil may be only 60–65% and, on soft soil, it may be as little as 40% (Brown et al., 1992). The rest is wasted, often destructively as slippage, compaction and frictional resistance. This stress causes soil aggregate destruction and in moist conditions, the remoulding of clays and the production of a compacted subsoil characteristic of Agrisols (Bridges and de Bakker, 1997 and Box 6).

The impact of tillage tools on the soil is the subject of much research. Soil disturbance and failure is studied empirically in laboratory soil bin tests. This work considers both applied force and resultant disturbance. Force is often applied as draught or horizontal force which, for a given speed, determines the power required. However, in tillage its vertical component is important, since this governs the penetration of the implement. But the process is difficult to model and workers continue to rely on empirical data produced for particular soils (Smith et al., 1989; Khatchatryan, 1963).

Box 6: The Agric Horizon or Plough Pan (FAO-UNESCO, 1988)

Many ploughed soils contain a zone of illuvially accumulated compacted soil and remoulded clays at the base of the plough layer. Illuviation results from the regular breakdown of soil aggregates during tillage. This liberates fines that eventually accumulate in the soil pores near the top of the highest undisturbed layer. Agricultural machinery is heavy and getting larger as time goes on. Inevitably, some of this weight loading is transmitted into the subsoil under the plough layer as raised soil densities.

Contd.

Contd. Box

The plough pan is a more particular feature. Ploughshares tend to remould the clays beneath them, reorienting the clay platelets parallel to the soil surface and packing them together as in a clay seal. This process, known as smearing, creates the 'plow-sole' or plough pan. The plough pan reduces water movement and allows the development of a perched water table and water-logging in the surface soils. It can also inhibit root growth because roots find it difficult to penetrate soil pores that are smaller than their root caps. Finally, the plough pan can also inhibit aeration of the lower profile. Consequently, crop yields can be severly reduced.

Smearing, of course, is a particular problem in clayey soils and wet conditions. Consequently, the timing of tillage is critical. Plough pans can also be difficult to eliminate. Measurements of plough pans in a sandy loam in Holland determined that the pans had a lower porosity (2 vs 8%), a higher bulk density (1.6 gm $\cdot$ cm^3) and a greater mechanical resistance than the soils above. However, rota-digging of the plough pan did not improve the hydraulic conductivity. Two years after the disturbance, the porosity remained higher but, since there were fewer large pore spaces, the hydraulic conductivity was much smaller (Kooistra et al., 1984). Many numerical models for the prediction of soil compaction under tillage have been developed. The deformation of soil by asymmetric tillage tools such as discs and mouldboard ploughs has been well studied, not least in the journal *Soil and Tillage Research*. Soil disturbance is modelled in each case in terms of the geometry of the cutting edge (Koolen and Kuipers, 1983; Khachatryan, 1963; Gyachev, 1961).

Porosity and permeability are often, but not necessarily, correlated. There is no functional reason why the pores of a mine spoil should interconnect. However, there are so many of them, during the early days after the creation of reclaimed land, that it is relatively easy for water flows to establish lines of connectivity. These connections may be challenged by the liberation of fines that follows from accelerated weathering. However, detailed studies of the hydrology of reclaimed surface coal mines suggest that there is hydrological connectivity through the dense compacted layers in many mine spoils (Kilmartin, 1994). This observation is supported by several studies wihich suggest that infiltration rates on clayey reclaimed soils may be two or three orders of magnitude smaller than on pre-mine soils in the same environment (Sharma and Carter, 1993). Infiltrometers established on the same sites studied by Kilmartin (1994) showed two patterns. Either infiltration rates were very low, often less than 0.1 mm $\cdot$ h^{-1}, or they were very fast. In one instance, water that disappeared into the infiltrometer reappeared as a spring some distance downslope.

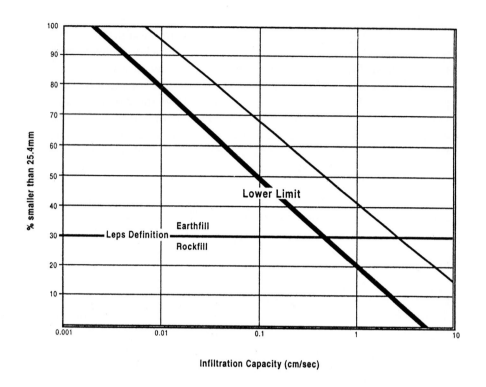

Infiltration Capacity (cm/sec)

Fig. 3: Infiltration into Mine spoils (after Dawson and Morgenstern, 1995, Vandre, 1995).

Recorded values of saturated hydraulic conductivity in mine spoils range from 10^{-4} to 10^{-6}, ranging down to 10^{-9} m · s $^{-1}$ for clayey spoils (Smith et al., 1995: 10). Vandre (1995) records conductivities of 1–10 cm · s^{-1} in spoils having less than 30% pass the 25-mm sieve and proposes a log-linear relationship between infiltration capacity and the proportion of the spoil passing a 25-mm sieve (Fig. 3). Smith et al. (1995) argue that this is equivalent to spoils in which less than 20% of the material passes the 2-mm sieve, where infiltration rates can also be high. Thus, many mine spoils are highly porous and permeable. High infiltration rates are recorded on spoils dominated by cobble and sand-size particles.

Poor water-holding capacity is a major problem for the reclamation of lignite mines in eastern Germany. In Wales, many of the oldest, loose tipped, deep-mine fan-ridge coal spoil-tips show a characteristic pattern of natural vegetation that also attests to rapid drainage. Water-loving *Juncus* spp. and sedges colonise the slope-foot. Drought-tolerant lichens (*Cladonia* spp.) are common on the dry upper surfaces alongside acid-tolerant ericaceous species such as Ling (*Calluna vulgaris*) (Haigh, 1978).

Soil Water

Soil water plays several critical roles in the soil. It is the main agency for the translocation of sediments and chemicals. It is the dynamic medium in the hydration and shrink-swell of clay minerals and, as such, implicit in the breakdown of soil aggregates and soil structures. Soil water has an important role in soil compaction. During loading, the pore water pressures decrease to a minimum and then increase again. This threshold of reversal, called the critical saturation pressure, is fairly constant for any given soil. It may mark the point where microaggregate structure begins to become degraded by compaction.

When a soil is submerged, a systematic change in its permeability follows. This is a three-phase process (Miyazaki et al., 1993; Allison, 1947). First, there is an initial decrease due to structural change, swelling and dispersal of soil aggregates followed by clogging of flow pathways. In impermeable and clayey soils, this decrease may last 10–20 days but in well-structured soils with stable aggregates, the impact may be short-lived and small. Second, there is an increase in flow due to the gradual removal of trapped air. This may dissolve into the percolating water or otherwise outgas as bubbles. Permeability reaches a maximum when all the trapped air is removed. Third, a gradual decrease in permeability occurs due to dispersion of aggregates under extended submergence—aided by microbial attack on the organic materials that bind the aggregates, and due to the clogging of soil pores with biological cells and their by-products of slimes and polysaccharides. There are two models for this mechanism of bacterial clogging in soils. The biofilm model suggests that the bacteria line the walls of soil pores in a meshwork of cells and polymers. The second, microcolony model involves the bacteria growing as colonies that adhere to the soil particle faces and which as they grow, consume the pore space (Miyazaki et al., 1993: 208). Gases produced by micro-organisms also contribute to the decrease in permeability (Poulovassilis, 1972).

A particular feature of unvegetated clayey mine spoils is the growth of a layer of algae at the soil surface, sometimes expanding into a full-scale cryptogamic soil crust. These algae hold water at the ground surface and may also reduce infiltration into the spoils beneath. The hydrology of these features has yet to be investigated.

Extended submergence also increases the activities of anaerobic bacteria and the reduction of ferric into ferrous iron that dissolves in the soil solution. Miyazaki et al. (1993: 201) detail the effects of a 40-day submergence on Kawazato light clay. Here, porosity increased from 54.5 to 67.8% and the pH rose from 5.4 to 6.7, but the saturated hydraulic conductivity declined from 4.9×10^4 cm^{-1} to 0.2×10^4 cm^{-1}.

It is suggested that the reason is the tendency for dissolved ferrous iron to hydrate or absorb to soil colloids, causing a reduction in effective pore space. This impact was also reported in other soils with low iron contents. Coal spoils, of course, tend to be very heavily loaded with iron and many other metals.

Soil waters are of three broad types. There is adsorbed water, which is bound to the soil particles, capillary water, which is held by the surface-tension meniscus and lastly, gravitational water, which is not bound to the soil at all and will drain freely under the influence of gravity alone. There are two threshold soil-water levels. At maximum hygroscopicity, the whole of the body of soil water is held by adsorption. At field capacity, the whole of the soil water is held by, in combination, adsorption and the surface-tension meniscus. Soils may be saturated by water, when all of their pore space is filled, but the balance of the water in the soil above field capacity is held by gravity alone.

The behaviour of soil water, unlike that of water in rivers, is governed by electrochemical processes. Water molecules are electrically charged dipoles (H^+–OH^-). These dipoles are attracted to charged surfaces, their own as well as those of the soil particles. This is the process of hydration (Verigo and Razumova, 1963).

Water is adsorbed to soil particles by the processes of molecular cohesion. At the edge of a soil particle crystal, the free electrical charges of the crystal lattice exert a field equivalent to several hundreds of atmospheres of pressure. This binds the first molecular layer of water to the crystal so tightly that translocation is possible only through vaporisation. This tightly adsorbed water has a density of 1.5 g · cm^3 and has the mechanical properties of a solid (Verigo and Razumova, 1963). However, the power of this field declines rapidly as the cube root of distance. So, attached to the layer of tightly bound dipoles oriented to the particle surface, there are layers of loosely bound water molecules. These are held at pressures of 40–50 atm. or less and have densities that decline outwards from 1.3 to 1.0 g · cm^3 (Kachinskii, 1958). This layer has a thickness of up to 100 molecular diameters on clays but it can be as little as 40 molecular diameters on quartz grains of 0.01–0.70 mm diameter (Rode, 1952). The maximum hygroscopicity of a soil, which is a measure of this bound water, ranges from 2% (sand) to 50% (clay) of the volume of soil water at field capacity and 0.5% (sand) to 10% (clay) by weight.

The second most important series of molecular attractions link the water dipoles to their neighbours in the liquid. Within this liquid, as within the crystal, most of the forces of molecular attraction are cancelled out. However, at the air/water interface, the water dipoles are more strongly attracted to other water molecules than to the gas molecules. The resultant of this excess electrical charge is directed into the liquid and its physical manifestation is the surface-tension meniscus that forms at the interface between water, air and soil. The tension in this film can be strong and, given the chance, will draw the liquid into a sphere.

If the charge on the surface of the soil particles is greater than that within the soil water that rests on the soil, then the soil water is pulled into the soil, causing wetting. The result is a concave surface-tension meniscus. If the charge is higher within the liquid, it will be convex. The pressure on a concave surface is higher than on a convex and most, but not all soils, are wettable, so capillary soil moisture exerts an additional negative pressure. The force is determined by the surface-tension energy of water and by the curvature of the surface-tension meniscus. The smaller the soil particles and pores, the less water between, the greater the importance of the meniscus. This negative pressure is the reason for capillary rise. These relationships are included in two classic formulae for the capillary transportation of water. In Jurin's formula:

$$h_c = 15/r \qquad \qquad \text{... (4)}$$

where: h_c is capillary rise and r is capillary interstice size in mm.
In Terzhagi and Peck's formula:

$$h_c = (c/e\, d_{10})\, 100 \qquad \qquad \text{... (5)}$$

where: c is a constant describing grain shape and texture, e the void ratio or total porosity, d_{10} effective particle size in mm, h_c the capillary rise in mm.

Capillary rise is greater in well-structured clayey soils than in sandy soils. However, poorly structured soils, like those on much reclaimed land, may allow little capillary rise at all.

As a soil is wetted, first water accumulates at the interstices where soil particles join. As more water accumulates, these water collars grow until they begin to interconnect. When the water reaches this funicular condition, water can move through the soil by hydrostatic pressure but the movement of air is not impeded. As more water is added, the soils become coated in a continuous film, which expands until the soil atmosphere becomes trapped in bubbles in the larger pores. During this process, the additional negative suction pressure of the soil declines from 50 to around 0.5 atmospheres.

As the negative suction pressure tends to zero, the dominant force holding water in the soil becomes gravity. Gravitational water has no electrochemical bond with the soil particles.

Thus, in a soil with closed capillaries and zero tension, the movements of soil water may be approximated as Darcy's Law, where flow is proportional to mass and pressures.

$$Q = k\, i\, f \qquad \qquad \text{... (6)}$$

where: Q is the quantity of translocated water in unit time, i the hydrostatic gradient or difference in water pressure at two points times the distance between them, f the cross-sectional area available for flow, k the infiltration coefficient—the height of a water column moving through the unit cross-section in unit time and directed perpendicular to the flow.

The infiltration coefficient is affected, of course, by many physical properties of the soil. These include pore size, pore volume and pore connectivity. The space available for gravitational transportation is affected by the amount held by adsorption and capillary tension. It is also affected by the speed of wetting, which affects the amount of trapped air and the degree to which soil aggregates are disrupted, and by the volumetric changes of expansive clay minerals.

Osmotic translocation results from solutes dissolved in the soil water. It is possible to determine the osmotic potential of a soil by enclosing it in a rigid membrane permeable only to pure water, then submerging it in pure water and exerting a pressure on the membrane sufficient to stop the process. Osmotic pressure (π) is therefore the difference between the osmotic pressure of the soil solution (p_1) minus that of pure water (p_2), (eqn 7). The chemical potential of water in a solution is affected by the osmotic pressure. The higher the osmotic pressure, the lower the chemical potential (Miyazaki et al., 1993). The amount of the decrease is called the osmotic potential.

$$\pi = (p_1) - (p_2) \qquad \qquad \text{... (7)}$$

In soils, two kind of osmotic processes operate. First, water interacts with exchangeable cations. These cations are attracted to the surfaces of colloidal soil particles and also to the water dipoles which surround the cation and divide cation and colloid, thereby creating an osmotic pressure. Second, sometimes there are major differences in the concentrations of ions in the soil. This is typical of mine spoils in which pockets or hot spots of particular chemicals often occur. Large-scale differences in the soil solutions cause a diffusion of solutes from high to low concentrations and the back-diffusion of water (Rode, 1952).

The osmotic potential of a soil is related to this water retentivity and its pore size, as well as the electrical properties of the soil's charged surfaces and the types/concentrations of ions in the soil solution. When the soil pores are larger than water molecules but smaller than the hydrated solute molecules, the soil acts exactly like a semi-permeable membrane. Sandy soils have pore sizes that are much larger than those of hydrated solute molecules, but sometimes this is not the case for clay soils. The water retentivity of a fine clay can be closely related to the osmotic potential. Similarly, its swelling can be due to osmotic potential as much as to the molecular forces associated with clay minerals and their hydration (Miyazaki, et al., 1993: 11).

Reclaimed lands are characterised by poor and unstable soil structures and relatively poor hydrological properties. Researchers from Serbia's Kolubara Basin report that all their lands suffer extremely high soil moisture deficits, albeit partly due to climatic factors (Institute of Forestry Belgrade, 1997: 51). In Serbia, as in Wales, clay soils showed the highest deficits. This is due to the absence of macropores to allow infiltration and wastage of much rainfall as runoff.

Soil Atmosphere

One of the key processes in soil aggregate breakdown (and the entrainment of soil particles by surface runoff during the wetting of dry soils) is the hydrostatic pressure built up by air bubbles trapped inside the wetting front. In laboratory studies with masonry sand, this can increase entrainment by as much as 21% (Badrashi et al., 1981). It is also a major force in the slaking of soil aggregates, while the escape of trapped air before a wetting front in an irrigated field is also capable of lifting soil particles from the surface and into the fluid stream. Entrapped air can also slow the flow of water through soils (Miyazaki et al., 1993). Studies by Suhr et al. (1984) suggest the infiltration of water into grass could be slowed by as much as 74%, declining to 10% after 28 weeks, thus making trapped air a major component in runoff generation.

The importance of entrapped air may be greater where there is a restrictive layer in the soil. This might be a traffic pan, a plough pan or the illuvial clay layer found in many clayey reclaimed lands. Suhr et al. (1984) found that air trapped above such a layer could halve the rate of water infiltration and, through increasing runoff, accelerate the final erosion rate by factors ranging from 1.8 to 5.4.

On reclaimed land, soil air also plays a major role in acid formation. The weathering of pyrites to sulphuric acid requires oxygen. Oxygen reaches the pyrite because the heat of pyrite oxidation (and reactions that remove gas from the spoil atmosphere), create a pressure difference in the soil atmosphere and compensating movements of soil air. The air also abets molecular diffusion caused by resulting differences in oxygen concentration. This process operates around 10,000 times faster in air than water (Curtis et al., 1988).

Finally, soil air provides the oxygen needed by the biological processes of the soil. The lower limit for the growth of agricultural crops is 10% by soil volume of air-filled pore (AFP) space (Erickson, 1982). Aerobic soil microbial processes seem to be maximised at 20% AFP but below this anaerobic processes increase rapidly (Linn and Doran, 1984).

Soil Loss Tolerance

In life, it is said happiness is someone who earns two dollars and spends one. Misery is someone who earns one dollar and spends two. So it is with soil losses on reclaimed land. The creation of a self-sustaining soil on reclaimed land depends on the achievement of a positive balance between soil gain through pedogenesis and soil loss through erosion.

The relationship is simple. After unit time (t), soil depth $D(t_i)$, is equal to the initial depth of soil $D(t_0)$, minus total soil removal due to water erosion, mass movement, wind, or biogenic processes (E), plus soil additions due to weathering and other soil-forming processes (P), plus further soil additions due to mass movement and the deposition of sediments mobilised by wind, water, or biogenic processes (A). So, for a stable soil:

$$D(t_i) = D(t_0) - E + P + A \qquad \qquad \text{... (8)}$$

Where E exceeds the sum of P and A, soil loss occurs and the soil is not sustainable. Where E is less than the sum of P and A, the soil accumulates. However, since sedimentation can damage plant growth and seedling emergence, the soil is better where the soil depth is dominated by the products of the soil-forming (P: pedogenetic) processes. Nikiforoff (1949) considers the distinction between accumulative and non-accumulative soils a major taxonomic distinction.

Unfortunately, studies of rates of soil formation on reclaimed lands are few and their results far from clear-cut (Hall et al., 1981: 34–35; Kukal, 1990). Even for natural soils, McCormick et al. (1982) consider that there is still insufficient data on the rate of development of regolith by rock weathering, although they suggest that a soil renewal rate of 1.1 $t \cdot ha \cdot yr^{-1}$ is a useful average. In fact, this lack of data is not unique to reclaimed land. It is a problem for all agricultural lands. On normal agricultural land, it is suggested that the rate of soil development may range from 8 $mm \cdot yr^{-1}$ in ideal circumstances, declining to 2.5 $mm \cdot yr^{-1}$ in more severe environments (Pimental et al., 1976). When a soil has a bulk density of 1.0 $g \cdot cm^3$, then $mm \cdot yr^{-1}$ times 10 converts to 2.5 to 8 $t \cdot ha \cdot yr^{-1}$.

Using some very doubtful data, Bennett (1939) considered that, in a permeable, medium textured loam on well-managed cropland, an A horizon can form at the rate of 25 mm in 30 years. McCormick (1982: 101) estimates this to be equivalent to a magical figure of 11.2 $t \cdot ha \cdot yr^{-1}$, the current United States Soil Conservation Service (S.C.S.) figure for maximum acceptable soil loss on arable farmland, the soil-loss tolerance.

Several workers have attempted to refine calculation of the soil-loss tolerance, defined as the maximum a soil can lose and maintain its productivity. The dimensions of a mean figure for soil-loss tolerance are much debated but the figure mentioned most commonly is 2 $t \cdot ha \cdot yr^{-1}$.

Rates of Soil Loss on Reclaimed Lands

Accelerated soil erosion is often proclaimed the major cause of soil degradation (cf. Hurni, 1996). It is more reasonably conceived as a symptom of soil degradation (Shaxson, 1995). Soil degradation increases surface runoff, the main cause of erosion.

A variety of erosion rates have been reported from reclaimed and orphan coal lands. They are summarised by Nicolau (1996: 75). In Catalonia, Porta et al. (1989) report losses of 119–507 t·ha·yr^{-1} on 23–43% slopes with 0–40% vegetation. In Australia, Day (1987) recorded soil losses from overburden of 9.4 t·ha·yr^{-1} from unvegetated overburden, 0.9 from vegetated overburden and 4.4% from topsoils on a 6% slope. In Ohio, over a very sandy and free-draining substrate, McKenzie and Studlick (1978) estimated soil losses of 0.16–0.4 t·ha·yr^{-1} from orphan surface coal lands.

Haigh (1988) used erosion pins to record ground losses from orphan strip-mine ridges at Utica, N. Illinois and plateau-type deep-mine coal-spoil heaps at Milfraen, near Blaenavon in Wales. Erosion pins measure ground elevation, not erosion. Changes in ground surface elevation are caused by erosion, but also by changes in soil density, soil moisture and soil organic accumulation. This means that ground losses in mm do not translate precisely as soil losses (Haigh and Sansom, 1999). At Milfraen, six-year ground losses from a 30–34 degree, mainly grassed NE-facing slope was 1.9 mm·yr^{-1}, from a grassed, warm, SW-facing slope 5.1 mm·yr^{-1} and from an unvegetated slope of neutral aspect (E) 6.6 mm·yr^{-1}. At Utica, un-vegetated, orphan strip-mine spoil banks showed ground losses between 90 and 270 mm between May 1978 and May 1983. In Wales, similar 6-year records from the 20–24 degree slopes of the former Pwll Du opencast mine, reclaimed 1947/8, showed ground losses of 3.6 mm·yr^{-1} from unvegetated and 2.2 mm · yr^{-1} from a grass-vegetated slope (Haigh, 1998). Since the bulk densities of the surface layers of these sites always exceed 1.2 g · cm^3 it may be assumed that these figures represent a soil loss very much greater than that normally offered as a soil-loss tolerance (Table 13).

However, results from Oklahoma and also from Pwll Du show that, with time, the early high erosion rates on reclaimed lands to ameliorate (Goodman and Haigh, 1981; Higgitt and Haigh, 1995). Part of the reason for this seems to be an adjustment of the topography towards a form whereby erosion is minimised (Haigh, 1999; Toy and Hadley, 1987).

The degree of erosion seems closely related to the volume of surface runoff. Haigh and Sansom (1999) studied erosion and runoff from coal lands near Pwll Du in South Wales, using a standard ORSTOM-Delta Lab Rainfall Simulator. This consisted of a motorised, computer-controlled, swinging nozzle mounted on a 4-metre high frame, which ensured that droplets accelerated to natural terminal velocity. The tests compared unvegetated spoils, naturally revegetated spoils, and grassed, topsoil-mantled spoils.

Erosion losses from the unvegetated plot were approximately a quarter (1:3.83) of those on the 15° vegetated test plots in which rainfall was applied at 240 mm·h^{-1} over 5 minutes. The ratio echoes those long-term records from erosion pins monitored on the same site (Haigh, 1988). Further tests at Blaenavon compared runoff on unvegetated, thinly vegetated slopes and full grassland as generated by tests using 330 mm·hr^{-1} over 5 minutes. The unvegetated plots converted 60% of the rainfall to runoff, the grassed plots converted 40% and the grassed and topsoiled Blaenant converted 13%. At Blaenant, test replication on the same wetted plot found 30% of the rain converted to runoff. This indicates the relatively high water-holding capacity of the topsoil (Nicolau in: Haigh and Sansom, 1999).

Similar tests conducted on the deep coal-mine spoils at Cadeby, Doncaster, England found that 5° slopes on newly laid, unvegetated, recently washed (and hence stabilised) coal spoils converted only 20% of rainfall to runoff. Older (reclaimed 1960s) unvegetated coal spoils, which had been affected by weathering and clay release, converted 25–35% of the rainfall to runoff. By contrast, adjacent spoils mantled with little topsoil but a near 100% grass turf converted less than 10% of the rainfall to runoff. Measurements of sediment yield indicated that soil losses for the same rain event were more than four times greater on unvegetated than partly vegetated sites. In sum, erosion can inhibit soil accumulation on spoils that are thinly vegetated (Haigh and Sansom, 1999).

Spoils mantled with a deep well-managed layer of applied topsoil absorbed most or all of the rain water applied in initial tests on both sites (Haigh and Sansom, 1998). However, not all of this rain water was held in the soil. Much was detected running near the junction between the manufactured topsoil and relatively impermeable mine-spoil subsoil below (Haigh, 1992). Similar results were found in laboratory tests that compared the ability of vegetated and unvegetated surface coal-mine spoils to transmit surface runoff (Haigh and Sansom, 1998 and this book).

At attempt was made to find a simpler way of determining soil losses using radioactive tracers (Higgitt and Haigh, 1995). De Jong and colleagues (1983) measured soil losses on agricultural slopes using a concentration of ^{137}C-s, radioactive Caesium, as their guide. This isotope was introduced into the atmosphere by the nuclear tests of the 1950s and 1960s, then washed by rain into the soil where it became sorbed to soil particles. The extent and character of the contamination is well documented (Jong et al., 1982). So, the isotope can be used as a radioactive tracer to monitor the movement of soil (Elliot et al., 1984; Ritchie et al., 1974). Using such evidence, De Jong et al. (1983) were able to calculate that over the past 20–25 years in Saskatchewan, Canada, the upper slopes of cultivated basins had lost 2–6 t·ha^{-1} and the lower slopes gained 2.5–8 t·ha^{-1}. Work at Pwll Du suggests that the technique's results correlate reasonably accurately with long-term records from erosion pins, although several sources of uncertainty remain to be resolved (Higgitt and Haigh, 1995). In the meantime, there is no easy technique for the direct rapid approximation of soil losses from reclaimed surface mine spoils, with the possible exception of rainfall simulation (Nicolau, 1996). Instead, many workers attempt to use the standard soil loss prediction equations for the evaluation of soil losses on reclaimed lands (Nicolau and Asensio this book).

Wind erosion is a well-known problem of operating surface coal mines. Mine spoils are prone to desiccation and, when unvegetated, may suffer wind erosion. However, studies are relatively few. It is reported to affect sandy spoil banks at Kolubara, Serbia (Institute of Forestry Belgrade, 1997) and the Annopol phosphorous mine spoils, Visutla, Poland (Repelewska, 1968).

II. MANAGING SOIL COMPACTION

The reclamation of a coal spoil mantled area involves a complete restructuring of the land surface. The new land is composed of new materials, surface-mine coal spoils and overburden, which have been moved into new landforms by heavy machinery, and drained by newly created artificial watercourses. The new landscape may be veneered with artificial topsoil, perhaps created from 'soil-forming materials', perhaps retrieved after storage from the pre-mining landscape, and probably enriched by chemical and organic additives.

Physical soil problems, especially soil compaction, are common problems on sites reclaimed after surface mining. Soil density has been called one of the best measures of land reclamation success (Ramsey, 1986). In Wales, it is a major player in the long- to mid-term degeneration of reclaimed coal land (Haigh, 1992). Among others, Pedersen et al. (1978), who compared newly reclaimed mine spoils with contiguous natural soils, also found that mine soils have a higher bulk density and coarser fragments. This despite the impacts of accelerated weathering, the extremely rapid breakdown of the newly exposed and slakable rock fragments in the spoil, that can transform a newly exposed spoil from open-textured cobbles to a dense clay-cobble mix within months (cf. Down, 1975; Taylor, 1987; Haigh and Sansom, 1999). This section examines the causes of soil compaction on surface-mine-disturbed lands and describes some techniques suggested as remedies.

Soil Quality and Soil Aggregates

The best-quality soils are found under forest and old pasture. These soils have a low bulk density and soil aggregates that are stable on immersion in water and that resist disruption when mechanically stressed. In silty and sandy loams, soil aggregates (a.k.a. peds) have high porosity. Even in heavy clay soils, the aggregates remain stable on wetting because cracks form in the same place each season. These macropores, supplemented by those created by earthworms and decomposing roots, ensure that the soils drain freely (Emerson, 1991: 905). The soils are fertile and have a high level of biological productivity.

Unfortunately, while everyone agrees that such soils are high quality, there is no single recognised measure of soil quality. The relationships between soil and plant productivity are complex. They vary with climate, soil type, vegetation type, moisture regime, soil structure, soil texture, soil chemistry, soil ecology and land management (Boone, 1986: 352). So, soil quality must be evaluated in terms of a number of interlinked 'favourable' soil properties, which Van Breemen (1993: 186) defines as any soil property that encourages increase in net primary production from higher plants on a plot with uniform vegetation.

Soil properties favouring biological productivity in the soil include many physical, chemical and biological attributes of the soil system, together with their interactions. However, certain of the physical attributes are both persistent and easy to measure. This has encouraged some to argue that, in the absence of chemical limitations, soil quality is best expressed in terms of the soil system's physical architecture (Shaxson, 1992, 1993).

This view suggests that the spaces in the soil are more important than the surrounding solid particles. This is where the important activities in the soil system take place: the movement of water and gases, the growth of roots and soil (micro) organisms. The inverse of soil pore space is soil density. Measuring soil density, therefore, measures the soil's capacity to provide a habitat for life. It is also much easier to measure and evaluate than the life in the soil *per se* (cf. Doran, 1996).

Soil Compaction: Causes

Soil compaction is a key problem in coal reclamation sites, especially in black coal mining areas. Soil compaction usually involves compression of the soil but, in fact, it is measured as increased soil density. The major causes of soil compaction on newly reconstructed reclaimed land are settling under gravity and trafficking/tillage. Additionally, it may be caused by natural soil-forming processes (such as the accelerated weathering of mines stones which may liberate large volumes of fine particles that choke the soil pores). Equally, it can result from die back of the soil biota, perhaps due to chemical toxicity, perhaps due to compaction itself. The history of the vegetation cover on a soil affects its interaggregate porosity and its capacity to resist compression (Bradshaw, 1997). Most soil degradation results from practices that allow a reduction in soil vitality (Doran, 1996). Since the soil organic system plays a major role in the stabilisation of soil aggregates and in the creation of soil voids, poorly managed soils suffer compaction (cf. Lynch and Bragg, 1985; Van Breemen, 1993). This will reduce the stability of soil aggregates and reduce the rate at which soil macropores are created (Haigh, 1992). In practice, while all of the processes are important, the last three—trafficking, accelerated weathering and low soil vitality—often working in combination—are the most critical.

Settlement

Reclaimed lands are constructed from loosened materials. After emplacement, they tend to settle under their own weight. Smith (1971), consequent to laboratory tests, suggested that the material in spoil-tips could be compressed to a 25% smaller volume. In northern England, four already established reclaimed opencast coal mines showed an average settlement of 1.5% during 14 years (Reed and Hughes, 1990). In Poland, Chwastek (1970) found that the bulk of settlement takes place within a few months of the site being laid and the amount is small. Kilkenny (1968) suggested that settlement could be estimated as 0.74% of the depth of fill per log cycle time. Thompson et al. (1986) suggest that if the fill is clayey, settlement can reach almost 7% of fill-height within 3–5 years of emplacement. However, numerical modelling by Naderian et al. (1996) predicts that, on average, settlement should amount to around 3% of the fill-height and that post-construction settlement should amount to no more than 1%. In their case study at Jeebropilly Colliey, Queensland, Australia, settlement amounted to an average of 64 mm over the first 2 years. They agree that settlement is a relatively short-lived phenomenon on reclaimed sites.

Trafficking and Wheelings

The first impression gained from a visit to the floor of an active surface mine is often of dust or deep mud. The incessant churning of heavy machinery grinds up the fragments of loose rock creating, in dry conditions dust that blows off-site and in wet conditions a thick layer of mud. These processes continue through reclamation. The spoils are repacked and graded by heavy machinery and, although care is taken to avoid over compaction of the surface and topsoil layers, layers of crushed and puddled mine stones are common features in the soil profiles of mine spoils. In sites near haul roads, they may dominate the soil profile.

Much reclaimed land is reclaimed by topsoiling, the application of a layer of topsoil, often soil stored from the pre-mine environment. Reinstating this soil is a major operation. Each hectare may require 300–500 mm of soil, which implies an earth-moving task of 3000–5000 t $\cdot$ ha^{-1}, 50–100 truckloads. Even if these trucks loose-tip onto the site, this implies a minimum 50–100 passes of a modest truck with just 6 wheels, creating ruts at least 250 or 300-mm wide, which could affect 500–750 m^2 $\cdot$ ha^{-1}, 5–7% by area. Some of the vehicles used in surface mining and reclamation are huge (3–18 m^3) and heavy (3.6–20 t). Large earth-moving equipment can exert ground pressure of 800 kPa, agricultural vehicles 100–250 kPa and grazing animals 64–170 kPa of static pressure (Ramsay, 1986; Soane, 1981). Felton (1992) found that the B horizons, beneath the reconstructed topsoils on such sites, became highly compacted, reaching densities of 1.6–1.9 g·cm^3. He argues that compaction of the topsoil and upper subsoil layers was subsequently relieved by tillage (Felton and Taraba, 1994).

To date, there are no reliable estimates of the degree or extent of trafficking experienced by the soil and subsoil layer on reclaimed land during reclamation. However, this is only the first stage of disturbance experienced by reclaimed land, much of which is subsequently subjected to agricultural treatment. In agricultural contexts, Giles (1983) found that, after cultivation, 11% of his North Dakota case study field was affected by tractor wheelings. However, this underestimates the true extent of the problem. Most tractor wheel ruts are reworked during sequential agricultural operations. In Scotland, seedbed preparation for spring barley involved fertiliser distribution, harrowing (twice), sowing and rolling. During these processes, 91% of the field area was affected by wheelings (Soane, 1981). The impact of this wheeling-induced compaction persists in the deeper layers of the soil (Sweigard and Escobar, 1989). It is added to compaction due to the movement of the plough through the soil, as mentioned earlier.

Tyre slippage causes compaction and structural damage in the soil. Sohne (1966) estimates that the efficiency of transmission between tractor wheels and the soil may be only 60–65% and, on soft soil, it may be as little as 40%. The extra energy is wasted, often destructively, as slippage, compaction and frictional resistance. Studies show that this tends to vary from 10% in good conditions to 20% in bad. Textbooks advise that slippage should be constrained between 10 and 15%. Field studies find that slip percentage varies widely between 13 and 39%. As a rule, slippage above 15% is visible as smeared tractor tread-prints. Where slippage wipes out these tread-prints, slippage is far too high (Koolen and Kuipers, 1983: 138).

Laboratory soil-bin studies by Smith et al. (1989; Smith, 1987) show that 80% of soil compaction occurs in front of the centre line of the axle. This compaction is more related to load than tyre characteristics, and the depth of maximum compaction below the centre of the wheel is a function of half the contact dimension. Maximum soil deformation occurs directly under the wheel and its amount is related to rut depth and soil strength. Smearing, of course, is a particular problem in clayey soils and wet conditions. Consequently, the timing of trafficking is critical. Compaction also affects plant growth. Deibert (1983) points out that increased soil density begins to inhibit plant root growth at densities between 1.3 and 1.8 g·cm^3. In North Dakota, yields of soya in tractor wheelings were 50% lower than those on adjacent ground, the plants in the wheel ruts were later maturing and the stand was 20% reduced (Deibert, 1983). Erbach et al. (1988) found that corn yields could be reduced by 30% on plots traversed by heavier wheeled vehicles. Average yields were 10.4, 9.3 and 8.7 t · ha^{-1} for untrafficked, track trafficked and wheel trafficked plots.

Trafficking, like tillage, intimately affects the structure of soil aggregates. Powers and Skidmore (1984) show that tillage significantly reduces soil structural stability and that the degradation is proportional to the degree of disturbance. Tillage and compaction increases the dry aggregate stability of soils but decrease the proportion of water-stable soil aggregates. Results from North Dakota confirmed that 100 years of cultivation increased soil bulk density, especially in the plough layer (Giles, 1983). Soil bulk density at 7.5 cm depth in the soil beneath undisturbed grassland was 0.8 g·cm^3 (1.0 at 15 cm, 1.3 at 30 cm), while that on a field cropped to sugar-beet was 0.9 g·cm^3 (1.2 at 15 cm, 1.3 at 30 cm). The soil density beneath a neighbouring track was 1.4 g·cm^3 throughout the profile (Giles, 1983). In sum, tillage damages the organic bonds in soils and disrupts short-range mechanical bonds. Since these dominate the soil aggregate structures on reclaimed soils, where organic activity is minimal, these soils tend to suffer major damage during tillage.

Vehicle tyres are only one of the ways in which trafficking engages the soil. In addition, there are track-laying machines. Reeves and Cooper (1960) suggest that ground pressures caused by these are about half those of wheeled vehicles at unit depth. Brown et al. (1992) found that the effect of trafficking was restricted to the top 125 mm of soil but that cores from beneath plots traversed by track-laying machinery had higher porosity and lower density than those crossed by wheeled vehicles. There are also agricultural devices such as ploughs, tines, plough shares, harrows and rollers, and also the trampling of animals—all of which have different soil mechanical properties (Koolen and Kuipers, 1983). All trafficking, whether by heavy machinery used in land reclamation or smaller machines used in agricultural processes, presses down soils, crushes soil aggregates, smears clays and in moist conditions, liquefies and disperses soils by churning (cf. Bragg, 1983; Sweigard and Escobar, 1989). Clay content and soil moisture conditions are known to be critical. It has been suggested that there is a linear increase in the resistance of a soil to tillage as the proportion of the soil smaller than 0.005 mm increases (Khatchatryan, 1985:15).

In Wales, much soil damage results from trafficking and trampling after reclamation has finished. Here, it is usual that a relatively loose, open-textured A-horizon of restored topsoil rests more or less directly above an impermeable layer of compacted mine spoils (Kilmartin, 1994). Trafficking with wheeled vehicles commonly cuts through the topsoil and the wheelings ground in the clayey subsoil. Where these wheelings run downslope, they provide a course for drainage and on steep slopes for gully incision (Haigh, 1992). Even the relatively small disturbances caused by cattle and· sheep can cause problems. In locations where vegetation regrowth is poor, the soil surface may be protected by an organically reinforced, cryptogamic soil crust. In wet conditions, the treading and turning motion of grazing animals is capable of punching through this layer, thereby exposing fresh and unprotected soil to erosion. On upland sites in Britain, reclaimed land is often managed as common grazing and on these sites maintenance may be low while grazing pressures may be high (Haigh, 1992).

Characteristics of Compacted Soils

Compacted soils have fewer pores, less water-stable soil aggregates, lower infiltration rates, reduced water-holding capacities and a greater capacity to resist root extension. Soil compaction causes an increase in particle-to-particle contact within the soil, often resulting in increasing soil resistance to root penetration and a reduction in number of macropores. This decrease in soil porosity and decrease in size of soil pores is associated with an increase in the soil's capacity to hold water to its particles and a decrease in its permeability to air and water. There is a corresponding increase in runoff and erosion (Kilmartin 2000, this book). Roots cannot penetrate soil pores smaller than their tips, so the replacement of macropores by micropores impedes root growth. Compaction also affects soil chemistry and nutrient availability. It encourages anaerobic rather than aerobic soil conditions and may foster unfavourable microbiological processes such as denitrification rather than nitrification (Sheptukhov et al., 1982). Because of these changes, the soil becomes less favourable for soil organisms and allows less growth of surface protecting vegetation. Compacted soils are less capable of sustaining life and of absorbing rainwater. Consequently, compared to better quality soils, they have reduced bioproductivity and exhibit more surface runoff and erosion. In sum, one way of reducing runoff and erosion problems is to reduce soil compaction (Haigh, 1998).

The most direct indicator of soil compaction is soil bulk density. Ramsay (1986) has suggested that soil bulk density may be the best single indicator of land reclamation success. He adds: 'The importance of bulk density as a measure of agricultural potential cannot be overestimated' (Ramsay, 1986: 31).

Soil Bulk Density Measurements

Soil bulk density is simply the dry weight of a soil per unit volume. Since the density of water is close to 1 $g \cdot cm^3$ the bulk density expresses a ratio of the mass of soil to that of an equivalent volume of water. Bulk density is measured by a variety of techniques. Many involve either direct measurement or measurements of displacement/replacement of the test material by a material of known density. Some of these techniques are very simple and easily replicated, such as hammering metal rings into the soil, collecting soil cores of known volume and dividing this volume by the collected soil's dry weight. Baize (1992, 1993) offers a critique of this and other methods.

The reference values for soil compaction are the bulk density values of natural soils. In nature, clay, clay loam and silt loam soils may have densities ranging from 1.00–1.60 $g \cdot cm^3$ and sands and sandy loams: 1.20–1.80 $g \cdot cm^3$ (Brady, 1984; Ramsay, 1986). However, the upper ranges of these natural values are unfavorable for plant growth. Agriculturists normally recognise optimum soil bulk density levels below 1.3 $g \cdot cm^3$ (Zhengqi Hu et al., 1993: 131). Pot studies by Nagpal et al. (1967) found that as soil bulk density decrease from 1.67–1.27 $g \cdot cm^3$ the dry matter productivity of wheat increased by 60%. Medvedov (1990: 66) reports optimum soil bulk densities ranging from 1.20–1.33 $g \cdot cm^3$ for winter wheat and small grain agricultural on various soil types in the Ukraine. These compare with equilibrium, normal, soil densities that range from 1.25 $g \cdot cm^3$ to 1.50 for chernozems and sod-podsolic soils respectively.

Hodgkinson (1989) compares soil densities on reclaimed lands with those recorded for natural soils of similar texture by the Soil Survey of England and Wales. Here, the mean bulk density of topsoil samples (clay content 25–30% and organic matter <2%) was 1.24 $g \cdot cm^3$ and for subsoils (clay content 27–33%) 1.38 $g \cdot cm^3$. Five years after restoration, the bulk density of the surface mined lands remained 18–23% (respectively) higher. Table 13 lists published bulk density values for surface coal-mine-disturbed lands in Britain and America. The density scores are, in general, higher than one would expect in similar natural soil profiles, particularly in the 30–50 cm depth layer and tend to remain thus (cf. Haigh, 1992).

Critical Soil Bulk Density

Increasing soil density creates mechanical impedance that can restrict root growth and penetration of soils (Russell, 1977; Deibart, 1983). The influence can begin to be felt at quite low soil densities. In some forest soils, densities above 1.2 $g \cdot cm^3$ restrict root growth (Ballard, 1981). In Australia, Reeves et al. (1984) found that spring wheat grown in an uncompacted soil of average bulk density 1.32 $g \cdot cm^3$ has 240 $g \cdot m^3$ of roots in the 0–200 mm layer while that grown in a soil compacted to 1.52 $g \cdot cm^3$ had only 155 $g \cdot m^3$.

The point at which root growth is effectively stopped is termed the **critical bulk density** (Barnhisel, 1988: 200). This varies with soil-water content, soil structure, soil texture, plant species and particularly in the case of mine spoils, the proportion of low density soil constituents such as coals, slag, or cinders. Jones (1983) suggested that the critical bulk densities for crop rooting are inversely related to soil clay and silt plus clay percentages. Roots may not penetrate heavy clays with bulk densities above 1.46 $g \cdot cm^3$ and sandy soils with densities above 1.75 $g \cdot cm^3$ (Veihmeyer and Hendrickson, 1948). In Britain, foresters suggest that the

Table 13: Bulk density of surface coal-mine-disturbed lands (after Haigh, 1995)

Site, Age, Texture (Source)	Depth (cm)	Dry Density (g.cm^3)
Maes Gwyn, S. Wales: 10–22 years/shales with sandstones (Bending et al., 1992)	00–20	1.54
	20–40	1.77
	40–60	1.83
	> 60	2.00
Pwll Du, South Wales: 45 years/unvegetated clay with shales and sandstones [9–18 samples] (Haigh and Sansom, 1999)	00–05	1.66
	05–10	1.77
	10–30	1.76
	30–50	1.67
Pwll Du, S. Wales: 45 years/grass (50% cover) on 5 cm topsoil over clay wth shales and sandstones [5–10 samples] (Haigh and Sansom, 1999).	00–05	1.55
	05–10	1.79
	10–30	1.77
	30–50	1.79
Blaenant, S. Wales: 20 years/grass (100% cover) with 100–150 mm topsoil over clayey spoils with shales and sandstones [14 samples] (Haigh and Sansom, 1999; Kilmartin, 1994]	00–05	1.39
	05–10	1.86
	10–30	1.83
	30–50	1.87
Cadeby Washings, Doncaster, England: 6 weeks/recently rewashed for coal extraction and relaid, unvegetated, deep mine coal spoils, pH 5.6–5.2 [1–4 samples] (Haigh and Sansom, 1998; Dyckhoff et al., 1993).	00–05	1.26
	05–10	1.69
	10–30	1.82
	30–50	1.91
Butterwell, Northumberland, U.K./ 1 year/clay loam/(King, 1988)	00–21	1.38
	21–51	1.64
	40–50	1.80
	51–82	1.68
	83–120	1.68
Acklington, Northumberland, UK/4 years/clay loam over shales and mudstones (Davies et al., 1992) (Densities of undisturbed topsoils 1.29 g.cm^3 undisturbed subsoils 1.76 g. cm^3).	00–25	1.49
	25–120	1.80
Rador, Northumberland, U.K./10 years/silty clay loam (Bragg, 1983; Bragg et al., 1984)	05–10	1.40–1.60
	15–20	1.57–1.67
	30–35	1.60–1.70
	40–45	1.68–1.75
Limekiln Lane, Shropshire, U.K./4 years/Sandy loam/(Hodgkinson et al., 1988; Hodgkinson, 1989) (Densities of local undisturbed soils range from 1.30–1.48 g. cm^3)	00–07	1.47
	08–15	1.57
	18–25	1.51
	27–34	1.80
	35–43	1.92
D. Johnston Mine, Glenrock, Wyoming /n.d. Sandy clay loam (Toy and Shay, 1987) (natural soils: 1.42 g.cm^3)	00–40	1.44

Contd.

Contd. **Table: 13**

Site, Age, Texture (Source)	Depth (cm)	Dry Density (g.cm^3)
River King 6, Randolph County and Eads Mine, Jefferson County, Illinois/n.d./silt clay loam (Indorante et al., 1981)		
(Densities of natural soils: 0–15 cm: 1.32–1.34 g.cm^3)	00–15	1.31–1.54
(Densities of natural soils: 15–30 cm: 1.46–1.47 g.cm^3)	15–30	1.53–1.75
(Densities of natural soils: 45–60 cm: 1.42–1.50 g.cm^3)	45–60	1.61–1.78
(Densities of natural soils: 60–75 cm: 1.50–1.55 g.cm^3)	60–75	1.60–1.74
Norris Mine, western Illinois/n.d./reinstated mollisol topsoil	00–30	1.7–1.9
over loess/glacial clay overburden (Indorante et al., 1981)	> 30	1.4–1.7
(Densities of natural subsoils 1.4–1.5)		
Gibraltar Mine of Peabody Coal, Kentucky, USA/n.d./reconstructed	01–07	1.36–1.40
Coarse silt, acid, aeric flauaquent	20–25	1.62–1.76
Soil for arable use/topsoil depth	50–56	1.71–1.80
200-mm/4 samples	81–87	1.60–1.90
Captain Mine, southern Illinois/n.d. reinstated topsoil from	00–30	1.4–1.9
alfisol/over coarse shovel spoil (Indorante et al., 1981)	730	1.7–1.9
(Densities of natural subsoils 1.4–1.7)		
Coopers Rock/Gladesville, West Virginia/70 to 130 years/clay loam (Smith et al., 1971) (Densities of natural soils: 0.90–1.13 g.cm^3)		1.39–15.8
Stellarton, N.W. Coal Reclamation, Nova Scotia/overburden (particle density 1.8 (range 1.6–2.1 g.cm^3, pH 6.5) (CANMET, 1977: 44)		1.1–1.4
Sydney, Cape Breton Is., Nova Scotia/overburden (particle density 2.1 (range 1.4–2.9 g·cm^3, pH 3.8) (CANMET, 1977: 45)		1.1–1.7
Minto Area, New Brunswick/overburden (particle density 2.5 (range 2.2–2.9 g·cm^3) pH 4.9 (CANMET, 1977: 46)		1.2–1.8
Bienfait—Estevan, Saskatchewan/overburden/pH 7.8 (CANMET, 1977: 47)		1.33–1.54
Kaiser Erikson, Michel Creek, British Columbia/overburden/(particle density 2.4 (range 2.34–2.52)/pH 6.0) (CANMET, 1977: 49)		1.73–1.99
Kaiser Lower C., Michel Creek, British Columbia/overburden/(particle density 2.05 (range 1.96–2.11 g.cm^3, pH 7.6) (CANMET, 1977: 50)		1.82
Kaiser McGillivray, Michel Creek, British Columbia/overburden/(particle density 2.08 (range 2.06–2.09, pH 7.5) (CANMET, 1977: 51)		1.25–1.38
Utrillas, Teruel, Spain/recently created (2–3 years), 18–20°,	00–50	1.38
surface coal mine spoil banks. Clay 25%, pH 8.3 (Nicolau, 1996)	> 100	1.43–1.51
7th September Mine, Pernik, Bulgaria/deep mine spoils	00–20	1.21
(particle density 2.73–2.75) (Kandev, 1987)	20–40	1.34
	40–60	1.35
	60–80	1.32
	80–100	1.38

Contd.

Contd. **Table 13:**

Site, Age, Texture (Source)	Depth (cm)	Dry Density (g.cm³)
Maxim Taban, Pernik, Bulgaria/briquette manufacturing	0–5	0.97
spoils/8 years old (particle density low)	15–20	1.35
	30–40	1.43

minimum standards for tree establishment on disturbed land should be a soil density < 1.5 g·cm^3 to 50 cm depth and < 1.7 g·cm^3 to 100 cm (Moffat and Bending, 1992: 2).

However, it is generally accepted that the upper limit for organic penetration of a soil layer does not exceed 1.8 g·cm^3 (cf. Ramsay, 1986: 31; Barnhisel, 1988:201, Zhengqi Hu et al., 1993: 131). Root impedance becomes a problem requiring concern and special treatment as soil bulk densities increase beyond 1.6 g·cm^3 (Verpraskas, 1988; Lal et al., 1989). Haigh (1995) has suggested the following standard for soil densities on reclaimed coal lands. The definition of 'successfully reclaimed' land should include the condition that soil bulk densities must not exceed 1.6 g·cm^3 (within 0–50 cm) and 1.8 g·cm^3 (within 100 cm) of the soil surface. Sites with soils that exceed these tolerances should be classed as substandard and 'in need of treatment'. This condition should obtain in checks conducted at reclamation and both 10 and 20 years subsequently.

Haigh further suggests that bulk density be evaluated by 5–cm depth increments through the soil profile using the small rings method. This involves pushing thin, sharp edged metal cylinders of known (100 cm^3) volume into the soil, recording the dry weight of the sample contained, dividing this by the volume and recording the result in grams per cubic centimetre (Haigh, 1995). This method has many problems, especially in stony soils, and results need to be based on the average of several tests (cf. Baize, 1993). However, the test is very easily conducted and its findings equally easily checked. It is also capable of detecting the thin bands of dense material that characterise 'reclaimed' site soil profiles.

Soil Strength: Cone Penetrometer Index

An alternative approach to measuring soil density is to measure soil strength. Field studies suggest that roots exert an axial, longitudinal pressure of around 0.9–1.5 MPa and that they cease to enlarge at a radial pressure of around 0.85 MPa (Eavis et al., 1969; Taylor and Ratcliff, 1969). Researchers seeking to measure soil capacity to constrain plant growth often use a soil strength index. This is usually based on penetrometer studies (Thompson et al., 1987; Zhengqi Hu et al., 1993: 131).

Busscher et al. (1987: 377) employed a 5-mm diameter flat head penetrometer and a resistance of 2 MPa as their index of root growth zero conditions for soils equilibrated at -100kPa soil-water potential. Most other workers prefer to employ a cone penetrometer. Many of these are based upon the device used by the United States Army Corps of Engineers to determine trafficability of soils. This cone penetrometer consists of a 30-degree circular cone with a 3.2-cm^3 base area. The Cone Index is measured as the force per base area required to drive this cone into the soil at the rate of 35 mm·s^{-1} (1829 mm·min^{-1}/72"·min^{-1}) (ASAE Standard 313.1, 1980).

There is a considerable range of error and variability in cone penetrometer measurements. At least 10–20 measurements should be collected for each data point (Smith, 1987: 87). Indeed, there are many problems with this approach. Not least among these is the fact that roots exert a different kind of pressure to a penetrometer. Cone penetrometers displace soil downwards, and also sideways, sometimes carrying a soil body in front. The consequence is that while field studies confirm that cone penetration results increase as biological productivity declines, the cone penetrometer indicates values much higher than those suggested by studies of the roots themselves. For example, Willat (1986) reports experiments involving the artificial compaction of a Scottish sandy clay loam (Stagnogley cambisol). Here, increasing cone resistance from 1.5–4.0 MPa did no more than halve the recorded root length of barley. Tollner and Verma (1984) have suggested that the accuracy of results from a cone penetrometer can be improved by lubrication of the cone. This reduces soil/metal friction.

Cone penetrometer readings are correlated with soil clay content, dry bulk density and soil moisture content (Ayers and Perumpral, 1982). Laboratory studies by Gerard et al. (1982) have shown that the Critical Cone Index, the Cone Index value in which root elongation is suppressed varies with clay content. More than this, Cone Index values vary directly with bulk density and inversely with soil moisture (Voorhees, 1983). Grimes et al. (1975) suggest that no one Cone Index value of even range of values is appropriate to the evaluation of root density at all depths. Root density tends to become more restricted at lower Cone Index values deeper in the soil. In consequence, Christensen et al. (1989) suggest that all Cone Index values must be referred to reference values of soil moisture and bulk density constructed at each soil depth. Ayers and Perumpal (1982: 1171) offer the following equation to link Cone Index values (CI) with dry bulk density (DD) and moisture content (MC) via four empirical constants related to soil texture and type (C_1–C_4):

$$CI = (C_1 + DD^{c4})/[C_2 + (MC - C_3)2] \qquad ... (10)$$

The statistically significant relationship provides a useful summary of the issue but it is of little practical utility.

Despite such problems, several workers have used the Cone Index to diagnose compaction problems and frame treatment strategies. Gerard et al. (1982) report that the Critical Cone Index values for cotton are approximately 6.5 MPa for a soil with 10% clay and 3.0 MPa for a soil with 50% clay. Field data from Verpraskas and Wagger (1989) suggest that effective root suppression can be achieved at mean Cone Index values of about 4.0 in a soil of 12% clay. However, they also report that, in soils with higher clay contents, the key variable is not soil strength but the presence of shrinkage cracks that create macropores allowing root penetration. Earlier workers have shown that cotton showed impeded root growth on soils with Cone Index values as low as of 2.8–2.1 MPa while 1.7 MPa was sufficient to impede the growth of maize in a coarse textured soil in Florida (Fiskell et al., 1968). It has also been suggested that, while the spherical expansion of a root system may be restricted at 1.6–2.4 kPa, cylindrical expansion may be restricted by pressures as low as 500–800 kPa (Tardieu, 1989: 153).

In sum, the recognition of soil strength parameters for reclaimed sites is more speculative than recognising limits for bulk density (Haigh, 1995). It has been suggested that a penetrometer index of 1.2 or less is optimal (Zhengqi Hu et al., 1993: 131). Further, a soil which, at an appropriate soil water content or potential, has a Cone Index value greater than 2.5 MPa (cf. Zhengqi Hu et al., 1993: 131) should be considered a cause for concern while soils with values greater then 7.0 MPa may require serious attention.

Ameliorating Soil Compaction

Although soil compaction may develop naturally on many reclaimed lands, due to the weathering of mine spoils, compaction during restoration should be avoided. Trafficking of the materials that lie close to the surface must be reduced to a minimum. Deep ripping and scarifying in advance of recultivation may cause a temporary alleviation of compaction but can also cause long-term damage to the soil by disrupting structural elements. Tillage may alleviate compaction in the surface horizons on reclaimed lands (Felton and Taraba, 1994). However, it may also cause severe damage to any soil with low structural stability and low vitality, so it is best kept to a minimum. It is often suggested that subsoiling of compacted layers in the spring may permit root growth and the amelioration of such layers (Daviers et al., 1992). Results from forestation are positive (Ashby, 1996; Wilson, 1985). However, this process also causes major structural damage to soils and is best undertaken infrequently.

Some interesting solutions to the problems of soil compaction emerge from ecological engineering. It has been argued that trees provide one route to alleviating the physical properties of soil. Forest fallowing is often advocated for the restoration of natural soil systems on reclaimed lands (Flege, this book). In Britain, great success has been achieved by sowing earthworms in compacted pasture lands (Scullion, 1992; Armstrong and Bragg, 1984). Care must be taken to select the forest or earthworm species best suited to the conditions in the new spoils. Species selection should be influenced by soil texture, chemistry, drainage and so forth. However, both practices seem to benefit from an initial loosening of the soil (Rushton, 1986; Wilson, 1985).

III. ACID MINE DRAINAGE AND ITS CONTROL

In many nations, the pH of reclaimed lands goes through a characteristic pattern. It begins at a respectable level, pH 5–7. Then, if there are reactive, usually pyritic rocks mixed into the spoils, that pH begins to plummet. Within a few months or years, it declines to pH 2–4. Eventually, decades later, the pH may begin to rise once again as the reactive materials in the spoils are either leached out or complexed by reactions with other components of the spoil and its organic system. In the mean time, the acidity causes serious problems for the revegetation of the reclaimed land. Vegetation can be destroyed by contact with these low pH materials and their leachate. However, these problems are dwarfed by the problems caused off-site by the leachate. It causes acid mine drainage, which is a huge environmental and legal headache for many mining companies in those many nations where coals or pyritic metals are mined.

Soil Acidity

The 1992 version of the French Referentiel Pedologique characterises pH levels below pH 3.5 as hyperacid and those between pH 3.5 and 4.2 as very acid (Baize, 1992). Brady (1984: 728) calls soils with pH levels between pH 5.0 and 4.5 very strongly acid, those below pH 4.5 extremely acid and those with a pH above 8.7 very basic.·

Low pH may have very negative impacts on land reclamation quality. However, different plants have developed different tolerances to soil acidity and, because of the physiological factors involved, it is difficult to generalise the limiting conditions implied by soil pH. Nevertheless, soil bacteria and actinomycetes do not thrive, while the oxidation/fixation of nitrogen is curtailed, in mineral soils of pH 5.5 and below. At pH 5.0 and below, there is a tendency for soil phosphates to become fixed and not plant available. It has been suggested that pH 3.0 is the threshold for forest recultivation (Strzyszcz, 1992). British foresters prefer that the minimum soil requirement for planting disturbed lands to amenity woodland is a soil pH within the range pH 3.5–8.5. (Moffat and Bending, 1992). In Wyoming, the suggested range is pH 5.0–9.0 (Ackerman, 1983). Haigh (1995) suggests that since reclaimed lands should not be allowed to develop hyperacidity, a pH < 3.0 (pH 3.5 minus 0.5—the maximum range of normal seasonal variation) might be offered as the minimum acceptable while perhaps pH 9 should be taken as the other extreme.

On reclaimed land, the traditional practice is to bury acid materials beneath a layer of topsoil. It is known that the depth of topsoil affects agricultural productivity, at least initially (Pole et al., 1979). The question arises, how deep should that burial be in order to avoid the problems due to acidity. Certainly, it should lie below the root zone. Thus, where such low pH materials exist, they should not be within 1.5 metres of the soil surface (cf. 2.0 metres recommended by Moffat and Bending, 1992). Even these depths are problematic. Smith (1993: 30) indicates the following circumstances that cause upward movements of contaminants buried in the soil. They are: 1) bulk movement of contaminated soil or fill as during construction or gardening; 2) increase in groundwater table levels or flooding; 3) action of soil organisms such as worms; 4) uptake, growth and decay processes of plants; 5) capillary rise of moisture in dry summer conditions; 6) vapour phase transfer; and 7) erosion, especially gully incision and mass movements. All these possibilities must be avoided. The best policy for dealing with acid waste may be to mix it with some neutralising agent (cf. Doyle, 1976) then bury it far below the reach of surface processes, including weathering.

Acid Mine Drainage

Acid mine drainage is a huge problem. In North America almost half the mine stones produced have acid potential (AGRA, 1995). In Silesia, Poland, acid-generating coal wastes are produced at the rate of 30 $30 \times 10^6 \text{t·yr}^{-1}$ (Bzowski, 1997). The problem is not restricted to coal-spoil deposits. In 1994, Natural Resources Canada examined the extent of acid-generating mine wastes at base metal mining operations. British Columbia, Saskatchewan, Manitoba, Ontario, Québec, New Brunswick, Newfoundland, Yukon and Northwest Territories had a total area of over 12,500 hectares of tailings and 740×10^6 t of waste mine rock from the last forty years of mining. The study assumed that an equal quantity of acidic tailings and waste rock would accumulate over the next twenty years. The United States Environmental Protection Agency's National Stream Survey of the mid-Atlantic and South-eastern USA found that 4,590 km (2%) of the total stream length was acidic due to AMD and that a further 5780 km were strongly impacted (Herlihy et al., 1990). This was broadly equal to the extent acidified from natural spring flows. A more recent report from Pennsylvania suggests that metal-loaded acid drainage from coal mines has degraded nearly 4200 km of streams and the cost to correct the problems will be $5 billion (1998). AMD is also an international problem (Wiggering, 1993; Singh and Bhatnagar, 1985; Costigan, et al., 1981). However, research seems to be led from the USA and Canada.

Acid mine drainage (AMD), sometimes styled acid rock drainage (ARD), is caused by the oxidation of sulphides in the mine stones exposed during mining. This oxidation begins immediately on exposure of the mine stones and is much enhanced by microbial processes (Bell and Finney, 1987). It results in the production of sulphuric acids and sulphates. The problem is compounded because the resultant acidic leachate mobilises other constituents in the mine spoils, not least metals, causing a further degradation of water quality (Table 14). Curiously, one of the first papers ever presented on the subject considered that AMD might be useful; the waters were shown to have a germicidal affect on the bacteria found in sewage (Dixon, 1910). However, most subsequent reports have been less positive and, by 1918, Sherlock was already discussing the responsibilities of American mine owners in the light of recent court judgements.

Acid drainage has been defined as water of pH < 6.0, in which total acidity exceeds total alkalinity (United States Code of Federal Regulations 30: Mineral Resources 1979 and seq.). However, worries about AMD tend to reflect a pH of water discharge that ranges between pH 4.5 and 2.0 (Bates and Jackson, 1980). In fact, the pH of AMD can often fall below 3. At this pH, iron and other metals such as zinc, copper, lead, arsenic and manganese are mobilised. The combination is toxic for most aquatic species. In many instances, streams receiving AMD or ARD are almost lifeless many miles downstream of the AMD source.

Part of the problem emerges on dilution. When the water pH rises above 3, ferric iron is precipitated and coats the stream bottom with an orange sludge of $Fe(OH)_3$ and related compounds. This sludge, which Americans call 'Yellow Boy', colours the stream water and smothers the organisms that live in and on the channel floor.

Acidity is the capacity of a water solution to neutralise basic or alkaline cations while alkalinity is the capacity of a water solution to neutralise acid solutions—a property largely attributable to the presence of the bicarbonate ion (HCO_3^-) but also carbonate CO_3^- and hydroxyl OH^- ions. Acid-base-accounting is the fundamental procedure used to evaluate the potential toxicity of coal-overburden materials. It consists of two measures: first, total or pyritic sulphur and second, the neutralisation potential-acid formation from pyrite oxidation and alkalinity due to the dissolution of carbonates and other basic minerals.

The acid generation process has three phases: initiation, propagation and termination (OSMRE, 1998). The initiation phase begins as soon as pyritic materials are exposed in an oxic environment but the amount of acid generated is small. During propagation, acid production increases rapidly. During termination, it declines. Peak acid loadings occur 5 to 10 years after mining. There follows a gradual decline extending beyond 50 years until acid release ceases (Ziemkiewicz and Meek, 1994). In laboratory studies, it has been shown that concentrations of soluble metals are one

Table 14: Typical USA ARD levels and permissble limits (Gazea et al., 1995)

.pH/Metals in mg.l	Appalachian coal-mine drainage (Phelps, 1987)	Coal-mine drainage (Hedlin et al., 1994)	AMD1–waste rock pile: Ignace, Ont. (Rao et al., 1992)	AMD2– tailings pond: Ignace, Ont. (Rao et al., 1992)	Limits for industrial effluents (USEPA, 1982)
.pH	1.4–7.0	2.6–6.3	–	2.2	6–9
Fe	1–10,000	1–473	1,630	2,360	3.5
Al	1–2,000	1–58	751	61	–
Zn	0–10	–	4,480	1,630	0.2–0.5
Mn	0–50	1–130	206	160	2

or two orders higher on the first two or three occasions a pyritic spoil is subject to leaching and that they decline rapidly thereafter (Doepker, 1994).

These trends in reaction rates can be offset or enhanced by the mass balance between acid and alkaline-producing minerals. If there is both a low pyrite and base content, drainage may be slightly or non-acid with low concentrations of metals. If there is low pyrite and high base content, drainage is alkaline with low concentrations of metals. If the spoils are high pyrite and low base content, drainage is acid with high concentrations of dissolved metals. Finally, if there is high pyrite and base content, drainage may be alkaline or acid with high concentrations of metals (Curtis et al., 1988). Naturally basic materials, such as limestone and hydrated lime, are often employed for the neutralisation of AMD (Curtis et al., 1988).

Several further factors affect the generation of ARD (Costigan et al., 1981). These include the physical characteristics of spoil substrates, especially their permeability. Porous pyritic sandstones tend to release their acid load rapidly while less permeable clayey mudstones release their acid load more slowly. The generation of AMD is also affected by mine-site hydrology (cf. Kilmartin 2000, this book). Saturation is one method used to control AMD generation and release. AMD release is affected by erosion that exposes pyrtic rocks to weathering (Curtis, 1971). Climate is also an important determinant of the rate and degree of flushing, which determines the concentration of acid in water leaving the soils.

Chemistry of AMD/ARD

In chemical theory, the main reactions between the oxidation of pyrites (FeS_2) and the production of acid (H^+) have been summarised as follows by Singh and Bhatnagar (1985). Different workers debate whether there is preferential release of sulphur or iron during oxidation; however, the balance of opinion is that the creation of a sulphur- rich surface is the first phase (Buckley and Woods, 1987; cf. Peters, 1986).

$$2FeS_2 + 7O_2 + 2H_2O \geq 2Fe^{2+} \geq 4SO_4^{2-} + 4H^+ \qquad \text{... (11)}$$

$$Fe^{2+} + \tfrac{1}{4}O_2 + H^+ \geq Fe^{3+} + \tfrac{1}{2}H_2O \qquad \text{... (12)}$$

$$Fe^{3+} + 3H_2O \geq Fe(OH)_3 + 3H^+ \qquad \text{... (13)}$$

$$FeS_2 + 14Fe^{3+} + 8H_2O \geq 15Fe^{2+} + 2SO_4^{2-} + 16H^+ \qquad \text{... (14)}$$

The net result is that each mole of FeS_2 generates 4 moles of acid. Equation (11) generates 4 moles of acid (eqn 13). Equation (14) comes into play when the ratio $Fe^{2+}/Fe^{3+} \geq 2$ (Singh and Bhatnagar, 1985). The speed of these reactions depends upon the character of the pyrites in the mine spoil. This occurs in four forms and some, notably fine-grained spherical particles of framboidal sulphur, are much more reactive than others.

Phelps (1987) summarises the same chemistry in a slightly different fashion and considers the second reaction (eqn 18) to be much faster.

$$2FeS_2 + 7O_2 + 2H_2O \geq 2FeSO_4^- + 2H_2SO_4 \qquad \text{... (15)}$$

$$4FeSO_4^- + 2H_2SO_4 + O_2 \geq Fe_2(SO_4)_3 + 2H_2O \qquad \text{... (16)}$$

$$Fe_2(SO_4)_3 + 6H_2O \geq Fe(OH)_3 + 3H_2SO_4 \qquad \text{... (17)}$$

$$FeS_2 + 7Fe_2(SO_4)_3 + H_2O \geq 15FeSO_4^- + 8H_2SO_4 \qquad \text{... (18)}$$

The literature contains many versions of the same reactions. However, the pattern is fairly clear—the oxidation of sulphur and pyritic compounds in mine spoils leads to the generation of a great amount of sulphuric acid. However, in the absence of micro-organisms, these reactions would occur relatively slowly and ARD might be a much smaller problem. The reality is that ARD becomes a problem mainly because the process is accelerated through the metabolic reactions of autotrophic bacteria, particularly their oxidation of Fe^{2+}. The volume of bacteria in mine spoils may be as high as 10^9 per ml (Singh and Bhatnagar, 1985). Many are conventional heterotrophs that use organic compounds to grow. However, it is not unusual to find even these bacterial cells and their remains encrusted with ferrihydrite and in association with other poorly ordered iron oxides (Ferris et al., 1989). Regardless of the type, bacteria can act as passive nucleation elements for iron oxide deposition in acidic sediments (Ferris et al., 1989; Konhauser, 1998). Iron oxide precipitation may occur as the by-product of the metabolism of the iron-oxidising bacteria. Bacteria excrete organic wastes into the environment that induce this mineralisation. However, in many cases, bacterial biomineralisation tends to be a two-phase and partly inorganic process. First, metal ions are electrostatically adsorbed onto anionic surfaces of the cell wall and surrounding organic polymers. Second, these sites provide nucleation sites for subsequent crystal growth. These crystal formations and, indeed, their governing processes are inorganic (Konhauser, 1998).

Two bacterial pathways of pyrite oxidation are described: direct and indirect (Singh and Bhatnagar, 1985). It is not known which is the more important to leaching. Direct requires contact between mineral and bacteria under anaerobic conditions. The bacteria adhere to the mineral surface and dissolve the metal by enzymatic attack of the surface. In indirect, ferric iron is the primary oxidant. The bacterium oxidises pyrite and reduces it to ferrous iron. This is oxidised by the bacteria and is restored as ferric iron (as sulphate) once again. The net result is summarised as equation (19).

$$FeS_2 + 3.5O_2 + H_2O = Fe^{2+} + 2SO_4^{2-} + 2H^+ \qquad \ldots (19)$$

During initial oxidation at neutral pH, the two rates of pyrite oxidation, biotic and abiotic are comparable. At pH values of 6 and above, bacterial activity is thought to be small compared to abiotic reaction rates. As the pH drops at around pH 4.5, the abiotic rate of oxidation decreases and the rate of bacterial oxidation increases. The bacterial actions remove constraints on pyrite weathering and allow the reactions to proceed rapidly. Below pH 3, a steady state cycle of bacterial oxidation of ferrous to ferric iron and the oxidation of pyrties by ferric iron is set up. Bacterial activity peaks in the range of pH 3-2.

ARD: Management and Control

Although ARD prediction is difficult, it is advisable that every mine operator evaluate the scale of any potential problem and also the extent of the costs that such treatment might impose (Skousen et al., 1990). These costs can be very high, especially if they have to be sustained for a long period (OSMRE, 1998). It is possible that treatment may have to be continued for the decades it may take for the discharge to decline to levels within current effluent standards (Curtis et al., 1988). Potentially, the burden may be enough to bankrupt a small mining company (Dreese and Bryant, 1971: OSMRE, 1998). America's Office of Surface Mining, Reclamation and Enforcement is currently working on producing a manual to assist with these problems through its Acid Drainage Technology Initiative (ADTI), a joint venture with industry, regional Government, academia, and other government agencies, that is seeking scientific solutions for AMD problems. The ADTI has two work groups: AMD prevention/remediation and AMD prediction.

Table 15: Acid-generating bacteria isolated from coal spoils (after Singh and Bhatnagar, 1985; Johnson, 1995)

Thiobacillus ferrooxidans:

This is a slow-growing, gram-negative, chemo-autotrophic acidophile that obtains energy from the oxidation of ferrous iron, elemental sulphur, several reduced sulphur compounds and sulphate. It uses atmospheric carbon dioxide as its sole carbon source. It tolerates pH 0.8–6.98 but requires some moisture (Kleinmann and Crerar, 1979; Twardowska, 1986). In the mineral processing industry, the use of T. ferrooxidans in metal liberation is growing rapidly.

Thiobacillus thiooxidans:

This bacterium thrives on ferrous iron, elemental sulphur and thiosulphate but is less efficient in acidification than *T. ferrooxidans* (Temple, 1952).

Ferrobacillus ferrooxidans

This bacterium thrives on ferrous iron, elemental sulphur and thiosulphate, and accelerates pyrite oxidation, it is said, by a factor of 8–13 (Rogoff et al., 1960).

Ferrobacillus sulphooxidans

First identified in Pennsylvanian mine waters, this bacteria derives energy by oxidising both elemental sulphur and ferrous iron (Kinsell, 1960).

Leptospirillum ferrooxidans

An iron-oxidising acidphillic mesophile found to be an important contributor to AMD at Iron Mountain, Ca., a heavily polluted former iron mine, (http://wisc.edu/news/this week 9/98; cf. Johnson, 1995: 46).

Also
Thiobacillus concretivorus
Thiobacillus neopalitanus
Thiobacillus thioparus

Other Possible Minor Players:
Metallogenium spp.
Acidophilium spp.

Representing a group of heterotrophic acidophilic bacteria that have a mainly passive role in AMD generation, they metabolise the organic substances that are potentially toxic to be autotrophic iron-oxidising bacteria (Johnson, 1995).

Sulpholobus acidocaldarius

This represents the exclusively archaean group of acidophilic and thermophilic bacteria that are usually isolated in hot springs (> 60°C) but occasionally from AMD. This species can couple both iron and sulphur reductions (Johnson, 1995: 48).

Prediction

Predicting acid generation from mine spoils remains difficult (Coastech Research Inc., 1989; OSMRE, 1998). The traditional approach involves acid/base accounting (Brady et al., 1994; Sobek et al., 1978). This quantitatively balances acid-generating pyrites against carbonates and other alkaline materials in the spoils. It has proved effective for the design of topsoils but unreliable as a predictor of drainage quality (Erickson and

Hedlin 1988). Other predictive methods have derived from simulated weathering tests. Many of these attempt to model cyclic wetting/drying and flushing of spoil piles (Renton et al., 1988; Scharer et al., 1991). Jaynes et al. (1984) attempted to adapt the general model of pyrite weathering of Cathles and Apps (1975) to the description of AMD. This uses first-order reaction kinetics, diffusion modelling and an activity model for bacterial action that is based on available energy in the substrate. The accuracy of these methods has yet to be adequately ground-truthed and the same is true of the current crop of computer models (OSMRE, 1998). The modellers try to incorporate the chemical reactions of acid generation, microbial catalysis and leaching (transport) of the weathering products into predictive programs. However, they generally lack verification and often include parameters that are difficult to measure or estimate. One promising model uses two composite variables, one for acid generation and a second for the leaching of weathering products (Rymer et al., 1990 Hart et al., 1991).

Prevention/Mitigation

ARD/AMD control has four foci: chemical inhibition of acid generation; isolation of the acid-generating rock, suppression of the bacterial catalysts of ARD reactions; and chemical or biological treatment of runoff. Many of the techniques that aim to prevent acid formation are based on the control of oxygen.

Oxygen reaches the pyrite by two mechanisms, convection and molecular diffusion. Convection is caused partly by the heat generated through pyrite oxidation and partly by the chemical reaction that removes gas from the spoil atmosphere. In either case, new oxygen is drawn into the spoil to fuel the reaction. Molecular diffusion operates wherever there is a difference in the oxygen concentration in a fluid. The process operates around 10,000 times faster in air than water, which is why soil moisture is sometimes touted as a barrier to acid formation.

Chemical Treatment with Alkaline Agents

The benefits of lime and other alkaline agents have long been recognised in AMD mitigation. However, the complex chemistry of spoil materials fosters varying levels of effectiveness in alkaline addition studies. Direct mixing and contact with pyritic materials appears most effective but an optimum lime-to-pyrite ratio remains unknown. Indirect treatments, such as alkaline recharge and borehole injection, have also yielded mixed results (Aljoe and Hawkins, 1991; Ladwig et al., 1985). Hence, treatment, as normally applied, involves the direct chemical neutralisation of acidity, followed by precipitation of iron and other suspended solids. Treatment systems include equipment for feeding the neutralising agent to the AMD; means for mixing the AMD and neutralising agent(s); procedures for ensuring iron oxidation; and settling ponds for removing iron, manganese, and other co-precipitates (OSMRE, 1998). The degree of treatment-system sophistication necessary depends on the effluent standards to be met, as well as on AMD chemistry, AMD quantity and AMD release period. Common AMD/ARD neutralising agents include limestone, hydrated lime, soda ash, caustic soda and ammonia.

Neutralisation by adding alkaline agents offers the benefits of removing acidity, of removing many metals, not least dissolved ferrous iron by precipitation, and of removing sulphates (Curtis et al., 1988). Its disadvantages include increased hardness of water, increased water content of dissolved solids, high runoff sulphate loading (often < 2000 $mg \cdot l^{-1}$), inability to reduce iron below 3–7 $mg \cdot l^{-1}$, and the creation of waste sludge along with the problem of its disposal.

HYDRATED LIME (CALCIUM HYDROXIDE)

Hydrated or slaked lime is the coal mining industry's neutralising agent of choice (OSMRE, 1998). It is easy and safe to use, effective and relatively inexpensive. However, high initial costs are incurred because of the size of the treatment plant. The lime is usually fed as slurry into AMD in a mixing chamber. If ferrous iron is low (> 50 $mg \cdot l^{-1}$), the water is treated to pH 6.5–8.0 and released. If it is high, the flow is aerated and the iron deposited as ferric hydroxide. The flow is then diverted to stilling tanks where metals (and when the calcium sulphate concentration rises above 4,000 $mg \cdot l^{-1}$ then gypsum) are precipitated as sludge (Curtis et al., 1988). One problem of the process is that it generates large volumes of sludge compared to the other most common agent, limestone (OSMRE, 1998).

$$Ca(OH)_2 + H_2SO_4 \geq CaSO_4 + 2H_2O \geq CaSO_4. 2H_2O \qquad ... (20)$$

$$Ca(OH)_2 + Fe(SO_4) \geq CaSO_4 + Fe(OH)_2 \qquad ... (21)$$

LIMESTONE (CALCIUM CARBONATE)

Limestone reacts in AMD as follows:

$$CaCO_3 + H_2SO_4 \geq H_2O + CO_2 \qquad ... (22)$$

$$3CaCO_3 + Fe_2(SO_4)_3 + 3H_2O \geq 3CaSO_4 + Fe(OH)_3 + 3CO_2 \qquad ... (23)$$

Limestone treatment costs are lower than for many other agents. It produces a dense and relatively low volume sludge. However, it is not very popular. Its disadvantages include the fact that its reaction with AMD produces carbon dioxide (eqns 22, 23) that buffers the reaction. This makes it difficult to raise pH above 6.0 and means that limestone is ineffective where acidity is above 50 $mg \cdot l^{-1}$ and in removing manganese. (Curtis et al., 1988). It is also ineffective in water high in ferrous iron or a high ferrous/ferric ratio. Efficiency is lost due to coating of the limestone particles with iron precipitates, sulphate, sediment and biological growths. This means that the effect of dumping limestone in a stream-bed may last no more than a month. In addition, in formal treatment plants, reaction times tend to be slow, the treatment system more complex than for hydrated lime and more dependent on the limestone characteristics. It should be high in calcium carbonate, low in magnesium, low in impurities and pulverised to a large surface area. Curtis et al. (1988) advocate the combined use of both limestone and hydrated lime. Limestone is used to raise pH 4.0–4.5, then hydrated lime used to finish the work. Bzowski (1997) used lime to enhance the buffering capacity of claystones present in Silesian coal-mine wastes; just 0.5% limestone in the surface layer enhanced the buffering capacity by nearly 20%.

Recently, researchers have begun to develop a treatment system that overcomes the problems caused by the tendency of limestone to become armoured with ferric hydroxide precipitates in oxic flowing waters. Beginning with Turner and McCoy (1990), various researchers have begun to develop treatment systems called anoxic limestone drains. These systems seek to preserve the value of limestone in passive AMD treatment by keeping the iron in ferrous form. Some positive results are reported but there is great variability in the effectiveness of these systems and the reasons for this are not yet clear. Hedlin et al, (1994) found that both of the systems they tested showed no sign of deterioration in the 18–30 months study. Both systems raised the pH and the alkalinity of the runoff. However, one system was 80% more effective than the other and the reason for this was not apparent (Hedin et al., 1994).

ANHYDROUS AMMONIA

Anhydrous ammonia is useful in treating AMD with a high ferrous iron and/or manganese content (Faulkner 1991).

$$NH_3 + H^+ \geq NH_4^+ \qquad \qquad ... (24)$$

The treatment system required for this reaction is simple, no more than a pressurised tank of anhydrous ammonia, a length of hose to discharge the ammonia to the ARD and a regulating valve. The only direct technical problems are: it can be lost to the atmosphere by diffusion or air-stripping and it generates a large volume of sludge (Curtis et al., 1988). Ammonia use costs little, although more than hydrated lime or limestone. However, it is dangerous to use because of its toxicity for fish and other aquatic life, and impacts on eutrophication and nitrification. The ammonia may reoxidise in streams to nitrate, thus eventually reacidifying the waters (eqn. 25).

$$NH_4^+ + 2O_2 \geq NO_3 + H_2O + 2H^+ \qquad \qquad ... (25)$$

Use of ammonia is generally not allowed in the United States and, where permitted, additional monitoring is required (OSMRE, 1998). Curtis et al. (1988) advise that it be used only where there is little possibility of runoff and the amount of AMD is small.

SODIUM CARBONATE

Sodium carbonate (soda ash) briquettes are especially useful for treating small AMD flows in remote areas. Simple treatment feeders have been developed for this purpose. The applied briquettes dissolve to neutralise the water.

$$Na_2CO_3 + 2H^+ \geq 2Na^+ + H_2O + CO_2 \qquad \qquad ... (26)$$

Major disadvantages are that proper control of pH is hard to achieve, especially in higher flows. In addition, the reagent cost is high relative to limestone and the sludge has poor settling properties. However, even short-term treatment had a positive effect on brook trout in affected streams in Pennsylvania (Skinner and Arnold, 1990).

SODIUM HYDROXIDE

Sodium hydroxide (caustic soda) is usually fed into the water by a small flume. The method is simple, requires no power sources and thus is especially effective for treating low flows in remote locations. Sodium hydroxide is often used to counter AMD with higher loadings of manganese. Sludge averaged 2% of the original volume treated in a test using 10% concentration of sodium hydroxide (Kennedy, 1972). Major disadvantages are high cost, the dangers of handling the chemical, poor sludge properties and freezing problems in cold weather.

Oxidation Inhibitors

Instead of countering acidity chemically, some agents are used to prevent development of acidity in the first instance. The next section considers some of the chemicals and techniques used to inhibit acid generation. These include the application of chemical inhibitors such as rock phosphate and electrochemical passivation, physical inhibitors such as cement, physical containment structures and submergence, and biochemical inhibitors such as SLS, ABS, hydrophobic fatty acid amines, antibiotics, disinfectants, and surfactants. It also briefly considers the potential of electrolytic remediation, which is the inverse of electrochemical passivation.

ROCK PHOSPHATE

Rock phosphate, a very widely used fertiliser, is also employed, not as an alkaline agent but as a pyrite oxidation inhibitor (Renton et al., 1988). Pyrite weathering produces free ferric iron, which oxidises additional pyrites establishing a self-propagating series of reactions (Peters, 1986). Rock phosphate dissolves in acid media to release highly reactive phosphate ions that combine with iron to form insoluble iron phosphate. The precipitation of insoluble iron phosphates breaks the cyclic reaction. Phosphate works in laboratory studies and, in one field study, produced a 70% reduction in acid load (Nyavor and Egiebor, 1995: OSMRE, 1998; Evangelou 1996). Application rates of 2 to 3% and complete mixing with the pyrites were required for effective treatment.

SURFACE SEALANTS AND ELECTROLYTIC REMEDIATION

Some researchers focus on the surface chemistry of pyrite and try to develop various types of sealers, coatings and inhibitors to halt acid production (OSMRE, 1998). This includes electrochemical passivation by coating with copper and silver ions (Lalvani and Shami, 1989). Other have used a block of massive sulphide graphite and variously: scrap iron, aluminium or zinc as the sacrificial anodes and acid leachate as electrolyte. The tests with iron raise the pH of the leachate (from 4.1 to 5.6) and reduced the redox potential from > 650 to < 300. Evangelou (1996) used a solution of H_2O_2 plus sodium acetate, with and without phosphate, which when buffered at pH 5.0 caused pyrite inhibition by building a surface coating of either Fe-OOH or $FePO_4$. The creation of these coatings consumed 5–10% of the pyrites. However, the Fe-OOH coating consumed more pyrite and was less effective than the phosphate coating in inhibiting oxidation.

CEMENT

Agglomeration is an AMD prevention technique that involves immobilising mine tailings in an aggregate, pellet or briquette that resists weathering. Cement is the binder of choice but it is expensive (Amaratunga, 1991; Thomas et al., 1989). Fly ash, bentonite and several other clays have also been studied in a similar context as well as for flow barriers (Skousen et al., 1987; Bowders and Chiado, 1990). The world is always seeking a new use for power station fly ash. Misra et al. (1996) suggest that combining mine tailings with a binder such as Portland cement and 20% fly ash results in increased aggregate stability and reduced leaching of metals.

Containment Structures

Various workers have adapted landfill technology to the treatment of reactive coal discards. This approach involves constructing a containment structure that isolates or encapsulates the pyritic material. Encapsulation can be extended to top, bottom and sides in an attempt to isolate the material from water and oxygen (Skousen et al., 1987). Some researchers use borehole injection to isolate buried pyritic material (OSMRE, 1998). Success demands high-grade civil engineering, to avoid leakage and to create a fully anoxic barrier that will also prevent the generation of acid. Clay liners, plastic, asphalt and some geotextiles have been deployed in this cause (Nicholson et al., 1989).

Physical barriers can be used in conjunction with, and sometimes be replaced by, barriers made from alkaline materials. One Canadian project used a composite soil layer to effect containment. The cover reduced temperatures and oxygen levels in the waste rock, demonstrating that the reaction was inhibited. However, experience suggests that acidification can rise into topsoil covers established on reactive spoils (Warburton et al., 1988) and seepage is often capable of destroying such covers (Haigh and Sansom, 1999).

Sometimes the aim is limited to avoiding acid leaching. In this case, workers may isolate pyritic material from water by placing it above the water table. Compaction and capping with clay or other materials are employed to reduce permeability. However, leakage is a problem and it proves difficult to keep the spoil materials dry (OSMRE, 1998). The alternative approach is to submerge the pyrites.

Flooding

Since ARD develops in an aerobic environment and becomes a problem when it is leached into water-courses, one solution is to keep the environment anaerobic and hide the spoils under stagnant waters (OSMRE, 1998). Oxygen diffuses very slowly and has limited solubility in water. However, a partial pressure of 1% is needed to stop pyrite oxidation and this may imply a saturated cover of more than 10 metres (Hammack and Watzlaf, 1990). However, tests at Solbec Cupra, Quebec suggest that 1.34 m may be sufficient and 0.7 m if the tailings are buried beneath an additional layer of sand (Mohamed et al., 1996). Submergence also provides a route to the elimination of surface erosion problems and creates a reducing environment in which soluble metals precipitate as sulphides and ammonia (Mohamed et al., 1996).

The OSMRE (1998) advise that this approach works best on gently sloped terrain, where groundwater gradients are low, the saturated zone is thick and the groundwater basin compartmentalised. Four strategies are employed: disposal into natural water bodies, disposal into artificial impoundments, disposal into flooded mine-workings and open-pits, and the building of a water cover on existing waste-management sites (Mohamed et al., 1996). Thus, Rahn (1992) recommends that the AMD from the anthracite fields of E. Pennsylvania should be suppressed by flooding the entire area. Dams should be built in the water gaps and the whole area, including its 700 million cubic meters of mining voids, should be submerged. The result would be a 140 km^2 reservoir at 410 m.a.s.l.

Rahn (1992: 50) argues that the water in the lakes would not be low quality. Once the mine voids are flooded, stagnant non-percolating conditions would prevail and the rate of weathering would be greatly retarded, as it is in flooded underground mines (cf. Leavitt, 1986). Disadvantages to this project include the cost of creating the reservoir and relocating local people, the time it would take to fill the basin (maybe 3 years), loss of the remaining coal in the area and loss of healthy ecosystems flooded along with the spoils.

Suppression of Bacteria

Although it is common practice to mix limestone etc. with the spoils reinstated on reclaimed lands in order to reduce AMD, most of the chemical treatments aim to treat the symptoms of the ARD problem rather than its source. The promise of biological remediation is that it can provide treatment at source (Nyavor et al., 1996). Techniques discussed in this section include the use of bacterial and biochemical inhibitors, bactericides and antibiotics.

Of course, these days the productive edge of research into the autotrophic bacteria that cause AMD concerns enlarging and exploiting their capacity to mobilise metals, rather than suppressing their activities. *Thiobacillus ferrooxidans* is used in bioleaching, a new method used to extract metals from low-grade ores. It is estimated that by the end of the millennium, more than US $3 billion worth of gold and US $8 billion worth of copper will be produced by this biotechnology.

BACTERIAL INHIBITORS

The role of bacteria in pyrite oxidation has been well documented (Kleinmann et al., 1981). In theory, it is possible to inhibit the work of these oxidising micro-organisms in high sulphur coal spoils and thus obtain a reduction in AMD. To be effective, inhibitors should be: non-toxic to non-target organisms, easily available, inexpensive and effective at low concentrations (Singh and Bhatnagar, 1985). Nyavor et al. (1996) used fatty acid amines to produce highly hydrophobic surfaces. This reduced both the chemical and biological oxidation of the pyrites, at least in laboratory conditions. Acetyl acetone and humic acids have also been the subject of experiments (Lalvani et al., 1990). Some other simple organic compounds that have been pressed into service to prevent the oxidation of iron are yeast extract, carbohydrate and carboxylic acid (Singh and Bhatnagar, 1985).

BACTERICIDES

If the AMD-generating bacteria cannot be slowed down, an alternative approach is to kill them. The application of bactericides has reduced acid loading in field experiments. Many compounds have been screened as selective bactericides. Kleinmann et al. (1981) employed detergent mixed with rubber to delay its release. However, the anionic surfactants sodium lauryl sulphate (SLS) and alkyl benzone sulphionate (ABS) have been found most effective and economical (Singh and Bhatnagar, 1985). Dugan (1987) conducted laboratory tests with an SLS and benzoic acid mix (1.1 mg per kg of spoil). The treatment is called 'relatively non-toxic' and works rapidly. Unfortunately, the effect is temporary. Acidity reappears within 2–5 weeks. The sulphur and iron-oxidising bacteria repopulate the spoil and catalyse the acid-producing reactions as soon as the bactericide is depleted. Doepker (1994) argues that the main impact of SLS is as a surfactant coating on the reactive minerals.

Various more general antibiotic agents (notably: novobiocin, oleandomycin, and aureomycin) and other toxins have also been applied against the AMD bacteria (Schearer et al., 1970). However, these antibiotics and other toxins all combine high expense with wide-spectrum toxicity.

Ecological Remediation

The frontline in AMD control involves techniques for ecobiological remediation. The most direct approach deploys biological predators of the bacteria to regulate their numbers. However, greater advances are being made using ecological agents that restrict the habitat of the acidophilic bateria or which encourage them to complex with the AMD. These schemes encourage the AMD and bacteria to co-precipitate harmlessely (Rao et al., 1992) or the AMD to convert into harmless chemicals.

BACTERIOPHAGES AND OTHER PREDATORS

It is unlikely that work with bacteriophages will ever solve AMD problems, although the potential of genetic engineering cannot be ignored. Years ago, Schearer et al. (1970) reported positive results from work with *Caulobacter* spp. In laboratory cultures, AMD bacteria proved vulnerable to major decreases under grazing pressure from protozoa.

CELLULOSE AND STRAW

Micro-organisms cause AMD but they are also capable of mitigating its impacts by neutralising pH and removing excess metals. Tuttle et al. (1969) reported an increase in pH and iron sulphide precipitation following the passage of acidified mine water through a wood debris dam. Outgassing of hydrogen sulphide may be another mechanism (King et al., 1974). Three equations are offered by Schindler et al., (1980) who believe that the deposition of iron sulphide is the more important mechanism:

$$2CH_2O + SO_4^{2-} + 2H^+ \geq 2CO_2 + H_2S + 2H_2O \qquad \text{... (27)}$$

$$CH_2O + 4FeOOH + 8H^+ \geq CO_2 + 4Fe^{2+} + 7H_2O \qquad \text{... (28)}$$

$$4Fe^{2+} H_2S \geq FeS + 2H^+ \qquad \ldots (29)$$

Adding extra organic matter to acidified lakes or streams helps negate acidity by sulphate reduction in sulphuric acid-contaminated waters. Brugam et al. (1995) in their study in southern Illinois raised pH from 3.1 to 6.5 at the expense of increased sulphide concentration and reduced oxygen. Cellulose substrates provide a physical support for the microbial AMD-mitigating bacteria and their biodegradation provides carbon to sustain them. In laboratory tests, some straw and hay reactors have failed because their decomposition rate was too slow to provide the organic acids needed by the AMD-mitigating bacteria. In fact, carbon proved the main limiting factor and some researchers advocate adding sugars to the water to facilitate AMD mitigation. However, ammonium also achieved a positive result. Bechard et al. (1994) recommend alfalfa as a suitable substrate. Others have recommended straw, peat, hay and mushroom compost.

SEWAGE SLUDGE

Sewage sludge is another kind of organic matter that has the same potential. The interactions between AMD and sewage sludge have long been discussed (Dixon, 1910). Like fly ash, sewage sludge is also another product in search of an effective means of disposal. It has been suggested that sludge is disinfected by mixing with AMD and concomitantly AMD may be amended with sewage sludge (King and Simmler, 1973). The general problem with the use of sludge in coal- and metal- mining contexts is that it tends to have high initial loadings of metals. There is a real risk of adding to the concentration of metals in affected waters and soils. Since many metals bioconcentrate, this enhances the risk of creating legally recognised contamination. However Rao et al. (1992) have had great success in encouraging the co-precipitation of metals and the contaminants found in municipal waste waters.

Bioremediation in Constructed Wetlands

The classic method for bioremediation of AMD is through the creation of a biochemical reactor in the form of a wetland or reed-bed. Constructed wetlands use soil- and water-born microbes associated with wetland plants to remove dissolved metals from mine drainage. Constructed wetland waste-water treatment systems can be defined as man-made, engineered wetland areas specifically designed for the purpose of treating waste-water by optimising the physical, chemical and biological processes that occur in natural wetland ecosystems. They have proven to be very useful in conventional waste-water treatment. US-EPA reports hint at BOD removal of 85%, and faecal coliform removal of 95%. However, their use against AMD is rather more speculative. Many of the details of these systems have yet to be worked out, not the least being how they actually work. Rao et al.

(1992) confirm the ancient maxim that AMD purifies contaminated water by coagulating and complexing with the organic matter in waste-water. The mechanism is certainly very important. Meanwhile, more than 400 such systems have been constructed in the eastern USA alone for the mitigation of AMD (Johnson, 1995). In one case study from Idaho Springs, Colorado, after passage through a wetland, the pH of AMD was raised from 2.9 to 6.0 and its loading of metals reduced by 94–99%. However, not all instances have been so successful (Wildeman and Loudon 1989, Wieder 1994).

Never the less, constructed wetlands are complex ecosystems that in ideal circumstances can raise the pH of AMD in flow by pH 3–5. They include both aerobic and anaerobic zones. They have high levels of variability and complexity (Johnson, 1995). Little is known about their microbial ecology. However, it is believed that metal uptake by plants and sedimentation through filtration is less important than microbial driven alkali-generating processes (Kalin et al., 1991). These include ammonification, denitrification, methanogenesis and the reduction of iron and sulphur. Ammonification has been identified as the major source of microbial alkali in AMD amended with organic matter. Sulphate reduction may only be important in newly established wetlands Vile and Wieder (1993) suggest that in established wetlands (pH > 6.0), reduction of ferric iron is the more important source of alkalinity. In addition, as living systems, they are prone to seasonal variations in efficiency with less going on in cold weather than warm, (OSMRE, 1998). Wetlands are also better in small-scale applications where throughput is, of the order of, a few literes per minute.

Although constructed wetlands are often offered as an inexpensive solution (Johnson, 1995), initial design and construction costs may be significant (OSMRE, 1998). There is also concern that these wetlands have a finite capacity for the accumulation of metals and, in time, become sources rather than sinks for contaminants (Wood and Shelley 1999: 232). However, wetland systems are essentially self-sustaining and require minimal maintenance. Recently, the USA has begun to issue guidance for wetland creation on reclaimed surface mines and on optimising wetland creation on coal-mined lands (Brooks and Gardner, 1994, 1995). This is based on research at the Corsica project, a constructed multicell wetland (Stark et al., 1994). A key problem is the sizing of wetlands. Traditional measures estimate the percentage of contaminant removed, which is a relative, so absolute measure of performance. The U.S. Burean of mines recommends area adjusted removal ($g \cdot m^{-2} day^{-1}$) for wetland sizing, but the technique fails to separate flow and concentration components (Hedlin et al. 1994). Tarutis et al. (1999) prefer that pollutant removal is described by first-order kinetics but concede that much more research is needed.

Meanwhile, some researchers are trying to improve upon nature by creating a biological treatment plant in which there are two bioreactors, one for the anaerobic reduction of ferric iron and one for aerobic reduction of sulphate, each containing targeted populations of micro-organisms (Johnson, 1995: 53). These include *T. ferrooxidans*, which is thus both cause and cure of AMD problems (Wildeman and Laudon, 1989).

Runoff and Drainage Control

Mine-spoil hydrology plays a crucial role in determining drainage quality and AMD (Kilmartin, this book). The hydrobiogeological controls of mine-water drainage quality need further investigation (OSMRE, 1998). Recent work by Kimball et al. (1994) found that, in their test site at St. Kevin Gulch, Colorado, while iron precipitates cover a kilometre of stream-bed, the processes that caused precipitation act in the first hundred metres. The dominant processes in the stream were sorption, precipitation and the formation of aqueous complexes. These were driven by the effects of metal loading and by dilution through acidic and non-acidic lateral inflows. Kimball et al. (1994) used Damkohler numbers to distinguish sites where removal processes are important. Damkohler (Da) numbers are the dimensionless ratio between the convective residence time and the characteristic reaction time, which is the reverse of the rate coefficients for precipitation and sedimentation. Where the Da number for the stream reach was greater than 0.1, removal processes became significant. Where Da exceeded 10, removal processes were dominant and reactions approached completion. In sum, the condition in a stream is a complex function of the material residence time and the speed of the reactions that can take place in that particular stream environment. Naturally, a variety of aerobic and anaerobic environments exist in each reach of a stream.

AMD release can also be regulated by water management techniques and by preventing the erosion, exposure and entrainment of reactive spoils (Haigh, chapter 5, this book). For example, at the Puentes Lignite Mine near La Coruna, Spain, concern focuses on 700 million m^3 of clayey mine spoils, which are being stored outside the mine on 1400 ha (Bueno et al. 1992). The area includes a former river-bed, now being converted to a culvert drain. Surface erosion rates from the spoil tip are high and so attempts are being made to contain and compartmentalise runoff-generating areas. The runoff is highly acidic (average pH 4.0, acidity ranges 20–0.2 meq·l^{-1} due to pyrite oxidation and is heavily loaded with suspended solids (average 1000–range 22,000–50 meq·l^{-1}. This water is collected in drains and diverted for treatment in a purification plant that reduces the suspended sediment load to 300 mg·l^{-1} or less and raises pH to 7.0.

However, strategies must be tailored to each individual site in the light of its geology, hydrology, mining method, spoil geochemistry and topographic context. The OSMRE (1998) list of runoff and erosion regulation strategies thus includes apparent contradictions (Table 16).

Table 16: Some suggested runoff control strategies for AMD regulation (after OSMRE, 1998)

1.	Divert drainage away from pyritic materials during and after mining.
2.	Divert AMD through alkaline materials, such as limestone riprap and anoxic limestone channels.
3.	Land-form so that ponding does not foster deep seepage into pyritic substrates.
4.	Drain site rapidly to reduce AMD concentration.
5.	Use under-drains to route water away from reactive mine spoils,
6.	Separate polluted waters so that the volume for treatment is less.

A Last Word on AMD

Acid mine drainage is one of the most serious problems affecting reclaimed land. It is also its greatest off-site impact. For this reason, a huge amount of research is directed towards the problem. However, although much useful research has been conducted in the understanding, prediction and prevention of acid mine drainage, there remains a feeling that not enough is yet known. AMD treatment and prevention techniques are likewise under development. But no universally effective technologies have been designed thus far. However, the tried and tested methods of chemical treatment with alkalines work as long as they are sustained, while the evolving areas of microbial ecological engineering and the use of constructed wetlands, hold promise for a self-sustaining future.

IV. MANAGING CONTAMINATION BY METALS

Chemical soil pollution may result from a number of causes. These include the misuse of fertilisers and plant protection products such as herbicides and pesticides, the release of organic pathogens and toxins, and accumulation of metals (Kambata-Pendias and Pendias, 1984; Ferguson, 1990). This section deals only with the last of these problems. However, many others merit consideration. These include organic compounds such as the phenols, sulphates, PCBs and hydrocarbon wastes (Smith, 1985).

Heavy metal contamination of reclaimed mine spoils may have any of a number of origins. The metals may be native to the spoils, due to the original chemical composition of the mine stones, or to the mineral processing technology that created them. They may also be alien additions, due to deposition from industrial activities, motor transport, or power generation emissions and effluents, or to substances applied to the site during soil amelioration in the course of recultivation (Hangyel and Benesoczky, 1992). Many common soil additives-fertilisers (Popova, 1991), especially those derived from rock phosphate, sewage sludge/cake, farm manure, especially pig manure, power station fly ash and composted refuse, are major sources of heavy metal contamination (Blum, 1987: 43).

Some heavy metal contaminants affect human and livestock health directly. Several, especially cadmium and lead, accumulate in the edible portions of some food and fodder crops (Poschenreider et al., 1989; Chernykh, 1991). These highly toxic metals may be ingested by eating food products that contain accumulations of the metals, by eating unwashed products that have metals as surface contaminants of soil and dust, or by the direct ingestion/inhalation of dust or soils. This last problem is particularly acute in young children's play areas and children suffering from pica are at greatest risk (ICRCL, 1987; 2.3). Metals may also leach into runoff and water supplies, especially under the influence of AMD (Mostaghimi et al., 1992). Additionally, several metal contaminants have the capacity to inhibit or prevent plant growth. In many cases, the level for phytotoxicity, as for zinc, is well below that which might harm humans. The main phytotoxic elements are zinc, nickel, copper and boron.

Special Problems of Metal Contaminantion on Reclaimed Land

The recongnition of heavy metal concentrations in the reclamation of coal-mine-disturbed lands has special significance. The argument runs as follows. Many mine spoils and topsoils, derived from mine spoil associated 'soil-forming materials', begin with a heavy loading of metals. This is especially true when the metals are concentrated by burning (Table 17). Around two-thirds of all British deep-mine spoil-tips have suffered spontaneous combustion and these materials are often found in spoils from surface mine's that reworked previously deep-mined areas.

The breakdown of mine spoils during extraction and by subsequent weathering, along with associated increases of micro-organic activity, may tend to increase the mobilisation of the heavy metal ions in the spoil/soil complex and possibly create local concentrations (Ford and Mitchell, 1992). This inherent contamination is supplemented by air pollution, which is

Table 17: Some trace elements in coal ashes (high-mean-low: mg·kg) (Wang and Sweigard, 1996)

Metal	Anthracite	Lignite and sub-bituminous coal	Power station fly ash
B (Boron)	130–90–63	1900–1020–320	No data
Ba (Barium)	1340–866–540	13,900–5027–550	10,850-1880–251
Cd (Cadmium)	No data	No data	16.9–12.0–6.4
Cr (Chromium)	395–304–210	140–54–11	651–247–37
Cu (Copper)	540–405–96	3020–655–58	1452–185–45
Mn (Manganese)	365–270–58	1030–688–310	41,800–11,800–1600
Ni (Nickel)	320–220–125	320–129–20	353–141–23
Pb (Lead)	120–81–41	165–60–20	2120–171–21
V (Vanadium)	310–248–210	250–125–20	652–272–95
Zn (Zinc)	1200–688–370	490–245–100	2880–449–27

often very heavy in mining areas and, more locally, by the movement of contaminated waters. These processes add an additional cocktail of heavy metals and other elements to the soil (Alloway, 1990). In areas of active heavy industry or power generation, this input may be considerable.

In practice, there is very little that can be done about either of the first two factors. Much 'reclaimed' coal land will contain high levels of heavy metals and lie close to, or beyond, the 'trigger' or threshold levels which governments use to recognise official 'contamination' (cf. ICRCL, 1987). Reclamation agencies may try to ensure that spoils that are contaminated are buried more deeply, beneath at least 1.5–2.0 metres of uncontaminated surface material, and/or that land which is, or will inevitably become, contaminated, is not used for food production, children's play areas etc. Coaling and environmental agencies may also work for cleaner coal combustion and the reduction of contaminated aerosol levels but, in practice, air pollution levels are outside their control.

Sources of Metal Contamination on Reclaimed Land

Reclaimed lands are often needlessly contaminated during the processes of 'recultivation'. Many rock phosphate fertilisers contain significant levels of cadmium (Cd), chromium (Cr) and vanadium (V). High levels of lead (Pb), manganese (Mn), and iron (Fe) are also common. Frequently, in agriculture the input from such sources may be equivalent to, or double that from other diffuse sources, such as air pollution (Blum, 1987: 44). Pig farming often involves feeds that contain high amounts of copper (Cu) that improve the processes of food conversion. Unfortunately, these compounds also end up in the liquid/semi-liquid pig manure slurries. Domestic sewage is routinely loaded with a wide range of heavy metals.

Kasatikov (1991) found that applications of sewage sludge to natural podsolic soils resulted in the accumulation of many potentially toxic elements (Pb, Zn, Cr, Ni, and Cu) along with micro elements including (Co, Mo, B and Mn) in the 0-40 cm of the soil profile. Many European and UNECE nations have set guidelines that attempt to limit the dosage of heavy metals applied to soils in this fashion (cf. Ewers, 1991; Alloway, 1990; USEPA, 1989). However, as Kasatikov (1991) complains, the basic factors affecting the transition of heavy metals from sewage sludge to soil are not sufficiently understood. The available models of metal transport are grossly inadequate, due in part to failure to take into account the complexity of microbial interactions with metals (Ford and Mitchell, 1992). So, 'the advisability of using sewage sludge as fertiliser is debatable' (Kasatikov, 1991). Some official sources call soil contamination by heavy metals, including contamination by radioactive elements, 'practically non-reversible' (Blum, 1987: 53). Ford and Mitchell (1992: 83) argue that the health risks from allowing high concentrations of heavy metals to migrate from soils into waters are unacceptable and worry that, with present technology, it is not possible to assess the extent of the hazard. In sum, in general, soil amendment with sewage sludge is probably not worth the risk involved.

Many heavy metal contaminants are found in higher concentrations in soils than in their parent materials. Many authors express concern about the potential for animals and plants to concentrate metals within their bodies. This allows the possibility that low levels of contamination in soils may lead to high, toxic levels of a contaminant in an organism. Concentration through the food chain is a well-documented problem in studies of pollution. Table 18 demonstrates how the concentration of heavy metals, especially cadmium and zinc, become enhanced in the tissues of earthworms living within the soil. Figures for the 'normal range of the metals in soils and plant tissues' are included as reference. The table illustrates, again, the hazards of sewage sludge application to soils. Further details of the environmental and health hazards imposed by high concentrations of particular metals are included in Table 19.

Table 18: Concentration of heavy metals in earthworms (after Beyer et al., 1982; Brady, 1984: 668–669)

Metal	Soil		Earthworm Tissues		Range in natural soils	Range in plant tissues
	Control	Sludge treated	Control	Sludge treated		
Cd	0.1	2.7	4.8	37	0.1–7.0	0.2–0.8
Zn	56	132	228	452	10–300	15–200
Cu	12	39	13	31	2–100	4–15
Ni	14	19	14	14	10–10,000	1
Pb	22	31	17	20	2–200	1.0–10

The issue of defining critical levels for land contamination is very complex. The emergence of toxic effects due to particular metals or combinations of metals is massively affected by the biology and chemistry of the soil together with its interactions with parent materials and land-use (Alloway, 1993). However, the reality is that such risks are best avoided. This is why many states and international organisations are working towards defining critical limits for the recognition of contamination (Ewers, 1991; Alloway, 1993). Some of these are detailed in Table 20. Some nations have also produced criteria and limits for the concentrations and loadings of heavy metals that may be applied by soil additives with sewage sludge and fly ash. These practices are not encouraged in the case of lands reclaimed from coal-mine spoils, which already have high metal loadings. Indeed, the application of any soil additive or fertiliser should be carefully evaluated with regard to its impact on the heavy metal content of the soil. Finally, it is suggested that metal content of soils should be

Table 19: Environmental significance of heavy metal contamination of soils (abstracted from Alloway, 1990; Brady, 1984)

Metal	Environmental origins and consequences of contamination by individual metals
As (Arsenic)	Extremely poisonous, common on reclaimed coal lands but not usually a major environmental problem. The uptake by plants is not usually great; radishes and grass accumulate most actively. Organic compounds are less toxic than inorganic. Animals may ingest as dust. Pollution often arises as a result of metal working, coal combustion or the application of pig/poultry manure/pesticides. Levels in uncontaminated soils usually range between 1–40 mg·kg^{-1}. (mean 10 mg · kg^{-1}). The European Community proposed maximum for soils treated with sewage sludge is 20 mg·kg^{-1}.
Cd (Cadmium)	Extremely poisonous, Cd has no essential biological function and is highly toxic for plants and animals in which it accumulates. Cd is accumulated by some vegetables, especially spinach and lettuce. Brassicas grown on cadmium-contaminated soils can accumulate high concentrations. Indeed, producing such crops has been evaluated as a method of eliminating Cd pollution. Cadmium uptake is especially high on acid soils. In humans, Cd accumulation causes kidney dysfunction. Normal intakes are 25–75 ng·day^{-1} and toxicity becomes a problem at 70 ng·day^{-1} according to the WHO. Cd contamination often results from soil treatment with sewage sludge. Overall, the relative contributions of anthropogenic sources is thought to be 2–5% sewage sludge, phosphate fertilisers 54–58% and atmospheric deposition 39–41%. The last, plus bioaccumulation, is why concentrations tend to be higher nearer the soil surface. The normal range of Cd in soils is 0.01–7.0 mg·kg^{-1}, mean 0.2–0.4. European Community proposed limits for the rate of Cd application are 0.15 kg·ha·yr^{-1} with a maximum permissible soil content of 3 mg·kg^{-1}. (Council of the European Community, 1987: *Environmental Pollution by Cadmium: Proposed Action Programme* COM 87: (165). Contaminated soils have loadings above 3–5 mg·kg^{-1}. The Cd half-life in soils ranges from 15–1100 yr.)
Cr (Chromium)	An essential micronutrient for animals and plants, at low doses Cr stimulates plant growth and prevents glucose intolerance in animals. At higher doses, Cr may be toxic and a carcinogen. Normally, levels of plant available Cr are very small and Cr is not easily mobilised. Cr accumulation tends to be associated with root crops. Chromium is often added to soil through the application of lime and fly ash, through coal combustion and the emissions of the iron and steel industry. It is associated with other metal concentrations in the surface layers of soils treated by sewage sludge. Normal concentrations in soil range from 0.3–10,000 mg·kg^{-1}. (mean 41–200 mg·kg^{-1}.). The European Community has suggested a limit of 150–250 mg·kg^{-1} for soils amended with sewage sludge.
Cu (Copper)	This essential micronutrient affects enzymatic activity in plants and animals. Toxicity is a problem in low pH conditions. Plants become affected at loadings above 20 mg·kg^{-1} dry matter. High doses may be

Contd.

Contd. **Table 19**

Metal	Environmental origins and consequences of contamination by individual metals
	toxic for animals and accumulations may develop in the liver. The recommended maximum dose is 2 mg·day^{-1}. Plants do not readily accumulate Cu so toxicity in animals develops as a result of the ingestion of soil (1–10% dry matter intake of grazing cattle and 30% sheep). Anthropogenic sources of Cu toxicity include mining, fly ash application and pig manure. The normal range of Cu in the soil is 25–40 mg·kg^{-1}. Critical loadings usually range from around 60 to 100 mg·kg^{-1}.
Hg (Mercury)	Moderately poisonous, with no essential biological function, Hg is toxic for animals. Methyl mercury, which forms naturally in the soil in association with the fulvic acid complex, has teratogenic, carcinogenic and mutagenic effects. Phytotoxicity is low and roots serve as a barrier to Hg uptake. Anthropogenic mercury reaches the soil through seed dressings, industrial emissions and the applications of sewage sludge. Mercury tends to concentrate in soil surface humus, especially in soils of lower pH.
Mn (Manganese)	This important plant nutrient is the trace element most often deficient in U.K. cereal production. Mn toxicity develops at high doses ranging from 80–5000 ppm (300–7000 mg·kg^{-1} in plant tissue dry matter). Sewage sludge may be heavily loaded with Mn and application concentrations in excees of 500–3000 mg·kg^{-1} and 0.02–1 kg.ha.yr^{-1} are not advised. The range of Mn in natural soils runs from 20–10,000 mg·kg^{-1}, but the normal level is less than 500 mg·kg^{-1}. Soil contamination thresholds, where recognised, usually begin at 1000–1500 mg·kg^{-1}.
Mo (Molybdenum)	Molybdenum is unusual among plant nutrients in that it exists in soil as an anion and shows increasing solubility with increasing pH. Soil contamination often results through soil amelioration by sewage sludge or fly ash. Mo affects plant nitrogen metabolism and high levels in herbage can cause molybdenosis, Mo-induced copper deficiency in livestock. Normal levels in British and American soils range from 0.08 to 40 ppm (mg·kg^{-1}) with a norm at 1–1.2 mg·kg^{-1}. Plant tissue levels of 0.03–0.15 mg·kg^{-1} are called adequate. Toxicity develops at 10–50 mg·kg^{-1}.
Ni (Nickel)	Moderately poisonous, Ni is an essential micronutrient of animals and plants. It affects enzymatic activity, iron intake and liver function. In large quantities, it may increase the risk of cancer. The main anthropogenic source of Ni are fly ash, smelting emissions, oil combustion or sewage sludge. Ni levels in natural soils range between 0.5 and 5000 mg·kg^{-1} but the normal range is 25–53 mg·kg^{-1}. European Community nations suggest that the maximum concentration in soils receiving sewage sludge is 30–75 mg·kg^{-1}.
Pb (lead)	Moderately poisonous, with no significant metabolic function in either plants or animals and high persistence in soils, lead has a tendency to long-term accumulation in the food chain. It accumulates

Contd.

Contd. **Table 19:**

Metal	Environmental origins and consequences of contamination by individual metals
	in crop plants, especially those grown on sewage sludge amended soils and it poisons animals through the ingestion of soil (nb. 1–10% dry matter intake of grazing cattle and 30% sheep). It impairs the mental development of children at levels far below those at which clinical symptoms are displayed. Soil is a sink for anthropogenic lead, which is generated by mining and smelting activities, vehicle exhausts and the application of sewage sludge. Natural Pb levels may be of the order of 10–30 mg·kg^{-1} but general, low-level contamination has raised this to 30–100 mg·kg^{-1}. An EC Directive advises against the use sewage sludge applications that add Pb in excess of 10 mg·ha·yr^{-1} over 10 years, or exceed a total soil loading of 50–100 mg·kg^{-1} (CEC 1982: *Council Directive on the Use of Sewage Sludge in Agriculture* 82: 527).
Sb (Antimony)	Antimony is non-essential to plants, but readily taken up by roots, and is normally found in greater concentrations in soil, where it has greater 'mobility' than in soil parent materials. Plant toxicity levels may be as low as 2–4 mg·kg^{-1}, but are more usually in the 5–10 mg·kg^{-1} range. Soil pollution is associated with coal spoils (U.K.: 110 mg·kg^{-1}, mean 3.3 mg·kg^{-1}), coal combustion and soil amendments such as fertilisers, sewage sludge (typical concentrations—USA: 2.6–44.4 mg·kg^{-1}, U.K. 15–19 mg·kg^{-1}) and fly ash (average 4.5 mg·kg^{-1}). Norway finds its highest Sb levels in humus, where it is associated with the deposition of aerosol. Normal levels of Sb in soils range from 0.3–9.5 mg·kg^{-1}.
Se (Selenium)	Selenium is an essential micronutrient that is toxic in large doses. In humans, deficiency begins at < 0.04 mg·kg^{-1} and toxicity at > 4.00 m·kg^{-1}. Organic selenium may be accumulated by plants and cause phytotoxicity but the impact is affected by pH and plant species. Normal levels in soils range from 0.5–5.0 mg·kg^{-1} but several countries prefer to limit Se concentrations in sewage sludge to below 10 mg·kg^{-1} and total soil loading to below 3–10 mg·kg^{-1}.
Sn (Tin)	An essential micronutrient, especially for ruminants, but toxic for higher plants and micro-organisms when present in large quantities (60–63 mg·kg^{-1} dry weight). Natural levels in soil range from 1–200 mg·kg^{-1} with a norm of 4 mg·kg^{-1}. It has been suggested that soil loadings above 50–300 mg·kg^{-1} should be avoided.
Tl (Thallium)	Thallium levels of 5–10 ppb are beneficial to plant growth. Levels greater than 3.5 ppm (mg·kg^{-1}) can destroy vegetation especially in soils of low fertility (low K), although not in clay loams. There is a world ban on pesticides that use thallium. Thallium affects photosynthesis and transpiration. Thallium toxicity in humans is reported from the North Rhine Westphalia region where urinary levels were up to 76.5 mg.l^{-1} versus the normal 0.8 mg.l due to the consumption of contaminated vegetables and meat products. The normal range of HNO_3–extractable thallium in British soils ranges from 0.03–0.40 with a peak in the Shipham anomaly of 0.99 ppm (mg·kg^{-1}). It has been suggested that the critical soil loading is 1 mg·kg^{-1}.

Contd.

Contd. **Table 19:**

Metal	Environmental origins and consequences of contamination by individual metals
U (Uranium)	The effects of radioactive elements in the environment are well documented. Ingesting uranium may cause cancers, teratogenic effects and genetic damage. Uranium is a major contaminant of some coal spoils, especially in eastern European regions like Ostrava-Karvina, of phosphate fertilisers and in areas affected by nuclear fallout. Uraniun is relatively immobile in soils. U that is added as an impurity in fertiliser has been found to remain in the plough layer of agricultural soils and the organic layer of grasslands. Levels in agricultural products and milk can become significant. The normal range of U in soils seems to be 0.79–11.0 $mg \cdot kg^{-1}$. Contamination is thought to involve concentrations greater than 100 $mg \cdot kg^{-1}$.
V (Vanadium)	This is an essential micronutrient for animals. The environmental threat posed by V is slight but at high levels of 2 $mg \cdot kg^{-1}$ it may inhibit plant growth in growing shoots. Coal combustion is the major source of V in the biosphere and it is often associated with fly-ash amendment of soils. Normal levels in soils range from 20–500 $mg \cdot kg^{-1}$ with a mean of 100 $mg \cdot kg^{-1}$ but extractable levels range from 0.03–26.0 $mg \cdot kg^{-1}$ in Scottish soils. Soil loadings of 50–100 $mg \cdot kg^{-1}$ are thought to be critical.
Zn (Zinc)	Zinc is an essential micronutrient for animals and higher plants. In humans, Zn affects enzymatic activity, growth and appetite. The recommended safe dietary intake is 15 $mg.day^{-1}$. Plants are Zn deficient if their Zn level is less than 10–20 $mg \cdot kg^{-1}$ dry matter and toxic if greater than 400 $mg \cdot kg^{-1}$. Zn is relatively insoluble in soils but availability increases with increasing pH. Zn becomes phytotoxic at the high levels associated with metal-workings, coal combustion, sewage sludge, manure/compost application and the use of agrochemicals. The normal background soil concentration of Zn is around 80 $mg \cdot kg^{-1}$. The European Community recommends that Zn concentrations in sewage sludge should not exceed 2500–4000 $mg \cdot kg^{-1}$, an application rate in excess of 30 $kg.ha^{-1}$ or a maximum total soil loading of 300 $mg \cdot kg^{-1}$.

evaluated as part of any routine post-project inspection of reclaimed land sites (cf. Munshower and Judy, 1987).

GENERAL CONCLUSIONS

Successful reclamation depends upon the creation of a self-sustaining rooting medium, a living, self-sustaining and self-regulating soil system. This implies creating a soil that is sustained by a healthy soil-organic control system (Doran 1996) and one that achieves a positive balance between soil growth and soil loss. The incorporation of organic materials,

Table 20: Standards and guidelines concerning metal contamination in soils

| Metal | Bowen (1979) | Alloway (1990) | Blum (1987) | Alloway (1990) | | | | Ewers (1991) | | Kloke (1980) | | Alloway (1990) | | | Poush-karov Instt. (1993) | Acker-man (1983) |
| | General | | Austria | UK (DOE) London (GLC) | | | | Swiss | | Germany | | Nether-lands | | | Bul-garia | Wyo-ming |
	Reference	Critical	Thres-hold	Thres-hold	Refer-ence	Caution	Critical	Water Soluble	Total	Reference	Criti-cal	Refe-rence	Caut-ion	Criti-cal	Criti-cal	Unsui-table
Ag	0.01–8.0	2.00														
As	0.1–4.0	20.0–50.0	20.0	10–40	0–30	30.0	50.0		0.8	2.0–20.0	20	20	30	50	25	
Cd	0.01–2.0	3.0–8.0	2.0	3–15	0–1	1.0	3.0	0.03		0.1–1.0	3	1	5	20	3	
Co	0.5–65.0	25.0–50.0	50.0						25	1.0–10.0	50	20	50	300	50	
Cr	5–1500	75–100	60.0	600–1000	0–100	100.0	200.0		75	2.0–50.0	100	100	250	800	100	
Cu	2–250	60–125	100.0	140–280	0–100	100.0	200.0	0.70	50	1.0–20.0	100	50	100	500	15–280	
Ga										0.5–10.0	10					
Hg	0.01–0.50	0.3–5.0	2.0	1–50	0–1	1.0	3.0		0.8	0.1–1.0	2	0.5	2	10		
Mn	20–10000	1500–3000			0–500	500.0	1,000.0									
Mo	0.1–40.0	2.0–10.0	10.0	35–70	0–20	20.0	50.0	0.20	5	1.0–5.0	5	10	100	200	5	> 1
Ni	2–750	100.00	60.0		0–500	500.0	1,000.0	1.00	50	2.0–50.0	50	50	100	500	50	
Pb	2–300	100.00	100.0	550–1500	0–30	30.0	50.0		50	0.1–20.0	100	50	150	600	20–80	> 10
Sb	0.2–10.0	5.0–10.0		60–500						0.1–0.5	5				5	
Se	0.1–5.0	5.0–10.0		3–6	0–1	1.0	3.0			0.1–5.0	10				10	
Sn	1–200									1.0–20.00	50	20	50	300		
Tl	0.1–0.8	1.00							1	0.1–5.0	1					> 2

Contd.

Table 20: *Contd.*

Metal	Bowen (1979) General		Blum (1987) Austria	Alloway (1990) UK (DOE) London (GLC)				Ewers (1991) Swiss		Kloke (1980) Germany		Alloway (1990) Netherlands			Poush-karov Instt. (1993) Bulgaria	Acker-man (1983) Wyoming
	Reference	Critical	Thres-hold	Thres-hold	Refer-ence	Caution	Critical	Water Soluble	Total	Reference	Critical	Refer-ence	Caut-ion	Criti-cal	Critical	Unsui-table
U	0.7–9.0				0–100	100.0	200.0			0.1–1.0	5					
V	3–500	50–100								10–100	50					
Zn	1–900	70–400	300.0	280–560	0–250	250.0	500.0	0.50	200	3–50	300	200	500	3000	20–370	

Sources: General: Bowen 1979, Alloway, 1990. *Austria:* Blum, 1987: determined with Agua Regia. *U.K.:* DOE in IRCRCL 1987. London (GLC); Alloway, 1990 (nb. UK extractable levels of Cu, Zn, Ni by EDTA extraction (0.05 M); Sb by HCl extraction (0.007 M) adjusted to pH 1.5). *Switzerland:* Ordinance on Soil Contaminants 1986 re: *Environmental Protection Law—Sections 33 and 39 (I),* total values indicate extraction with concentrated HNO$_3$, water soluble indcates extraction with aqueous NaNO$_3$. *Germany:* Kloke, A. 1980: *Orienterungsdaten fur tolierbare Gesamtgehalte einiger Elemente in Kulturboden. Miteilingen des Verbandes Deutscher Landwirtschaftlicher untersunchungs,- und Forschungshallten Heft 1–3.* Determined with Agua Regia. *Netherlands: Leidsrad Bodemsanering. Afintering 4, Nov, 1988, Staasuitgeverij's Gravenhage, Bulgaria:* Poushkarov Institute, Sofia, 1993, range values indicate variations from low to high pH sites. *Wyoming:* Ackermann, 1983. See also: Ewers; 1991; Ferguson, 1990; Giesler, G. 1987. *Contaminated Land in the EEC. Fredrickshafen: Dornier System.* (*Note: data are not necessarily based on standardised extraction or sample collection techniques and are tabulated merely as a prelude to further investigation*).

including inoculation with micro-organisms and earthworms, may have beneficial effects and may help 'kick-start' the soil biological system. However, great care must be taken not to use soil treatments that increase the heavy metal loading of the surface soils. Attention must also be paid to preventing the development of acidity and the prevention of acid drainage, which can destroy vegetation, mobilise and spread contaminants into the environment. The physical and chemical deficiencies of overburden and spoils produced by coal mining may be remedied in a number of ways. It is common practice for a mine operator to save and restore topsoils from the pre-mining environment. However, wiser contractors also ensure that the more favourable soil-forming materials— those least inclined to toxicity or breakdown on weathering—are restored closer to the surface while those of greater toxicity or more prone to breakdown are more deeply buried (Gentcheva-Kostadinova and Haigh, 1988). Finally, it must be recognised that problems in the soil and water systems on reclaimed lands neither begin instantaneously when such are created, nor disappear quickly. Many, e.g. acidity build up over 5–10 years and may take several decades to resolve. In the meantime, the quality of the soil on reclaimed lands depends upon continuous monitoring, care, repair and maintenance.

References

Ackerman, W. 1983, *Topsoil and overburden*. Wyoming Department of Environment Quality, Land Quality Division, Wyoming Draft Guidelines: 1.

ADAS (Agricultural Development and Advisory Service), 1990. *Agricultural Land Classfication: Physical Characteristics and Soil Forming Materials Survey*. A Report to the Public Inquiry into the Proposed Pwell Du Opencast Coal Site, Blaenavon, Gwent. Gwent Country Hall, Cwmbran: 1–13.

Agassi, M. and Ben-hur, M. 1992: Stabilising steep slopes with soil conditioners and plants. *Soil Technology* 5: 249–256.

Aljoe W.W. and Hawkins, J.W. 1991, Hydrologic characterization and in-situ neutralization of acidic mine pools in abandoned underground coal mines. *Second Int. Conf. Abatement of Acidic Drainage, Montreal, Canada, Proc.* 1: 69–90.

AGRA, 1995. *Minesite Reclamation Planning: Issues and Solutions*. AGRA Earth and Environmental Calgary, Canada.

Allison, L.E. 1947. Effect of micro-organisms on the permeability of a soil under prolonged submergence. *Soil Science* 63: 439–450.

Alloway, B.J. (ed.). 1990. *Heavy Metals in Soils*. Blackie, Glasgow, 339 pp.

Amaratunga, M.L. 1991. A novel concept for the safe disposal of acid generating tailings by agglomeration and encapsulated alkaline and bacterial additives, pp. 389–400. In: R.W. Smith and M. Misra (ed.). *Mineral Bioprocessing*. Warrendale, Pa.

American Society of Agricultural Engineers, 1980. *Standard S313.1*. Agricultural Engineers Yearbook. St. Josephs, Michigan.

Anderson, D.W. 1977. Early stages of soil formation on glacial till mine spoils in a semi-arid climate. *Geoderma* 19: 11–19.

Armstrong, M.J. and Bragg, N.C. 1984: Soil physical parameters and earthworm populations associated with opencast coal working and land restoration. *Agriculture, Ecosystems and Environment* 11: 131–143.

Ashby, W.C. 1996. Red oak and black walnut growth increased with mine soil ripping. *Int. J. Surface Mining, Reclamation and Environment* 10(3): 113–116.

Aveyard, J.W. 1983. Soil erosion and productivity research in New South Wales to 1982. *J. Soil Conservation* (N.S.W.) 25, 258–268.

Ayers, P.D. and Perumpral, J.V. 1982: Moisture and density effect on cone index. *Amer. Soc. Agric. Eng., Trans.,* 25(5): 1169–1172.

Azzam, R.A.I. 1980. Agricultural polymers—polyacrylamide preparation, application and prospects in soil conditioning. *Comm. Soil Sci. Plant Analysis* 11(8): 767–834.

Badrashi, B. Jarrett, A.R. and Hoover, J.R. 1981. The role of escaped air on erosion in sand. *Amer. Soc. Agri. Eng., Paper* 81–2002: 18 pp.

Bagautdinov, F.V. 1996. Turnover of humus components of grey forest soil and of typical chernozem during long-term humification of carbon-labelled plant residues. *Eurasian Soil Science* 28(2): 106–115.

Baize, D. (ed) 1992: *Referential Pedologique* AFES/INRA, Paris 222 pp.

Baize, D. (ed.). 1993. *Soil Science Analyses: A Guide to Current Use.* J. Wiley/INRA, Chichester, 192 pp.

Ballard, T.M. 1981: Physical properties and physical behaviour of forest soils, pp. 113–120: In: P.E. Heilman, H.W. Anderson and D.W. Baumgartner (eds.). *Forest Soils of the Douglas-Fir Region Pullman, Washington, USA.* Coop. Ext. Serv., Washington State Univ.

Banov, M. Filtcheva, E. and Hristov, B. 1989. Humosonatravane i kachestven sostav na humusa pri recultivirani zemi. *Pochvohnamie i Agrochimia* (Sofia) 4: 3–9.

Barnhisel, R.I 1988. Correction of physical limitations to reclamation, pp. 191–211. In: L.R. Hossner (ed.). *Reclamation of Surface-Mined Lands,* I. CRC Press, Inc., Boca Raton, Florida, 212 pp.

Bates, R.L. and Jackson, J.A. (eds.). 1980. *Glossary of Geology* (2e). Amer. Geol. Inst., Falls Church, Va., 749 pp.

Bechard, G., Yamazaki, H., Douglas Gould, W. and Bedard, P. 1994. Use of cellulose substrates for the microbial treatment of acid mine drainage. *J. Environ. Quality* 23: 111–116.

Bell, A. and Finnery, K.D. 1987. *Mines and Mill Wastewater Treatment.* Environment Canada: Minerals, Mining and Metallurgical Process Division, Ottawa Report EPS 3/HA.

Bell, A.V. Riley, M.D. and Yanful, E.K. 1996. Evaluation of a composite soil cover to control acid waste rock pile drainage. *Can. Bur. Mines, Bull.,* 88: 41–48.

Bending, N.A.D., Moffat, A.J. and Roberts, C.J. 1992. Site factors affecting tree response on restored opencast ground in the South Wales Coalfield. *Land Reclamation,* 91. Elsevier, Amsterdam, 400 pp.

Bennett, H.H. 1939. *Soil Conservation.* McGraw Hill, New York.

Berezin, P.W., Voronin, Ad, and Shein, Ye. V. 1984. An energetic approach to the quantitative evaluation of soil structure. *Soviet Soil Science* 15(5): 103–109.

Berkovitch. I., Manckerman, M. and Potter, N.M. 1959. The shale breakdown problem in coal washing: Part 1: Assessing the breakdown of shales in water. *J. Inst. Fuel,* Dec. 1959, pp. 579–589.

Beyer, W.N., Chaney, R.L. and Mulhern, B.M. 1982: Heavy metal concentrations in earthworms from soils amended with sewage sludge. *J. Environ. Quality* 11: 381–385.

Biro, B., Voros, I., Kovespechy, K. and Szegi, J. 1993. Symbiont effect of rhizobium bacteria and vesicular arbuscular mycorrhizal fungi on *Pisum sativum* cultivation in recultivated mine spoils. *Geomicrobiology J.* 11(3-4): 275–284.

Bishop, A.W. 1973. The stabilty of tips and spoils heaps. *Quart. J. Eng. Geol.,* 6: 335–376.

Blackwell, P.S. 1979: A method of predicting dry bulk density changes of field soils beneath the wheels of agricultural vehicles. Ph.D. thesis, Univ. Edinburgh, Edinburgh (unpubl.).

Blum, W. 1987. *Soil Conservation Problems in Europe.* Council of Europe, Strasbourg. 60 pp.

Boone, F.R. 1986: Towards soil compaction limits for crop growth. *Netherlands J. Agric. Sci.,* 34: 349–360.

Bowders, J. and Chiado, E. 1990. Engineering evaluation of waste phosphatic clay for producing low permeability barriers. *Mining and Reclamation Conference and Exhibition (West Virginia University), Proc.,* 1: 11–18.

Bowen, H.J.M. 1979. *Environmental Geochemistry of the Elements.* Academic Press, London.

Bradshaw, A.D. 1997. Restoration of mined lands using natural processes. *Ecol. Eng.,* 8: 225–269.

Brandsma, R.T. 1996. Soil conditioner effects on soil structure and crop performance. Ph. D. thesis, Univ. Wolverhampton, School of Applied Sciences, 277 pp (unpubl).

Brandsma, R.T., Fullen, M.A. and Hocking, T.J. 1996. The contribution of an anionic soil conditioner to soil conservation. *Int. Erosion Control Assoc., Proc.,* 27: 467–479.

Brady, K.B.C. 1984. *The Nature and Properties of Soils* (9th ed.). Macmillan, New York, 737 pp.

Brady, K.B.C., Perry, E.F., Beam, R.L. et al. 1994. Evaluation of acid-base accounting to predict the quality of drainage at surface coal mines in Pennsylvania, USA. *US Bureau of Mines, Spec., Publ.* SP-06A-94: 138–147.

Bragg, N.C. 1983. The study of soil development on restored opencast coal sites. *ADAS Land and Water Service Report* RD/FE/9, 4 pp.

Bragg, N.C., Arrowsmith, R. and Rands, J.G. 1984. The re-examination of 'Radar' Opencast Coal Drainage Experiment. *ADAS Land and Water Service Report* RD/FE/19, 8 pp.

Brechtel, H.-M. 1992. Impact of acid deposition caused by air pollution in central Europe, pp. 52–63. *In:* J. Krecek and M.J. Haigh (eds.). *Environmental Regeneration in Headwaters* (Proc. 2nd Int. Conf. Headwater Control). ENVCO, Prague, 368 pp.

Bridges, E.M. and de Bakker, H. 1989. Soil as an artefact: the human impact on the soil resource. *The Land* 1(3): 197–215.

British Standards Institution, 1975. *Methods of Test for Soils for Engineering Purposes: BS 1377:* London: BSI: 144 pp.

Broad, K.F. 1979. Tree planting on man-made sites in Wales. *Forestry Commission Occasional Paper* 3, Edinburgh.

Brooks, R.P. and Gardner, T.W. 1994. *Optimizing Wetlands Creation on Coal Mined Lands* Washington: US Dept. Interior, Office of Surface Mining, Wash. Final Report. ER9404.

Brooks, R.P. and Gardner, T.W. 1995. *Handbook for Wetland Creation on Reclaimed Surface Mines.* U.S. Dept. Interior, Office of Surface Mining, Wash, Final Report: ER9503.

Brown, H.J., Cruse, R.M., Erbach, D.C. and Melvin, S.W 1992. Tractive device effects on soil physical properties. *Soil and Tillage Research* 22: 41–53.

Browing, G.M., Parish, C.L. and Glass, J. 1947. A method for determining the use and limitations of rotations and conservation practices in the control of erosion in Iowa. *Agron, J.* 39(4): 65–73.

Brugam, R.B., Gastineau, J. and Ratcliff, E. 1995. The neutralisation of acidic coal-mine lakes by additions of natural organic matter: a mesocosm test. *Hydrobiologia* 316: 153–159.

Buckely, A.N. and Woods, R. 1987. The surface oxidaton of pyrite. *Applied Surface Science* 27: 437–452.

Bueno, A.G., Caballero, C.V., Vasquez, F.M. and Martinez, C.M. 1992: Water management and control measures adopted in the reclamation surfaces of the Puentes Lignite Mine (La Coruna, Spain). *U.N. Econ. Comm. Europe, Comm. Energy, Working Party on Coal, Symp. Opencast Coal Mining and Environment (Nottingham, U.K.)* ENERGY/WP.1/SEM. 2/R.22: 13 pp.

Busscher, W.J., Spivey, L.D and R.B. Campbell. 1987. Estimation of soil strength properties for critical rooting conditions. *Soil and Tillage Research* 9: 377–386.

Bzowski, Z. 1997. Changes of buffer action in the carboniferous clay-stones located in limestone waste on hard coal heaps. *Int. J. Surface Mining, Reclamation and Environment* 11: 79–82.

CANMET. 1977: Reclamation by vegetation—mine waste description and case histories. *CANMET (Ottawa, Canada: Energy, Mines and Resources) Pit Slope Manual* 10(1/1): 1–51.

Cathles, L.M. and Apps. J.A. 1975. A model of the dump leaching process that incorporates oxygen balance, heat balance and air convection. *Metal Transactions* 68: 617–624 (in Jaynes et al., 1984).

Coastech Research Inc. 1989. Investigation of prediction techniques for acid mine drainage. DSS File No. 30SQ.23440-7-9178. Report to Canada Center for Mineral and Energy Technology for Mine Environment Neutral Drainage (MEND) Program. MEND Secretariat, Ottawa, Ontario, Canada.

Chernykh, N.A. 1991. Alteration of the concentrations of certain elements in plants by heavy metals in the soil. *Agrokhimiya* 1991/3: 68–76 (tr. *Eurasian Soil Science* 23, 1991: 45–53).

Chino, M. et al., 1992. Behavior of zinc and copper in soil with long term application of sewage sludges. *Soil Sci. Plant Nutrition* 38 (1): 159–167.

Chorley, R.J. and Kennedy, B.A. 1971. *Physical Geography: A Systems Approach.* Prentice Hall, London, pp. 210–216.

Christensen, N.B., Session J.B., and Barnes, P.L. 1989: A method for analyzing penetration resistance data. *Soil and Tillage Research* 13: 83–91.

Chwastek, J. 1970. Wplw czynnikow gorniczo-geologicznych na formy zwalowisk. *Czasopismo Geograficzne* 4: 409–425.

Constantinesco, I., 1977: Soil conservation for developing countries. *FAO Soils Bull.* 30: 80 pp.

Cook, K. 1982 Soil loss: a question of values. *J. Soil Water Cons.,* 37: 89–92. 603–604.

Coppin, N.J. and Bradshaw, A.D. 1982. *Quarry Restoration.* Mining Journal Books, London.

Costigan, P.A., Bradshaw, A.D. and Gemmel, R.P. 1981 and seq. The reclamation of acidic colliery spoil, I-IV. *J. Appl. Ecol.,* 18(3): 865–878, 879–887, 19(1) 1982: 193–201, 21(1) 1984: 377–385.

Curtis, W.R. 1971. Strip-mining, erosion and sedimentation. *Amer. Assoc. Agric. Eng. Trans.,* 14: 434–436.

Curtis, W.R. Dyer, K.L. and Williams G.P. 1988. *A Manual for the Training of Reclamation Inspectors in the Fundamentals of Hydrology.* Soil and Water Conservation Society/Office of Surface Mining and Enforcement/US Dept. Agriculture, Northeast Forest Experiment Station, Ankeny, Iowa, 178 pp..

Daddo, R.L. and Warrington, G.E. 1983. *Growth Limiting Soil Bulk Densities as Influenced by Soil Texture.* Fort Collins, Co. USDA, Forest Service—Watershed Development Group, Report WSDG-TN-0005.

Dang Xhi, Wan Guojiang, Haigh, M.J. and Watts, S.F. 1994; Coal mine spoil-water interaction (I): release mechanism of clay minerals. *Chinese Bull. Sci.,* 39(24): 2053–2055.

Daniel, M. 1993. *European Coal Prospects to 2010.* IEA Coal Research (London), Perspective IEAPER/05: 28 pp.

Davies, R., Younger, A. and Chapman, R. 1992. Water availability in a restored soil. *Soil Use and Management* 8(2): 67–73.

Davy, T.G. 1979. An industry view of the problems of dealing with land reclamation legislation. *Can. Land Rec. Assoc. Proc.* 4: 277–283.

Dawson, R.F. and Morgenstern, N.R. 1995. *Liquefaction Flowslides in Rocky Mountain Coal-mines, Phase 3: Summary Report.* Ottawa: Canmet SSC File: XSG23440-3-9135/00/A.

Dawson, R.F., Martin, R.L. and Cavers, D.S. 1995. *Review of Long Term Geotechnical Stability of Mine Spoil Piles.* Calgary, Agra Earth and Environmental Report CG25038, for British Columbia Ministry of Energy, Mines and Resources, 84 pp.

Day, D.G. 1987. Surface coal mining and the geomorphic environment in eastern Australia, pp. 359–380. In: V. Gardiner V (ed.). *International Geomorphology I.* Chichester, J. Wiley.

De Boodt, M. 1979. Soil conditioning for better management. *Outlook on Agriculture* 10: 63–70.

De Boodt, M. 1990. Applications of polymeric substances as physical soil conditioners, pp. 517–556. In: De Boodt, M. et al. (eds.). *Soil Colloids and their Associations in Soil Aggregates.* NATO-ASI Series B, vol. 215. Plenum Press, New York.

De Boodt, M. 1992. Synthetic polymers as soil conditioners: thirty-five years of experimentation, pp. 137–164. In: H.J.W. Verplancke et al. (eds) *Water Saving Techniques for Plant Growth.* NATO-ASI series E, vol. 217. Kluwer Academic, Dordrecht.

De Jong, E. Begg, C.B.M. and Kachenoski, R.G. 1983. Estimates of soil erosion and deposition for some Saskatchewan soils. *Can. J. Soil. Sci.,* 63: 607–617.

De Jong E, Villar, H. and Bettany J.R. 1982. Preliminary investigations on the use of 137's to estimate erosion in Saskatchewan. *Can J. Soil Sci.* 62: 673–683.

Deibert, E.J. 1983. Compaction, hydrological processes and soil erosion in loamy sands in east Shropshire, England. *Soils and Tillage Research* 6(1): 17–29.

Dimitrov, K. 1992. Restoration, using natural zeolites (clinoptilolites), of spoil heaps in the area of Maritsa East opencast mine. *UNECE Comm. Energy, Working Party on Coal, Symp. Opencast Coal Mining and Environment (Nottingham, U.K.).* ENERGY/WP.1/ SEM.2/R.21:2 pp.

Dixon, S.G. 1910. Germicidal effect of mine waters and tannery wastes. *Eng. Rec.* 61(16): 533–534.

Dobrzanski, B., Dechnik, I. Debiki, R. and Lipiec, J. 1982. Suitability of some soil conditioners and waste products for decreasing soil susceptibility of erosion. *Polish Soil Sci.* 15 (2): 155–160.

Doepker, R.D. 1994. Laboratory detremination of parameters influencing metal dissolution from sulphidic waste rock. *Int. J. Surface Mining, Reclamation and Environment* 8(2): 55–64.

Doran, J.W. 1996. Soil health and sustainability. *Advances in Agronomy* 16: 1–54.

Doubleday, G.P. 1969. The assessment of colliery spoil as a soil forming material. *North of England Soils Discussion Group, Proc.,* 6: 5–13.

Down, C.G. 1975. Soil development on colliery waste tips in relation to age. *J. Appl. Ecol.,* 12: 617–624.

Doyle, W.S. 1976. *Deep Coal Mining Waste Disposal Technology.* Noyes Data Corp. Environ. Tech. Rev. (Park Ridge, New Jersey) 29: 392 pp.

Dreese, G.R. and Bryant, H.L. 1971. Costs and effects of a water quality program for a small strip mining company. *US Army Eng. Inst. Water Resources, IWR-Report* 71–77: 150 pp.

Dugan, P.R. 1987. Prevention of formation of acid drainage from high-sulfur coal refuse by inhibition of iron and sulfur oxidizing organisms. I/II. *Biotech. Bioengi,* 29: 41–54.

Dyckhoff, C.J., Fowler, M.B., Haigh, M.J. and Watts, S.F. 1993. *Geochemistry of Mine Spoils at the Site of the Proposed Earth Centre Ecological Parklands, Denaby Main, S. Yorks.* Oxford Brookes, Univ., Land Reclamation Collective/Soils Research Group, Research and Consultancy Report EC1: 74 pp.

Eavis, B.W., Ratliff, L.F. and Taylor, H.M. 1969. Use of a deadload technique to determine axial root pressure. *Agronomy. J.,* 61: 640–643.

Elliot, G.L., Campbell, B.L. and Lougharan, R.J. 1984. Correlation of erosion and erodibility assessments using Caesium 137. *J. Soil Cons.* (NSW) 40: 24–29.

Emerson, W.W. 1967. Classification of soil aggregates. *Australian J. Soil Res.,* 5: 47–57.

Emerson, W.W. 1991. Structural decline of soils: assessment and prevention. *Australian J. Soil Res.,* 29: 905–921.

Erbach, D.C., Melvin, S.W. and Cruse, R.M. 1988. Effects of tractor tracks during secondary tillage on corn production. *Amer. Soci. Agric. Eng. Paper* 88–1614: 15 pp.

Erickson, A.E. 1982. Tillage effects on soil aeration. *Predicting Tillage Effects on Soil Physical Properties and Processes.* Amer. Soci. Agron., Spec. Publ. 44: 91–105.

Erickson, P.M. and Hedlin, R. 1988. Evaluation of overburden analytical methods as a means to predict postmining coal mine drainage quality. *US Bureau of Mines, Information Circular* 9183, pp. 11–19.

Erickson, P.M., Kleinmann, R.L.P. and Onysko, S.J., 1985., Control of acid mine drainage by application of bactericidal materials. In: Control of Acid Mine Drainage—Proceedings of a Technology Transfer Seminar: *US. Bureau of Mines, Information Circular* 9027, p. 25–34.

Evangelou, V.P. 1996. Pyrite oxidation inhibition in coal-waste by PO_4 and H_2O_2 pH buffered pretreatment. *Int. J. Surface Mining, Reclamation and Environment* 10(3): 135–142.

Ewers, U. 1991. Standards, guidelines and legislative regulations concerning metals and their compounds, pp 706–710. In: E. Merian (ed) *Metals and their Compounds in the Environment.* VCH, Weinheim, Germany, 860 pp.

Fadygas, K. 1992. Reclamation of spoil heaps at opencast coal mines by tree planting. *UN Econ. Comm. Europe, Commit. Energy, Working Party on Coal, Symp. Opencast Coal Mining and Environment (Nottingham, U.K.).* ENERGY/WP.1/SEM.2/R. 15: 2 pp.

Fairfax, J.A. 1985. *Completion Report for Blaenant Opencast Site.* Cardiff: Welsh Office Agricultural Department (WOAD), Land and Water Service, 3 pp.

FAO-UNESCO 1988. *Soil Map of the World: Revised Legend.* Rome FAO: World Soil Resources Report 60.

Faulkner, B. (ed.). 1991. *Handbook for Use of Ammonia in Treating Mine Waters.* West Virginia Mining and Reclamation Assoc., Chaerleston, W.V.

Feda, J. 1992. *Creep of Soils and Related Phenomena.* Academia (CSAV), Prague, 422 pp.

Felton, G.K. 1992. Soil hydraulic properties of reclaimed prime farmland. *Amer. Assoc. Agricul. Eng., Trans.,* 35 (3): 871–877.

Felton, G.K. and Taraba, J.L. 1994. A laboratory examination of organic matter degradation in a B horizon soil from post-mining reconstructed prime farmland. *Int. J. Surface Mining, Reclamation and Environment* 8(4): 153–158.

Ferguson, J.E. 1990. The heavy elements. *Chemistry, Environmental Impact and Health Effects.* Pergamon, Oxford pp. 340–355.

Ferris, F.G., Tazaki, K. and Fyfe, W.S. 1989. Iron oxides in acid mine drainage environments and their association with bacteria. *Chem. Geol.* 74: 321–329.

Filtcheva, E., Chakalov, K. and Popova, T. 1998. Model approach to improve soil quality using organo-zeolite compost. *J. Balkan Ecol.* 1: 72–80.

Filtcheva, E. Gentcheva-Kostadinova, Sv., Noustorova, M. and Haigh, M.J. 1988. Forestation as an ecological approach for improving surface coal-mine spoils I: Impact on organic matter accumulation. *J. Balkan Ecol.* 1: 47–55.

Fiskell, J.G.A., Carlisle, V.W., Kashirad, R. and Hutton, C.E. 1968. Effect of soil strength on root penetration in coarse textured soils. *Int. Cong. Soil Science, Trans.,* 9(1): 793–802.

Flege, A. 2000. Forest recultivation of coal-mined land: problems and prospects. *In*: M.J. Haigh (ed.). *Reclaimed Land: Erosion Control, Soils and Ecology.* A.A. Balkema Rotterdam, Oxford and IBH Pub., New Delhi. Land Reconstruction and Management, vol. 1.

Ford, T. and Mitchell, R., 1992: Microbial transport of toxic metals, pp. 83–101 *In*: R. Mitchell (ed.). *Environmental Microbiology.* J. Wiley & Sons, New York, 403 pp.

Gazea, B., Adama, K. and Kontopoulos, M. 1995. A review of passive systems for the treatment of acid drainage. *Minerals Engineering* 9(1): 23–47.

Gentcheva, Sv. 1994. Klasifikatitsiya na antropogennite pochvi. *Lesobudska Misul* (Sofia) I/94: 87–95.

Gentcheva, Sv. 1995. *Klasifikatsiya I Iyakoi Okobenosti na Antropogennite Pochvi.* Sofia: Ministervo na Obrazovanieto Naukata I Tehnologiite, Vistch Lesotehnicheski Institut, Avtoreferat na Disertatsiya, 56 pp.

Gentcheva-Kostadinova, Sv. and Haigh, M.J. 1988. Land reclamation and afforestation research on the coal-mine disturbed lands of Bulgaria. *Land Use Policy* 5: 94–102.

Gentcheva, Sv. and Hubenov, G. 1988. Processa vivetrivaniya i pochvoobazobaniya na sybstratah obrazovannih na promishlennih otvalah pi dobrich yglya [Weathering and soil forming processes on mine spoil substrata]. *Ninth Int. Symp. Landscape Reclamation of Lands Disturbed by Industrial Activities. PRB-GDR-PRP-SRR, Collected Reports* 1. Kompolt—Dyondyosh, Agric. Univ. Kompolt, 230 pp.

Gerard, C.J., Sexton, P. and Shaw, G. 1982. Physical factors influencing soil strength and root growth. *Agron. J.* 74: 875–879.

Gerrard, A.J. 1981. *Soils and Landforms.* Allen and Unwin, London, 219 pp.

Giles, J.F. 1983. Soil compaction and crop growth. *North Dakota Farm Research* 41(1): 34–35.

Gilley, J.E. Gee, G.W., Bauer, A., Willis, W.O. and Young, R.A. 1977. Runoff and erosion characteristics or surface-mined sites in Western North Dakota. *American Society of Agricultural Engineers, Transactions* 5: 697–704.

Goodman, J. and Haigh, M.J. 1981. Slope evolution on abandoned spoil banks in eastern Oklahoma. *Phys. Geog,* 2(2): 160–173.

Grimes, D.W., Miller, R.J. and Wiley, P.L. 1975. Cotton and corn root development in two field soils of different strength characteristics. *Agron. J.* 67: 519–523.

Gyachev, L.V. 1961. (1985). *Theory of Surfaces of Plow Bottoms.* Amerind, New Delhi, 294 pp (Trans. from Russian).

Haigh, M.J. 1978. *Evolution of Slopes on Artificial Landforms, Blaenavon, UK.* Univ. Chicago, Chicago Dept. Geography Research Papers 183: 292 pp.

Haigh, M.J. 1988. Slope evolution on coalmine disturbed land, pp. 3–13. In: Balasubramaniam A.S., Chandra S., Bergado D.T. and Prinya Nutalaya (eds.). *Environmental Geotechnics and Problematic Soils and Rocks.* A.A. Balkema, Rotterdam, 640 pp.

Haigh, M.J. 1992. Problems in the reclamation of coal-mine disturbed lands in Wales. *Int. J. Surface Mining, Environment and Reclamation* 6(1): 31–37.

Haigh, M.J. 1993. Surface mining and the environment in Europe. *Int. J. Surface Mining, Environment and Reclamation* 7(3): 91–104.

Haigh, M.J. 1995. Surface mining in the South Wales environment, pp. 675–682. In: Singhal, R.K., Mehotra A., Hadjigeorgiou J. and Poulin R. (eds.). *Mine Planning and Equipment Selection '95* A.A. Balkema, Rotterdam, 1117 pp.

Haigh, M.J. 1998. Promoting better land husbandry in the reclamation of surface coal-mined land: towards soil quality standards for reclaimed coal lands. *Advances in GeoEcology* 31: 767–779.

Haigh, M.J. 2000. Erosion control: principles and some technical options. In: M.J. Haigh (ed.). *Reclaimed Land: Erosion Control. Soils and Ecology.* A.A. Balkema Rotterdam Oxford and IBH, New Delhi. Land Reconstruction and Management, vol. 1.

Haigh, M.J. and Blake, L. 1989. Soil degradation on reclaimed surface mined coal-land in South Wales: an hypothesis, pp. 12–19, In: M. Penkov (ed.). *Melioratsi Pochv*—Int. Sci. Conf. Soil Melioration, Proc., Sofia, Bulgaria: Higher Institute of Architecture and Civil Engineering, 318 pp.

Haigh, M.J. and Sansom, B. 1999. Soil compaction, runoff and erosion on reclaimed coal-lands (UK). *Int. J. Surface Mining, Reclamation and Environment* 13 (in press).

Haigh, M.J. and Gentcheva-Kostadinova, Sv. 2000. Case Study: Forestation controls erosion on coal-briquette spoil banks, Pernik, Bulgaria. In: Haigh M.J. (ed.) *Reclaimed Land: Erosion Control, Soils and Ecology.* A.A. Balkema, Rotterdam/Oxford and IBH, New Delhi. Land Reconstruction and Management, 1: 281–290.

Haley, G. and O'Keefe, J. 1994. Contaminated Land—What Implications for the City? S.G. Berwin & Co., London-Brussels, 14 pp.

Hall, G.F., Daniels, R.B. and Foss, J.E. 1981: Rates of soil formation and renewal in the USA, pp. 23–40. In: *Determinants of Soils Loss Tolerance.* Amer. Soc. Agron., Spec. Publ. 45: 153 pp.

Hallberg, G.R., Wollenhaupt, N.C. and G.A. Miller. 1978. A century of soil development in spoil derived from loess in Iowa. *Soil Soc. Amer. J.* 42: 339–343.

Hammack R.W. and Watzlaf, G.R. 1990. The effect of oxygen on pyrite oxidation. *Mining and Reclamation Conf. Exhibition, W.Va. Univ., Proc.,* 1: 257–264.

Hangyel, L. and Benesoczky, J. 1992. Research on the effects of municipal effluents on the reclamation of spoil heaps at opencast coal mines. *UN Econ. Comm. Europe, Committ. Energy, Working Party on Coal, Symp. Opencast Coal Mining and Environment (Nottingham, U.K).* ENERGY/WP.1/SEM.2/R.17: 8 pp.

Harris, J.A. and Birch, P. 1989. Soil microbial activity during coalmine restorations. *Soil Use and Management* 5(4): 155–160.

Harris, J.A., Birch, P. and Short, K.C. 1989. Changes in the microbial community and physicochemical characteristics of topsoils stockpiled during opencast mining. *Soil Use and Management* 5(4): 161–165.

Harris, J.A., Hunter, D., Birch, P. and Short K.C. 1987. Vesicular arbuscular mycorrhizal populations in stored topsoils. *British Mycological Soc., Trans.,* 89: 600–603.

Hart, W.M., Batarseh, K., Swaney G.P. and Stiller, A.H. 1991. A rigorous model to predict the AMD production rate of mine waste rock. *Second Int. Conf. Abatement of Acidic Drainage, Montreal, Canada, Proc.,* 2: 257–270.

Hattori, H. 1992. Influence of heavy metals on soil microbial activities. *Soil Sci. Plant Nutrition* 38(1): 93–100.

Hawkins, A.B. and Pinches, G.M. 1992. Engineering description of mudrocks. *Quar. J. Eng. Geol.* 25: 17–30.

Hays, O.E. and Clark, N. 1941. *Cropping Systems that Help Control Erosion* Univ. Wisconsin, Agric. Exper. Station, State Soil Conservation Committee Bulletin 452, Madison, W.

Herlily, A.T., Kaufmann, P.R. and Mitch, M.E. 1990. Regional estimates of acid mine drainage impact on streams in the Mid-Atlantic and Southeastern United States. *Water Air and Soil Pollution* 50 (2): 91–107.

Hedlin, R.S., Nairn, R.W. and Kleinmann, R.L.P. 1994. Passive Treatment of Coal-mine Drainage. *US Bureau of Mines Information Circular* 9389: 35 pp.

Higgitt, D.L. and Haigh, M.J. 1995. Rapid appraisal of erosion rates on mining spoils using caesium-137 measurements. *Int. J. Surface Mining, Reclamation and Environment* 9(3): 141–148.

Hillel, D. 1982. *Introduction to Soil Physics.* Academic Press, New York, 364 pp.

Hodgkinson, R.A. 1989. The drainage of restored opencast coal sites. *Soil Use and Management* 5(4): 149.

Hodgkinson, R.A., Bragg, N.C. and Arrowsmith, R. 1988. Surface and subsurface drainage treatments for restored opencast coal sites, pp. 8–9. In: *Ten Years of Research—What Next?* British Coal Opencast Executive, Mansfield.

Hodgson, J.M. 1979. *Soil Survey Field Handbook.* Soil Survey of England and Wales, Technical Monograph 5: 1–99.

Hole, F.D. 1964. Earthworms and the development of coprogenous A1 horizons in forest soils of Wisconsin. *Soil Sci. Soc. Amer Proc.,* 28, 426–430.

Hood, R. and Moffat, A. 1995. Reclamation of opencast coal-spoil using alder. *NERC* (Natural Environmental Research Council, Swindon, UK), *News*, October 1995: 12–15.

Horn, R. 1989. Strength of structured soils due to loading: a review of processes on macro and micro-scale: European aspects, pp. 9–22. In: W.E. Larson, G.R. Blake, R.R. Allmaras, W.B Voorhees and S.C. Gupta (eds.). *Mechanics and Related Processes in Structured Agricultural Soils.* Kluwer, Dordrecht, NATO Advanced Science Institute Series E: 172: 273 pp.

Hruska, J. and Kram, P. 1992. Monitoring the Lysina catchment with high aluminium concentration in runoff, pp. 111–116. In: J. Krecek and M.J. Haigh (eds.). *Environmental Regeneration in Headwaters* ENVCO, Prague, 368 pp.

Humi, H. 1991. President's report—the soil as a living resource. *World Assoc. Soil Water Conserv. Newsletter* 7(2): 1.

Humi, H. 1996. *This Precious Earth.* ISCO/Univ. Berne, Berne, 96 pp.

ICRCL. 1987. *Guidance on the Assessment and Redevelopment of Contaminated Land.* London: Department of Environment, Inter-Departmental Committee on the Redevelopment of Contaminated Land ICRCL 59/83 (2nd ed.): 19 pp.

IECA. 1994. *Erosion and Sediment Control* (video). Steamboat Springs, Co.: ASCE/Int. Erosion Control Assoc. 25 min.

Indorante, S.J., Jansen, I., and Boast, C.W. 1981. Surface mining and reclamation: initial changes in soil character. *J. Soil Water Conse.,* 36(6): 347–351.

Insam, H. and Domsch, K.H. 1988. Relationship between soil organic carbon and microbial biomass on chronosequences of reclamation sites. *Microbial Ecology* 15: 177–188.

Institute of Forestry, Belgrade. 1997. *Recultivation by Afforestation of Mine-spoil Banks of Opencast lignite Mine, 'Kolubara'.* Republic of Serbia, Ministry of Environmental Protection, Belgrade. 152 pp.

Jaynes, D.B., Rogowski, A.S. and Pionke, H.B. 1984. Acid mine drainage from reclaimed coal-strip mines, I: Model description. *Water Resources Research* 20(2): 233–242.

Jaynes, D.B., Pionke, H.B. and Rogowski, A.S. 1984. Acid mine drainage from reclaimed coal-strip mines, II: Simulation results of model. *Water Resources Research* 20(2): 243–250.

Jobling, J. and Stevens, F.R.W. 1980. *Establishment of Trees on Regarded Colliery Spoil Heaps.* Forestry Commission, Occasional Paper, Edinburgh.

Johnson, D.B. 1995. Acidophilic microbial communities: candidates for bioremediation of acidic mine effluents. *Int. Biodeterioration and Biodegradation* 1995: 41–58.

Jones, C.A. 1983. Effect of soil texture on critical bulk densities for root growth. *Soil Sci. Soc. Amer. J.* 47: 1208–1214.

Kachinskii, N.A. 1958. *Mechanical and Microaggregate Composition of Soil* (Mekhannicheski i mikroagregatnyi sostav pochvy metody ego izucheniya). Israel Program for Scientific Translations, Jersualem, 1611 (1966): 134 pp. (trans. from Russian).

Kalin, M., Cairns, J. and McCready, R. 1991. Ecological engineering methods for acid mine drainage treatment of coal wastes. *Resource Conservation and Recycling* 5: 265–275.

Kambata-Pendias, A. and Pendias, H. 1984. *Trace Elements in Soils and Plants.* CRC Press, Boca Raton, Fla., 380 pp.

Kandev, T. 1987. Recultivation of anthropogenic landscapes in the Pernik Basin. *Annuaire de l'Université de Sofia, Kliment Ohridski, Faculté de Geology et Geographie* 76(2): 141–150.

Kapelkina, L.P. 1992. Special features of the recultivation of damaged land in areas of atmospheric pollution. *UN Comm. Europe, Comm. Energy, Working Party on Coal, Symp. Openscast Coal Mining and Environment (Nottingham. U.K.).* ENERGY/WP.1/SEM/2R.24: 2 pp.

Kasatikov, V.A. 1991. Effect of municipal sewage sludge on soil microelement composition. *Pochvovedeniye* 1991/9: 41–49 (tr. *Eurasian Soil Science* 24 (UDC 631.41), 1992: 11–20.

Kennedy, J.L. 1972. Sodium hydroxide treatment of acid mine drainage. *US Environmental Protection Agency, Crown Mine Drainage Field Site Report:* 6 pp.

Khachatryan, Kh. A. 1963. *Rabota Pochvoobrabatyvaushchikh Orudi v Usloviyakh Gornogo Reliefa.* Armgosizdat, Erevan, 250 pp.

Khachatryan, Kh. A. 1963 (1985). *Operation of Soil Working Implements in Hilly Regions.* USDA/Amerind Publ. New Delhi, 231 pp. (translated from Russian).

Kilkenny, W. 1968. *A Study of the Settlement of Restored Opencast Coal Sites and Their Suitability for Building Development.* Univ. Newscastle-on-Tyne, Dept. Civil Eng. Bull. 38.

Kilmartin, M.P. 1989. Hydrology of reclaimed surface coal-mined land: a review. *Int. J. Surface Mining* 3(2): 71–83.

Kilmartin, M.P. 1994. Runoff generation and soils on reclaimed land, Blaenant, South Wales. Ph.D. thesis, Oxford Brookes Univ., Oxford, U.K.: 384 pp. (unpubl.).

Kilmartin, M.P. 2000. Hydrological management of reclaimed opencast coalmine sites. In: Haigh M.J. (ed.). *Reclaimed Land: Erosion Control, Soils and Ecology.* A.A. Balkema, Rotterdam/Oxford and IBH, New Delhi, Land Reconstruction and Management, 1: 137–158.

Kilmartin, M.P. and Haigh, M.J. 1997. Initial statistical comparisons between natural and reclaimed soils in the South Wales coalfield. *Lesobudska Misul* (Sofia) (in press).

Kimball, B.A., Broshears, R.E., Bencala, K.E. and McKnight, D.M. 1994. Coupling of hydrologic transport and chemical reactions in a stream affected by acid mine drainage. *Environ. Sci. Tech.,* 28: 2065–2073.

King., D.L. and Simmler, J.J. 1973. *Organic Wastes as a Means of Accelerating the Recovery of Acid Strip Mine Lakes. Completion Report* USDI, Office of Water Resources Research. Univ. Missouri, Water Resources Centre, Columbia, MO, 65 pp.

King. D.L., Simmler, J.J., Decker, C.S. and Ogg, C.W. 1974. Acid strip-mine lake recovery. *J. Water Pollution Control Federation* 46: 2301–2316.

King, J.A. 1988: Some physical features of soil after opencast mining. *Soil Use and Management* 4(1): 23.

Kinsell, N.A. 1960. New sulphur oxidising bacterium. *J. Bacteriology* 60: 628–632.

Klecka, W.R. 1975. Discriminant analysis. pp. 434–467. In: *Statistical Package for the Social Sciences.* McGraw Hill, New York.

Kleinmann, R.L.P. and Crerar, D.A. 1979. Thiobacillus ferrooxidans and the formation of acidity in simulated coal mine environments: *Geomicrobiology J.* 1(4): 373–388.

Kleinmann, R.L.P., Crerar, D.A. and Pacelli, R.R., 1981. Biogechemistry of acid mine drainage and a method to control acid formation: *Mining Engineering* 33: 300–305.

Komissaroff, D. 1979. Gimicheskaiapriroda gummosovik veshtests molodyith pochv tehnogeniih eluviev y okisleniih uglei Kuzbasa y ihvesaimodeistvil s mineralami. pp. 212–258. In: *Procvoobrazaovanii y Technogennih Landshaftah*. Nauka, Novosibirsk,

Konhauser, K.O. 1998. Diversity of bacterial iron mineralization. *Earth Science Reviews* 43(3–4): 91–121.

Kononova, M.M. 1966. *Soil Organic Matter* (2e). Pergamon, New York, 544 pp.

Kooistra, K.J., Bouma, J., Boersma, O.H. and Jager, A. 1984. Physical and morphological characteristics of undisturbed and disturbed plough pans in a sandy loam soil. *Soil and Tillage Research* 4(4): 405–417.

Koolen, A.J. and Kuipers, H. 1983. *Agricultural Soil Mechanics*. Springer Verlag, Berlin, 235 pp.

Koorevaar, P., Menelik, G. and Dirksen, C. 1983. *Elements of Soil Physics*. Elsevier, Amsterdam, Developments in Soil Physics 13: 228 pp.

Krauss, H.A. and Allmaras, R.R. 1982. Technology masks the effects of soil erosion on wheat yeilds–a case study in Whiteman County, Washington, In: *Determinants of Soil Loss Tolerance*. Amer. Soc. Agron. Publ. 45: 75–86.

Kukal, Z. 1990. *The Rate of Geological Processes*. Academia (CSAV), Pragye.

Kwaad, F.J.P.M. 1991. Summer and winter regimes of runoff generation and soil erosion on cultivated loess soils (Netherlands) *Earth Surface Processes and Landforms* 16(7): 653–662.

Ladwig, K., Erickson, P. and Kleinmann, R. 1985. Alkaline injection: an overview of recent work. *US Bureau of Mines Information Circular* 9027.

Lal, R., 1984. Mechanised tillage system effects on soil erosion from and alfisol in watersheds cropped to maize. *Soil and Tillage Research* 4(4): 329–348.

Lal, R., Hall, G.F. and Miller, F.P. 1989: Soil degradation I: basic processes. *Land Degradation and Rehabilitation* 1(1): 51–69.

Lalvani, S.B. and Shami, M. 1989. Passivation of pyrite oxidation with metal cations. *J. Materials Science* 33: 300–305.

Lalvani. S.B., Deneve, B.A. and Weston, A. (1990). Passivation of pyrite due to surface treatment. *Fuels* 69: 1567–1569.

Larionov, A.K. 1971 (1982). *Methods of Studying Soil Structure*. (Metody Issledovaniya Structury Grantov) Amerind Publ., New Delhi, 193 pp. (trans from Russian)

Leavitt, B.R. 1986. Federal groundwater legislation significant to coal mining. *Amer. Mining Congress J.* 72(6): 13 et seq.

Ledlin, M. and Pedersen, K. 1996. The environmental impact of mine wastes—roles of microorganisms and their significance in treatment of mine wastes. *Earth Science Reviews* 41: 67–108.

Levy. G.J. 1996. Soil stabilisers, pp. 267–299. In: Agassi, M. (ed). *Soil Erosion, Conservation and Rehabilitation*. Marcel Dekker, New York.

Lewis, R. 1993. Civil liability for contaminated land. *Land Contamination and Reclamation* 1(1): 6–8.

Linn. D.M. and Doran, J.W. 1984. Effect of water-filled pore-space on carbon dioxide and nitrous oxide. *Soil Sci. Soci. Amer. J.* 48: 1267–1272.

Lynch, J.M. and Bragg, E. 1985. Microorganisms and soil aggregate stability. *Advances in Soil Science* 2: 133–166.

McCormick, D.E. Young, K.K. and Kimberlin, L.W. 1982. Current criteria for determining soil loss tolerance (Chapter 9) In: *Determinants of Soil Loss Tolerance*. Amer. Soc. Agron. Spec. Publ. 45, 94–111.

Mckenzie, G.D. and Studlick, J.R.J. 1978. Determination of spoil bank erosion rates in Ohio by using interbank sediment rates. *Geology* 6: 499–502.

Medvedev, V.V. 1990: Variability of the optimal soil density and its causes. *Pochvovdeniye* 1990(5): 20–30 (tr. *Eurasian Soil Science* 21, 1991: 65–75).

Merriwether, R. and Sobek, A. 1978. Physical and chemical properties of overburdens, spoils, mine-wastes and new soils. In: *Reclamation of Drastically Disturbed Lands*. ASA/SCSA/SSSA, Madison, Wi pp. 149–172.

Mishustin, E.N. and Mirsoeva, V.A. 1968. Spore forming bacteria in the soils of the USSR. pp. 459–473. In: T.R.G. Grey and D. Parkinson (eds.) *Ecology of Soil Bacteria*. Liverpool Univ. Press, Liverpool.

Misra, M., Yang, K. and Mehta, R.K. 1996. Application of fly ash in the agglomeration of reactive mine tailings. *J. Hazardous Materials* 51: 181–192.

Miyazaki, T., Hasgawa, S. and Kasubuchi, T. 1993. *Water Flow in Soils*. Marcell Dekker, New York, 290 pp.

Moffat, A. and Bending, N. 1992: Physical site evaluation for community woodland establishment. *Forestry Commission Research Division (Wrecclesham, U.K.); Research Information Note* 216: 3 pp.

Mohamed, A.M.O., Young, R.N., Caporuscio, F. and Li, R. 1996. Flooding of a mine tailing site: suspension of solids, impact and prevention. *Int. J. Surface Mining, Reclamation and Environment* 10(3): 117–126.

Moldenhauer, W.C. and Onstad, C.A. 1975. Achieving specified soil loss levels. *J. Soil and Water Conservation* 30: 166–168.

Monier, G. and Goss, M.G. 1987. *Soil Compaction and Regeneration*. A.A. Balkema, Rotterdam, 170 pp.

Mostaghimi, S., Younas, T.M. and Tim, U.S. 1992: Effects of sludge and chemical fertilizer application on runoff water quality. *Water Resources Bulletin* 28(3): 545–552.

Mostaghini, S., Gidley, T.M., Dillaha, T.A. and Cooke, R.A. 1994: Effectiveness of different approaches for controlling sediment and nutrient losses from eroded land. *J. Soil Water Conservation* 49(6): 615–620.

Munshower, F.F. and Judy, C. 1987. Resource perspectives and the surface coal mine regulatory system, pp. 19–39. In: Hossner, L.R. (ed.). *Reclamation of Surface-Mined Lands* (1) CRC Press, Inc., Boca Raton, Florida, 212 pp.

Naderian, A.R., Williams, D.J. and Clark, I.H. 1996. Numerical modelling of settlement in back-filled open-cut coal mines. *Int. J. Surface Mining, Reclamation and Environment* 10(1): 25–29.

Nagpal, N.K. Kathavate, Y.V. and Sen, A. 1967. Effect of compaction on Delhi soil on the yield of wheat and its uptake of common plant nutrients. *Indian J. Agron.*, 12: 375–378.

Nakariakov, A.V. and Trofimov, S. 1979. O molodih pochv formirovanih nat otvalah otrabotannik rosipei v podzone juzhnoi taigi sre Dnego Urala. pp. 59–106. In: *Pocvoobrazazovanii y Technogennih Landshaftah*. Nauka, Novosibirsk.

Neurerer, H. and Genead, A. 1991. Potentiality for soil erosion control and improving plant production in arid zones. *Die Bodenkulture* 42: 295–306.

Nicolson, V.R., Gilham, R.W., Cherry, J.A. and Reardon, E.J. 1989. Reduction of acid generation in mine tailings through the use of moisture retaining cover layers as oxygen barriers. *Can. Geotech.* 26: 108 et seq.

Nicolau, J.M. 1996. Effects of topsoiling on erosion rates and processes in coalmine spoil banks in Utrillas, Teruel (Spain). *Int. J. Surface Mining, Reclamation and Environment* 10(2): 73–78.

Nicolova, R., Georgiev, N. and Christov. B. (1992): Biostimulator from Bulgarian lignites and its application in agricultural site recultivation. *UN Econ. Comm. Europe, Comm. Energy, Working Party on Coal, Symp. Opencast Coal Mining and Environment (Nottingham, U.K.)*. ENERGY/WP.1/SEM.2/R.38: 2 pp.

Nielsen, G.A., Ruhe, R.V., Fenton, T.E. and Ledesma, L.L. 1975. Missouri River history, floodplain construction and soil formation in southwestern Iowa. *Iowa Agric. Exper. Station, Res. Bull.* 580; 738–791.

Nikiforoff, C.C. 1949. Weathering and soil evolution. *Soil Science* 67: 219–230.

Noustorova, M., Gencheva-Kostadinova, Sv., Filtcheva, E. and Haigh, M.J. 1998: Forestation as an ecological approach for improving surface coal-mine spoils: II: Impact on the microbial action. *J. Balkan Ecology* 1: 56–60.

Nyavor, K. and Egiebor, N.O. 1995. Control of pyrite oxidation by phosphate coating. *Science of the Total Environment* 162: 225–237.

Nyavor, K., Egiebor, N.O. and Fedorak, P.M. 1996. Suppression of microbial pyrite oxidation by fatty acid amine treatment. *Science of the Total Environment* 182: 75–83.

OSMRE. 1992. *Overburden Strength-Durability Classification for Surface Coal-mining*. Office of Surface Mining Reclamation and Enforcement, Technical Report, Wash., DC.

OSMRE. 1998. Acid Mine Drainage. Office of Surface Mining Reclamation and Enforcement, [*http:/www.osmre.gov/amd.htm*] Washington, DC.

Pedersen, T.A. Rogonski, A.G. and R. Pennock, Jnr. 1978. Comparison of some properties of minesoils and contiguous natural soils. *Interagency Energy/Environment R and D Program Report,* E.P.A.—600/7: 78–162.

Peters, E. 1986. Leaching of sulphides, pp. 445–462. In: P. Somasundaran (ed.). *Advances in Mineral Science, Arbiter Symposium*. SME-AIME, New Orleans, La.

Phelps, L.B. 1987. Acid mine drainage, pp. 131–140. In: S.K. Majumdar, F.J. Brenner and E.W. Miller (eds.). *Environmental Consequences of Energy Production*. Pennsylvania Academy of Sciences Easton, Pa.

Pimental, D., Terhune, E.C., Dyso-Hudson, R., Rochgereu, S., Samis, R., Smith, E.A., Denman, D., Reifschneider, D., and Shepard, M. 1976. Land degradation, effects on food and energy resources. *Science* 194: 149–155.

Pionke, H.B. and Rogowski, A.S. 1979. How effective is the deep placement of acid spoil materials? *Can. Land Reclamation Assoc., Proc.* 4: 87–104.

Plyusnin, I.I. 1964. *Reclamative Soil Science* (tr. Meliorativnoe Pochvovdenie). Foreign Languages Publ. House, Moscow, 398 pp.

Pole, R., Bauer, A., Zimmerman, L. and Melsted, S. 1979. Effects of topsoil thickness on spoil banks on wheat and corn yields in North Dakota, *Can. Land Reclamation Assoc., Proc.,* 4: 139–155.

Popova, A.A. 1991: Effect of mineral and organic fertilisers on the status of heavy metals in soils. *Agrokhimiya* 1991 (3): 62–67 (tr. *Eurasian Soil Science* 23, 1991: 38–44).

Porta, J., Poch, R. and Boixadera, J. 1989. Land evaluation and erosion control practices on mined soils in N.E. Spain. *Soil Technology Series* 1: 189–206.

Poschenreider, C., Gunse, B. and Barcelo, J. 1989: Influence of cadmium on water relation, stomatal resistance and abscissic acid content in expanding bean leaves. *Plant Physiology* 90: 1365–1371.

Potter, K.N., Carter, F.S. and Doll, E.C. 1988: Physical properties of constructed and undisturbed soils. *Soil Sci. Soc. Amer. J.* 52: 14.

Poulovassilis, A. 1972. The changeability of hydraulic conductivity of saturated soil samples. *Soil Science* 113(29): 81–87.

Powers, D.H. and Skidmore, E.L. 1984: Soil structure as influenced by simulated tillage. *Soil Sci. Soc. Amer. J.* 48: 879–884.

Quastel, J.H. 1954. Soil conditioners. *Ann. Rev. Plant Physiology* 5: 75–92.

Rahn, P.H. 1992. A method to mitigate acid mine drainage in the Shamokin area, Pennsylvania, U.S.A. *Environmental Geology and Water Sciences* 19(1): 47–53.

Ramsay, W.J.H. 1986: Bulk soil handling for quarry restoration. *Soil Use and Management* 2(1): 30–39.

Rao, S.R., Gehr, R. Riendeau, M., Lu, D. and Finch, J.A. 1992. Acid mine drainage as a coagulant. *Minerals Eng.* 5(9): 1011–1020.

Reed, S.M. and Hughes, D.B. 1990. Long term settlement of opencast mine backfills—case studies from the North East of England. *Third Int. Sympo. Reclamation, Treatment and utilisation of Coal Mining Wastes, Proc.:* 313–323.

Reeves, C.A. and Cooper, A.W. 1960. Stress distribution in soils under tractor loads. *Agric. Eng.* 41: 20–21, 31.

Reeves, T.J., Haines, P.J. and Coventry, D.R. 1984: Growth of wheat and clover on soil artificially compacted at various depths. *Plant and Soil* 80: 135–138.

Remezov, N.P. and Pogrebynak, P.S. 1965. *Forest Soil Science*. (tr. Lesnoe Pochvovdenie). Israel Program for Scientifc Translations, Jerusalem, 5177 (1969): 261 pp.

Renton, J.J., Rymer, T.E. and Stiller, A.H. 1988. A laboratory procedure to evaluate the acid producing potential of coal associated rocks. *Mining Science and Technology* 7: 227–235.

Renton, J.J., Stiller, A.H. and Rymer, T.E., 1988. The use of phosphate materials as ameliorants for acid mine drainage: *US Bureau of Mines Information Circular* 9182, 1: 67–75.

Repelewska, J. 1968. Procesy erozyjne na zwalach kopalnianych. *Czasopismo Geograficzne* 39(1): 31–43.

Ritchie, J.C., Spradberry, J.A. and McHenry, J.R. 1974. Estimating soil erosion from the redistribution of fallout 137-Cs. *Soil Sci. Soc. Amer., Proc.* 38, 137–139.

Rode, A.A. 1952 *Soil Moisture* (tr. Pochvennaya Vlaga). Israel Programme for Scientific Translations (1967), Jerusalem, 260 pp.

Rogoff, M.H., Silverman, M.P. and Wender, I. 1960. The elimination of sulphur from coal by microbial action, pp. 25–36. In: Amer. Chem. Soc., Division of Gas and Fuel Chemistry, Conference Preprints.

Rogowski, A.S. and Jacobi, E.L. 1979. Monitoring water movement through strip-mine spoil profiles. *Amer. Soc. Agric. Eng. Trans.*, 22: 104–109.

Rushton, S.P. 1986. The effects of soil compaction on *Lumbricus terrestris* and its possible implications for populations on land reclaimed after surface coal-mining. *Pedobiologia* 29: 85–90.

Russell, R.S. 1977: *Plant Root Systems*. McGraw Hill, London, 298 pp.

Rymer, T., Stiller, A., Hart W. and Renton, J. 1990. Some aspects of SSPE/PSM modelling for quantitative assessment of disturbed hydrologic systems. *Mining and Reclamation Conf. Exhibition (W.Va Univ.) Proc.*, 1: 61–68.

Schafer, W.M., Nielsen, G.A. and Nettleton, W.D. 1980. Minesoil genesis and morphology in a spoil chronosequence in Montana. *Soil Sci. Soc. Amer. J.* 44: 802–808.

Schamp. N., Huylebroek, J. and Sadones, M. 1975. Adhesion and absorption phenomena in soil conditioning, pp. 13–24. In: S. Stewart (ed.). *Soil Conditioners*. Madison, WI.: Soil Sci. Soc. Amer., Spec. Publ. 7.

Scharer, J.M., Garga, V., Smith, R. and Halbert, B.E. 1991. Use of steady state models for assessing acid generation in pyritic mine tailings. *Second Int. Conf. Abatement of Acidic Drainage, (Montreal, Canada), Proc.*, 2: 211–230.

Schearer, R.E., Everson, W.A., Mausteller, J.W. and Zimmerer, R.P. 1970. Characteristics of viable antibacterial agents used to inhibit acid-producing bacteria in mine-wastes. *Coal Mine Drainage Research, Third Symposium, (Pittsburg, Pa.), Preprints:* 188–199.

Schelp, G.S., Chesworth, W. and Spiers, G. 1995. The amelioration of acid mine drainage by an in situ electrochemical method. I/II. *Applied Geochemistry* 10: 705–732.

Schindler, D., Wagemann, R., Coo, R.B., Rusczinski, T. and Prokopsich, J. 1980. Experimental acidification of Lake 233, Experimental Lakes Area. *Can. J. Fish. Aqua. Sci.* 37: 342–354.

Scullion, J. 1992. Re-establishing life in restored topsoils. *Land Degradation and Rehabilitation* 3(3): 161–168.

Scullion, J. and Mohammed, A.R.A. 1986. Field drainage experiments and design of former opencast coal mining land. *J. Agric. Sci.* 107(3): 521–528.

Scullion, J., Mohammed, A.R.A. and Ramshaw, G.A. 1991. The effect of cultivation on structure in soils affected by opencast mining of coal. *J. Sci. Food Agric.*, 55: 327–339.

Sequi, P. 1978. Soil structure—an outlook. *Agrochimica* 22: 403–425.

Serebriakov, V.N. and Sharshovets, G.A. 1992. Restoration of the fertility of land damaged by opencast workings. *UN Econ. Comm. Europe, Comm. Energy, Working Party on Coal, Symp. Opencast Coal Mining and the Environment (Nottingham, U.K.).* ENERGY/WP.1/SEM.2/R.45: 2 pp.

Sharma, P.P. and Carter, F.S. 1993. Infiltration and soil water distribution of pre- and post-mine soil profiles. *North Dakota State University: Mine-Land Reclamation Research Review, Proc.*, 2–16.

Shaxson, T.F. 1992. Erosion, soil architecture and crop yields. *J. Soil Water Conservation* 47(6): 433.

Shaxson, T.F. 1993. Organic materials and soil fertility. *Enable* (Newsletter, Association for Better Land Husbandry) 1: 2–3.

Shaxson, T.F. 1995: Principles of good land husbandry. *Enable* (Newsletter, Association for Better Land Husbandry) 5: 4–13.

Sheptukhov, V.N., Voronin, A.I. and Shipliov, M.A. 1982. Bulk density of the soil and its productivity (in Russian). *Agrokhimiya* 1982(8): 91–100 (English trans: UDC 631.431.1:631.452—Scripta Technica).

Sherlock, C.G. 1918. The pollution of streams. *Eng. Mining. J.* 106: 861.

Silburn, D.M. and Crow, F.R. 1984. Soil properties of surface mined lands. *Amer. Soc. Agric. Eng. Trans.*, 27: 825.

Singh, G. and Bhatnagar, M. 1985. Bacterial formation of acid mine drainage: causes and control. *J. Sci. Ind. Res.*, 44: 478–485.

Skinner, W.D. and Arnold, D.E. 1990. Short-term biotic response before and during the treatment of acid mine-drainage with sodium carbonate. *Hydrobiologia* 199: 229–235.

Skousen, J.G., Sencindiver, J.C. and Smith, R.M. 1987. *Review of Procedures for Surface Mining and Reclamation in areas with Acid-producing Materials.* W. Va. Univ. Energy and Water Research Center and West Virginia Mining and Reclamation Association, Morganstown, Va, 39 pp.

Skousen, J.G., Politan, K., Hilton T. and Meek, A. 1990. Acid mine drainage treatment systems: chemicals and costs. *Green Lands* (West Virginia Mining and Reclamation Association, Charleston) 20(4): 31–37.

Slick, B.M. and Curtis, W.R. 1985. A guide for the use of organic materials as mulches in the reclamation of coal minesoils in the Eastern United States. *US Dept. Agriculture, Forest Service, Northeastern Station, General Technical Report* NE-98: 144 pp.

Smith, D.D. 1941. Interpretation of soil conservation data for field use. *Agric. Eng.*, 22, 173–175.

Smith, D.L.O. 1987. Measurement, interpretation and modelling of soil compaction. *Soil Use and Management* 3(3): 87–93.

Smith, D.L.O., Goodwin, R.J. and Spoor, G. 1989. Modelling soil disturbance due to tillage and traffic pp. 121–136. In: W.E. Larson, G.R. Blake, R.R. Allmaras, W.B. Voorhees and S.C. Gupta (eds.). *Mechanics and Related Processes in Structured Agricultural Soils.* Kluwer, Dordrecht, NATO Advanced Science Institute Series E: 172: 273 pp.

Smith, G.N. 1971. *Elements of Soil Mechanics for Civil and Mining Engineers* (2e). Crosby Lockwood, London, 341 pp.

Smith, L., Lopez, D., Beckie, R., Morin, K., Dawson, R. and Price, W. 1995. *Hydrogeology of Waste Rock Dumps.* Calgary: Agra Earth and Environmental: Final Report to the Department of Natural Resources (Canada, Contract 23440-4-1317/01-SQ: 130 pp.

Smith, M.A. 1985. Contaminated land: reclamation and treatment. *NATO—Challenges of Modern Society* 8: 407–417. Plenum Press, New York.

Smith, M. 1993. Dealing with contaminated ground conditions. *Land Contamination and Reclamation* 1(1): 22–36.

Smith, R.M., Tryon, E.H., and Tyner, E.H. 1971. Soil development on mine spoil. *W. Va. Univ. Agricultural Experiment Station, Bulletin* 604T: 1.

Smyth, C.R. 1997. Native grass, sedge and legume establishment and grass-legume competition at a coal-mine in the Rocky Mountains, British Columbia. *Int. J. Surface Mining, Reclamation and Environment* 11(2): 105–113.

Soane, B.D. 1981. Soil degradation attributable to compaction by wheels and its control, pp. 27–46. In: D. Boels, D.B. Davies and A.E. Johnston (eds.). *Soil Degradation.* Balkema, Rotterdam.

Sobek, A.A., Schuller, W.A., Freeman, J.R. and Smith, R.M. 1978. Field and laboratory methods applicable to overburden and minesoils. *USEPA Report* No. 600/2-78-054: 203 pp.

Sohne, W.H. fur 1966. Characteristics of tillage tools. *Satrych fur Grundforbattring* (Oslo) 1: 31–48.

Sohnitzer, H. et al., 1969. Free radicals in soil humic compounds. *Soil Science* 108: 6.

Soil Survey Staff. 1994. *Keys to Soil Taxonomy* (6e). Washington: US Dept. Agric. Soil Conservation Service, Agriculture Handbook 436.

Sojka, R.F. and Lentz, R.D. 1996. Polyacrylamide in furrow irrigation: an erosion control breakthrough, pp. 183–189. In: *Int. Erosion Control Assoc., First European Conf. Trade Exposition* (Barcelona), *Lecture Book* 1: 189 pp.

Stark, L.R., Brooks, R.P., Webster, H.J., Unz, R.F. and Ulrich, T. 1994. *Treatment of Mine Drainage by a Constructed Multi-cell Wetland: the Corsica Project.* Bureau of Mining and Reclamation, and Pennsylvania Department of Environmental Resources. Final Report ER9411, Washington, D.C.

Steila, D. 1976. *The Geography of Soils.* Prentice Hall, Englewood Cliffs, 222 pp.

Stewart, V.I., Scullion, J., Salih, R.O. and Al-Bakri, K.H. 1989. Earthworms and soil structure rehabilitation in subsoils and in topsoils. *Biol. Agric. Hortic.,* 5: 325–337.

Strohm, W.E., Bragg, G.H., and Ziegler, T.W. 1978. *Design and Construction of Compacted Shales Embankments.* Federal Highway Administration Report: FMWA-RD-78-141-v 5. Wash., DC.

Strzyszcz, Z. 1992. Biological reclamation of phytotoxic Tertiary formations composing overburden of the brown coal mine in Zar region. *UN Econ. Comm. Europe, Comm. Engery, Working Party on Coal, Symp. Opencast Coal Mining and the Environment (Nottingham UK).* ENERGY/WP.1/SEM.2/R.6: 2 pp.

Suhr, J.L., Jarrett, A.R. and Hoover, J.R. 1984. The effect of soil air entrapment on erosion. *Amer. Soc. Agric. Eng., Trans.,* 27(1): 93–98.

Sweigard, R.J. and Escobar, E. 1989. A field investigation into the effectiveness of equipment alternatives in reducing subsoil compaction. *Mining Science and Technology* 8: 313–320.

Szegi, J. and Voros, T. 1992. Research on the vesicular-arbuscular endomycorrhizas during the recultivation of mining spoils in Hungary and Poland. *UN Econ. Comm. Europe, Comm. Energy, Working Party on Coal, Symp. Opencast Coal Mining and the Environment (Nottingham, U.K).* ENERGY/WP.1/SEM.2/R.14: 1 pp.

Taranov, A. et al. 1979. Partziliarnaia struktura fitotzenoza y neodnomodnost molodiih pochv tehnogennih landshaftah. pp. 19–58. In: *Pocvoobrazazovanii y Technogennih Landshaftah.* Nauka, Novosibirsk.

Tardieu, F. 1989. Root system responses to soil structural properties: micro and macro-scale, pp. 153–172. In: W.E. Larson, G.R. Blake, R.R. Allmaras, W.B. Voorhees and S.C. Gupta (eds.). *Mechanics and Related Processes in Structured Agricultural Soils.* Kluwer, Dordrecht. NATO Advanced Science Institute E-172–273 pp.

Tarutis, W.J. J., Stark, L.R., and Williams, F.M. 1999. Sizing and performance estimation of coal mine drainage wetlands. *Ecol. Eng.* 12: 353–372.

Tate, R.L. III. 1985. Microorganisms, ecosystem disturbance and soil formation, pp. 1–33. In: R.L. Tate III and D.A. Klein (eds.). *Soil Reclamation Processes: Microbiological Analyses and Applications.* Marcel Dekker, Basel, 345 pp.

Taylor, H.M. and Ratcliff, L.F. 1969. Root elongation of cotton and peanuts as a function of soil strength and soil water content. *Soil Science* 108(2): 113–119.

Taylor, R.K. 1974. Colliery spoil heap materials—time dependent changes. *Ground Engineering,* July '74: 24–27.

Taylor, RK. 1987. Colliery wastes in the United Kingdom—findings and implications 1966–1985, pp. 435–446. In: *Environmental Geotechnics.* A.A. Balkema, Rotterdam.

Taylor, R.K. 1988. Coal Measures mudrocks: composition, classification and weathering processes. *Quart. J. Eng. Geol.,* 21: 85–89.

Taylor, R.K. and Spears, D.A. 1970. The breakdown of British coal measure rocks. *Int. J. Rock Mech. Mining Sci.,* 7: 481–501.

Temple, K.L. 1952. Investigations on coal-mine drainage. *W. Va. Agric. Res. Station, Indus. Res. Fellowship 11, Interim Rep.* 19: 13 pp.

Theng, B.K.G. 1982. Clay-polymer interactions; summary and perspectives. *Clays and Clay Minerals* 30: 1–10.

Thomas, D. and Jansen, I. 1985. Soil development in coalmine spoils. *J. Soil Water Conservation* 40: 439–443.

Thomas, M.D.A., Kettle, R.J. and Morton, J.A. 1989. The oxidation of pyrite in cement stabilised colliery shale. *Quart. J. Eng. Geol.,* 22(3): 207–218.

Thompson, L.M. 1952. *Soils and Soil Fertility (le).* McGraw Hill, New York.

Thompson, P.J., Jansen, I. and Hooks, C.L. 1987. Penetrometer resistance and bulk density as parameters for predicting root system performance in mine soils. *Soil Sci. Soc. Amer. J.* 51: 1288–1293.

Thompson, S., Scott, J.D., Seogo, D.C. and Schultz, T.M. 1986. Testing of model footings on reclaimed land, Wabamun, Alberta. *Can. Geotech. J.*, 23: 541–547.

Tollner, E.W. and Verma, B.P. 1984. Modified cone penetrometer for measuring soil mechanical impedance. *Amer. Soc. Agric. Eng., Trans.*, 27: 31–336.

Toy, T.J. and Hadley, R.F. 1987. *Geomorphology and Reclamation of Disturbed Lands.* Academic Press, Orlando, Fla. 480 pp.

Toy T.J. and Shay, D. 1987. Comparison of some soil properties on natural and reclaimed hillslopes. *Soil Science* 143(4): 264–277.

Turner, D. and McCoy, D. 1990. Anoxic alkaline drain treatment system, a low cost acid mine drainage treatment alternative, pp. 73–75. In: D.H. Graves (ed.). *Nat. Symp. Mining,* (Lexington), *Proc.* Univ. Kentucky, Lexington, Ky.

Tuttle, J.H., Dugan, P.R. and Randles, C.I. 1969. Microbial dissimilatory sulphur cycle in acid mine water. *J. Bacteriology* 97: 594–602.

Twardowska, I. 1986. The role of *Thiobacillus ferrooxidans* in pyrite oxidation in colliery spoil tips. I/II. *Acta Microbiologica Polonica* 35(3–4): 291–298/36(1–2): 101–107.

USEPA. 1982. *Development Document for Final Effluent Limitations Guidelines.* US Environmental Protection Agency EPA 440/1–82/057. Washington, D.C.

USEPA. 1989. Standards for the disposal of sewage sludge: US Environmental Protection Agency Proposed Rules. *Federal Register* 54: 5746–5902.

Van Breemen, N. 1993. Soils as biotic constructs favouring net primary productivity. *Geoderma* 57: 183–211.

Van Doren, C.A. and L.J. Bartelli 1956. A method of forecasting soil loss. *Agric. Eng.* 37(5), 335–341.

Vandre, B. 1995. Contribution to Smith et al. 1995: 11–14.

Varela, A. et al. 1993. Chemical and physical properties of opencast lignite minespoils. *Soil Science* 156(3): 193–205.

Veihmeyer, F.J. and Hendrickson, A.H. 1948. Soil density and root penetration. *Soil Sci.* 65: 478–493.

Verigo, S.A. and Razumova, L.A. 1963. *Soil Moisture and Its Significance in Agriculture* (tr. Pochvennaya vlaga i ee znachenie v sel'skolkhozyaistevennom proizvodstve). Israel Program for Scientific Translations 1689 (1966), Jerusalem, 234 pp.

Verpraskas, M.J. 1988. Bulk density values diagnostic of restricted root growth in coarse textured soils. *Soil Sci. Soc. Amer. J.*, 52: 1117–1121.

Verpraskas, M.J. and Wagger, G. 1989. Cone index values diagnostic of subsoiling in corn root growth. *Soil Sci. Soc. Amer. J.*, 53: 1499–1505.

Vile, M.A. and Wieder, R.K. 1993. Alkalinity generation by Fe(III) reduction versus sulphate reduction in wetlands constructed for acid mine drainage treatment. *Water, Air and Soil Pollution* 9: 425–441.

Visser, S. 1985. Management of microbial processes in surface-mined land reclamation in western Canada, pp. 203–241. In: R.L. Tate III and D.A. Klein (eds.). *Soil Reclamation Processes: Microbiological Analyses and Applications.* Marcel Dekker, Basel, 345 pp.

Vogel, W.G. 1987. *A Manual for Training Reclamation Inspectors in the Fundamentals of Soils and Revegetation.* Soil and Water Conservation Society/Office of Surface Mining and Enforcement/US Dept. Agric. Northest Forest Experiment Station SWCS Ankeny, Iowa, 178 pp.

Volk, T. 1998. *Gaia's Body: Toward a Physiology of Earth.* Copernicus (Springer Verlag), New York, 269 pp.

Voorhees, W.B. 1983. Relative effectiveness of tillage and natural forces in alleviating wheel induced soil compaction. *Soil Sci. Soc. Amer. J.*, 47: 129–133.

Vos, W.J. and Stortelder, A. 1992. *Vanishing Tuscan Landscapes: Landscape Ecology of a SubMediterranean-Montane Area (Solano Basin, Tuscany, Italy).* Pudoc Scientific, Wageningen, Netherlands, 404 pp.

Wallace, A. and Wallace, G.A. 1990. Soil and crop improvement with water-soluble polymers. *Soil Tech.*, 3: 1–8.

Wallace, A., Wallace, G.A. and Cha, J.W. 1986. Mechanisms involved in soil conditioning by polymers. *Soil Science* 141: 381–386.

Walsh, J.P. 1985. Soil and overburden management in western surface coal mine reclamation—findings of a study conducted for the Congress of the United States—Office of Technology Assessment. *Amer. Soc. Surface Mining and Reclamation, Proc.*, 2: 257–264.

Wang, D. and Sweigard, R.J. 1996. Characterisation of fly ash and bottom ash from a coal-fired power station. *Int. J. Surface Mining, Reclamation and Environment* 10(4): 181–186.

Warburton, D.B., Klimstra, W.B. and Spitzkeit, J. 1988. Stabilizing coal-mine tailings by direct revegation. *Int. Erosion Control, Proc.* 19: 139–147.

Ward, A.D., Wells, A.G. and Phillips, R.E. 1983. Infiltration through reconstructed surface mined spoils and soils. *Amer. Soc. Agric. Eng., Trans.*, 5: 821–829.

Wieder, R.K. 1994. Changes in iron (III)/iron (II) in effluent from constructed acid-mine drainage wetlands. *J. Environ. Quality* 23: 730–738.

Wiggering, H. 1993. Sulphide oxidation—an environmental problem within colliery spoil dumps. *Environ. Geol.*, 22(2): 99–105.

Wildeman, T.R. and Laudon, L.S. 1989. Use of wetlands for treatment of environmental problems in mining: non-coal-mining applications. *Constructed Wetlands for Wastewater Treatment*, pp. 550–557. Lewis Publishers, Boca Raton, FLA.

Willat, S.T. 1986. Root growth of winter barley in a soil compacted by the passage of tractors. *Soil and Tillage Research* 7: 41–50.

Williamson, J.C. and Johnson, D.B. 1991. Microbiology of soils at opencast coal-sites II. *J. Soil Science* 42: 9–15.

Williamson, N.A., Johnson, M.S. and Bradshaw, A.D. 1982. *Mine Wastes Reclamation.* Mining Journal Books, London 103 pp.

Wilson, K. 1985. A guide to the reclamation of mineral workings for forestry. *Forestry Commission (Edinburgh), Research and Development Paper* 141: 56 pp.

Wischmeier, W.H. and Smith, D.D. 1978. Predicting rainfall erosion losses; a guide to conservation planning. *US Dept. Agriculture, Agricultural Handbook 537:* 58 pp.

Witsell, L.E. and Hobbs, J.A. 1965. Soil compaction effects on field plant growth. *Agron. J.*, 57: 534–537.

Wood, T.S. and Shelley, M.L. 1999. A dynamic model of bioavailability of metals in constructed wetland sediments *Ecol. Eng.* 12: 231–252.

Zheleva, E. and Haigh, M.J. 2000. Case Study: Vegetation, erosion and soil development on lignite spoil banks: Maritsa-Iztok, Bulgaria. In: M.J. Haigh (ed) *Reclaimed Land: Erosion Control, Soils and Ecology.* A.A. Balkema Rotterdam/Oxford and IBH, New Delhi. *Land Reconstruction and Management*, 1: 275–280.

Zhengqi Hu, Caudle, R.D. and Chong, S.K. 1993. Evaluation of farmland reclamation effectiveness based on reclaimed mine soil properties. *Int. J. Surface Mining and Reclamation* 6: 129–135.

Ziemkiewicz P.F. and Meek F.A. 1994. Long term behaviour of acid forming rock: results of 11-year field studies. pp. 49–56. In: *Proc. Int. Land Reclamation and Mining Drainage Conf. and Third Inte. Conf. Abatement of Acid Mine Drainage.* Pittsburgh, Pennsylvania.

10

Case Study: Vegetation, Erosion and Soil Development on Lignite Spoil Banks: Maritsa-Iztok, Bulgaria

Elena Zheleva and Martin J. Haigh

Abstract

Recultivating the steep terrace risers of lignite spoil banks with black locust (Robinia pseudoacacia L.) rapidly converted soil loss into soil gain. It also allowed deeper penetration of soil organic matter than recultivation with grass.

INTRODUCTION

The lignite fields of Maritsa-Iztok provide the most important energy resource of Bulgaria (Malakov, 1993). This study concerns the reclamation of spoil banks associated with the Troyan power generation complex of the Maritsa-Iztok Economic Association, near Stara Zagora, south-central Bulgaria (cf. Fig. 1, Haigh and Gentcheva-Kostadinova 2000, this book). Here, lignites are excavated in very large surface mines and are fed directly by conveyor belt to three large power stations. The mine complex cuts through prime agricultural land at the rate of around 1 km every year.

Prior to mining, the humic layers of the fertile soils are removed and stored. When the mining front has passed, the pit is refilled and the land is reshaped as a series of terraces with near-horizontal benches separated by steep (20 degree) risers. During reclamation, care is taken to bury toxic and infertile materials and to leave materials with favourable soil-forming properties at the surface. This policy was adopted after some early attempts at reclamation foundered due to poor quality and expansive clays being exposed at the soil surface. After land forming, stored topsoils are

Land Reconstruction and Management Vol. 1, 2000, pp 275–279.
ISSN 1389-2541
ISBN 90 5410 793 6
A.A. Balkema, Rotterdam, The Netherlands

laid on the terrace benches and these areas are recultivated for agricultural production. There is not enough soil for the restoration of the steep risers, so these are reclaimed by direct forestation. In recent years, the severe and protracted economic crisis, which has followed the transition from Socialism, has restricted funding available for this forestation. This situation provides the context for this project.

ENVIRONMENTAL CONDITIONS

The climatic and hydrological context of Maritsa-Iztok is similar to that of Pernic, described in Haigh and Gentcheva-Kostadinova (2000, this book, Table 1). However, it is warmer, drier and more prone to late-summer soil water deficit.

Table 1: Climate characteristics at Maritsa-Iztok and Pernik

Climate/Coalfield	Pernik	Maritsa-Iztok
Mean annual temperature	8–10°	12.6°
Mean annual precipitation	550–650 mm	542 mm
Wettest months	May–June	May–June
Driest months	August–September	August–September
Warmest month	July: 37°	August: 42.4°

GEOLOGY AND SOIL CHARACTERISTICS

Maritsa-Iztok's lignites are found in Pliocene deposits 70 m thick. The overburden lithologies, which are stripped in layers, are: clay, clayey sands, sands, and lignite coal. Some clays, which are toxic, are buried well away from the soil surface (Gentcheva-Kostadinova and Haigh, 1988). The final spoil banks are composed of clays with some clayey sands. The clay percentage is typically 50% and ranges upwards to 70–74%. The spoils tend to develop relatively high bulk densities of 1.65 up to 1.80 $g.cm^3$ against specific gravities of 2.36 up to 2.85 $g.cm^3$ (Gentcheva-Kostadinova et al., 1994). The high densities inhibit root hair penetration and make forestation more difficult. Trees develop mainly horizontal root systems, which may penetrate no deeper than 25–40 cm. However, unlike the briquette spoils at Pernik (Haigh and Gentcheva-Kostadinova, this book), the Maritsa-Iztok clays readily undergo natural vegetative colonisation. This natural vegetation can be employed as a green manure and assists subsequent recultivation through sideration (Zheleva, 1988).

EXPERIMENT PLAN

Normally, these lignite-spoil terrace risers have been forested with dense plantings of the leguminous, nitrogen-fixing, black locust/false acacia (*Robinia pseudoacacia*). Unfortunately, severe national economic stress has meant that, in this region, while many spoil banks created before 1989 have been forested, many of those created since the 'Blue Revolution' have not.

This unfortunate circumstance has created an opportunity to demonstrate the importance of forestation for erosion control and soil development. Erosion is monitored on two neighbouring spoil banks. One, created around 1989, is densely forested with black locust. The second, created after 1991, remained untreated. In 1993, when this study commenced, it existed in an almost totally unvegetated state and was suffering severe rill and interrill erosion that was also causing sediment pollution damage to cultivated fields on the adjacent terrace bench. However, as the experiment progressed, this slope developed a dense herbaceous ground vegetation cover, surface erosion was replaced and soil pipe development took place.

Ground surface changes were measured by means of erosion pins. As at Pernik, these were laid out in rows at 5-metre intervals down the slope profile (cf. Haigh and Gentcheva-Kostadinova 2000, this book). Results were recorded for the whole slope up to the break of slope at the start of the terrace bench where slope-foot deposition begins.

RESULTS

Preliminary ground retreat scores from the two slopes are strikingly different. Fig. 1 shows results mapped down the mainly rectilinear terrace riser from crest to slope-foot. The data incorporates records from April 1993, April 1994 and September 1994.

Results from the forested slope, away from the slope-crest (05–metres) are mainly negative. This indicates that there has been net ground advance in the first 17 months of the study. The main reason is litter accumulation, but there may also be changes in the packing density of the spoils.

By contrast, on the initially unvegetated slope, away from the slope-crest (00 and 05-metres), apart from an anomalous result at the 40-metres grid row, there is net ground loss on the untreated slope. This ground loss increases with distance downslope.

Visible evidence suggests that the bulk of the ground loss may be due to soil erosion. However, a proportion may be due to increases in soil compaction.

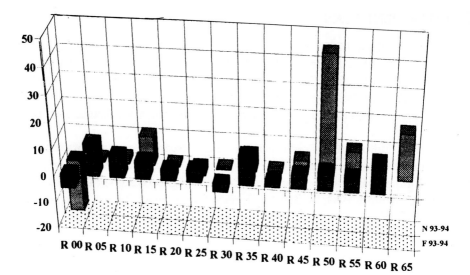

Fig. 1. Preliminary ground retreat scores (1993–1994), Maritsa-Iztok.

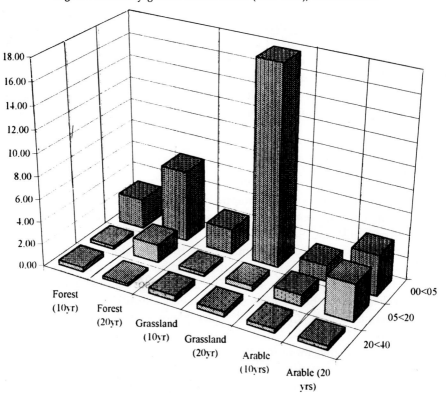

Fig. 2. Organic matter accumulation under different land-uses; lignite spoil banks, Maritsa-Iztok.

DISCUSSION

The moral of this tale is that forested slopes suffer much less erosion than unvegetated slopes and, ground advance rather than soil loss. At Maritsa-Iztok, colleagues have compared the percent organic matter of sites reclaimed as forest, grassland and for agriculture, 10 and 20 years after reclamation (Fig. 2 and Banov et al., 1998, 1993, 1995). In these spoils of pH 7.4–7.8, high carbonate (1.0–6.9%), and high K (1.9–2.4%), results indicate that, over 20 years, grassland generates the greatest organic matter accumulation at the soil surface and that forest enriches the soil surface layer more than arable agriculture. The artificial topsoil, despite the huge cost of its creation, fares little better than the site recultivated without topsoil. However, in the absence of tillage and fertiliser input, only forest enriches the profile at 05–20 cm depths and, experience suggests, only the tree roots are likely to influence progressively greater depths.

CONCLUSION

Forest-fallowing surface coal-mine-disturbed land with appropriate tree species accelerates organic accumulation in the soil, reduces rilling and sheet erosion, and quickly transforms soil loss into soil gain, even on steeply sloping sites.

Acknowledgements

The authors thank the British Council, Sofia and the Earthwatch Institute, Oxford, England, for supporting this project.

References

Banov, M., Filtcheva, E. and Hristov, B. 1989. Humosonatrvane i kachestven sostav na humusa pri recultivirani zemi. *Pochvoznanie i Agrochimia* 4: 3–9.

Banov, M., Hristov, B., Filtcheva, E. and Georgiev, B. 1993. *Humus Accumulation and its Quality Composition in Reclaimed Lands.* Poushkarov Institute of Soil Science, Sofia, p. 9.

Banov, M., Hristov, B., Filtcheva, E. and Georgiev, B. 1995. Humus accumulation and its quality composition in reclaimed lands. pp. 279–285. *Sbornik Nauchni Dokladi, Jubileina Nauchna Conferentsiya 125 Godini BAN i 65 Godini Institut za Gorata.* Institut za Gorata i Bulgarska Acadamiiya na Naukite, Sofia.

Gentcheva-Kostadinova, Sv. and Haigh, M.J. 1988. Land reclamation and afforestation research on the coal-mine disturbed lands of Bulgaria. *Land Use Policy* 5(1): 94–102.

Gentcheva-Kostadinova, Sv., Zheleva, E., Petrova, R. and Haigh, M.J. 1994. Soil constraints affecting the forest-biological recultivation of coal-mine spoil banks in Bulgaria. *Int. J. Surface Mining, Reclamation and Environment* 8(2): 47–54.

Haigh, M.J. and Gentcheva-Kostadinova, Sv. 2000. Case study: forestation controls erosion on coalbriquette spoil banks, Pernik, Bulgaria. In: M.J. Haigh (ed.). *Reclaimed Land: Erosion Control, Soils and Ecology.* A.A. Balkema Rotterdam/Oxford and IBH, New Delhi. *Land Reconstruction and Management,* 1: pp 281–290.

Malakov, P. 1993. Country Report: Bulgarian Delegation. *United Nations Economic Commission for Europe, Committee on Energy, Working Party on Coal, Workshop on Environmental Regulations in Opencast Mining Under Market Conditions (Most, Czech Republic) Nov. 9–11, 1993,* (see: UNECE Secretariat 1994: Outcome of the Workshop on the Development of Environmental Regulations in Opencast Coal Mining under Market Conditions (Most, Czech Republic, Nov. 1993). ENERGY/WP.1/R.31: 14 pp.

Zheleva, E. 1988. Lesnaya rekultivatsia mineralnim uadobreniem i sideratskia otvalov pouchennih otkritoyia dobrichey [Forestry reclamation with mineral fertilization and sideration on waste tips from surface lignite mining]. pp. 82–96. In: J. Bender (ed.). *Vivotrivchine i Pochoobvoozrazovatelinie Processa na Otvalah Tekhnogenikh,* Polskiya Akademiya Nauk, Konin.

11

Case Study: Forestation Controls Erosion on Coal-briquette Spoil Banks, Pernik, Bulgaria

Martin, J. Haigh and Sv. Gentcheva-Kostadinova

Abstract

Forested coal-briquette spoil banks at Pernik suffer much less ground loss than unvegetated slopes. Even in the absence of ground vegetation and litter, the rate was reduced by a factor of 4 (48.5 vs 12.1 mm between 1988–1994) on 16–17° slope test plots. On a forested crest site, negative ground loss was recorded indicating soil development or at least decompaction. Soil bulk density decreased (from 0.97 to >1.3 $g \cdot cm^3$ in the 0-5-cm layer), soil pH moderated from pH < 3.0 to pH > 4.0, and the organic content increased. By contrast, an unvegetated test plot, protected by mechanical means, showed no soil development, and where its contour wattle protection broke down, it suffered as much interrill erosion as an untreated, unvegetated control plot. Despite this, the contour wattles proved nearly as efficient as forest in preventing gully incision.

INTRODUCTION

This case study from the Republic of Bulgaria focuses on the problems of erosion control on lands created from coal-mining wastes. Most of Europe's industrial landscapes include large areas of such land (Haigh, 1993). In Bulgaria, coal spoil banks cover around 16,000 hectares, including 12,700 hectares that have not yet been successfully reclaimed (Malakov, 1993). Erosion is a problem throughout Bulgaria and nearly 60% of its 'low productivity' lands are severely degraded (Onchev and Minev, 1988;

Land Reconstruction and Management Vol. 1, 2000, pp 281–290.
ISSN 1389-2541
ISBN 90 5410 793 6
A.A. Balkema, Rotterdam, The Netherlands

Onchev, 1988). Coal land, especially steep slopes, is particularly prone to accelerated erosion and damage can be severe.

This project aims to evaluate the potential of forest fallowing for successful long-term erosion control and environmental improvement on steep slopes. It is based in the Pernik coal basin, west of Sofia, which produces brown coals by both deep and surface coal mining (Fig. 1). Pernik ranks with the oldest of Bulgaria's coal-mining areas and is provided with some of its better quality Oligocene brown coals (Ivanov et al., 1973). Production peaked in the 1960s and has since declined. Until 1975, some of Pernik's surface-mined coals were enhanced by conversion to coal-briquettes. These spoils generated by briquette manufacture have proved intractable to land reclamation treatment and are the subject of this study.

Local Context

Bulgaria has a continental climate with an average annual precipitation of 606 mm·yr^{-1}. Recorded R-factors range from around 48 to 130 (Onchev and Kolchakov, 1988). Most rainfall is recorded in May and June. The warmest months are July and August. The two driest months are August and September.

This situation is a recipe for severe late-summer soil moisture deficits (Table 1).

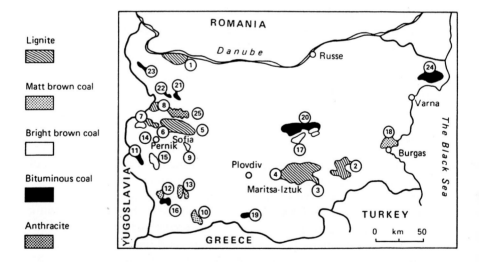

Fig. 1: Coal basins of Bulgaria
1. Lom. 2. Elhovo. 3. Maritsa-Iztok. 4. Maritsa-Zapad. 5. Sofia. 6. Beli Brjag. 7. Aldomirovci. 8. Stanjanci. 9. Cukurovo. 10. Goze Delcev. 11. Kjustendil. 12. Oranovo. 13. Razlog. 14. Pernik. 15. Bobovdol. 16. Pirin. 17. Nikolaevo. 18. Burgas. 19 Smoljan. 20. Balkan. 21. Gorno Ozirovo. 22. Gorna Luka. 23. Belogradcik. 24. Dobrudza 25. Svoge.

Table 1: Climate of Pernik

Climate	Pernik
Mean annual temperature	8–10°
Mean annual precipitation	550–650 mm
Wettest months	May–June: 71–77 mm
Driest months	August–September: 39 and 34 mm
Warmest month	July 37°

PERNIK RESEARCH PROJECT

Traditionally, Pernik was Bulgaria's major brown coal-producing area (Ivanov et al., 1973). However, in recent decades its significance has declined relative to lignite production in the Maritsa-Iztok Economic Association (Zheleva and Haigh 2000, this book). At Pernik, production peaked in the 1960s and has since declined.

Until 1975, a part of Pernik's surface-mined coal production was enhanced by conversion to coal-briquettes. This process generated a large volume of dark-coloured, fine-textured, and not very fertile ash-spoils (Gentcheva-Kostadinova et al., 1994). These include coal shales, marls and a high proportion of low-density coal cinders. The briquette spoils contain high levels of sulphate and hydrocarbon ions. The pH often strays into hyperacidity: pH 2.2–3.2 is typical. Salinisation of the soil solution is also a serious problem.

Experiment Site

Most of Pernik's briquette spoils were dumped as low mounds by loose-tipping from an aerial bucketway. However, adjacent to the northern peripheral road of Pernik., they have been recompacted as a 16–20° slope roadside batter. This south-facing batter has resisted natural vegetative colonisation. This may be due to the very high summer ground surface temperatures and very poor water-holding capacities of the spoils. Anyway, there is little or no grass cover.

Despite this, between 1972–96, one segment of the slope was successfully planted with *Betula pendula* L., supported by *Pinus nigra* Arn. at the slope-crest. Apparently, these deep-rooting trees were able to survive by tapping reserves of moisture deeper in the spoils (Gentcheva-Kostadinova, 1994).

Adjacent sections of the slope were not forested and have remained totally devoid of vegetation. However, between 1978 and 1979, part of this untreated area was defended mechanically by contour wattles. The whole site has suffered moderately severe rill/gully and interrill erosion with channels achieving depths of more than a metre on the unprotected slope.

Experiment Design

The condition of this batter provided an ideal field laboratory for demonstrating the role of trees in erosion control. The site displayed three adjacent segments of a single mine-spoil batter, all created at identical times, from identical material, by identical means and all entirely devoid of ground-surface vegetation. The three differed only in their reclamation treatment. One had been provided with a tree cover. One had been protected mechanically by contour wattles. One had been left entirely alone and thus could serve as control.

In September 1988, these three adjacent, south-facing (TB 170°) slope profiles were instrumented with erosion pins. Rows of pins were installed at 5-metre intervals down each test slope from crest to slope-foot. One erosion pin grid, the control, was laid out across the unvegetated segment. This control plot was set onto an entirely unvegetated, 16.6° (average slope) profile. Some 50 metres distant, a second grid was installed down a 17.6° slope profile that ran through the 1–2 metre spaced, 2–3 metre tall Betula plantation. Between these two sites, on a 19.2° slope that had been terraced by contour wattles (30 cm high) spaced at 5-metre intervals, a single terrace was instrumented with erosion pins.

MEASUREMENT OF EROSION (GROUND LOSS)

Erosion is interpreted here as year-on-year change in ground surface elevation as measured by means of erosion pins (Haigh, 1977; Haigh and Sansom, 1998). An erosion pin is simply a metal rod (length > 600 mm, diameter < 7 mm) hammered into the spoil to a depth where it is beyond disturbance by frost and incidental trampling by humans or grazing animals. A small length of the erosion pin, usually around 25 mm, is left exposed at the ground surface. The upper surface of the erosion pin is counted a fixed bench-mark against which changes in height of the surrounding soil may be registered.

Each erosion pin score is calculated from two measurements—exposure to the right and exposure to the left. Ground retreat across each slope zone is reported as the average change in exposure from a row of between three and five erosion pins located on the same part of the slope profile. The averaging process helps eliminate small scale variations in rates of erosion, such as those associated with individual rills or stones. Individual results are replicable to within 1 mm.

Data are collected in September and April and compared year on year. The reason for this is that erosion pins do not measure erosion. They measure changes in ground surface elevation, which can be affected by changes in soil moisture content, by frost action and by changes in soil density. In fact, changes in the altitude of the ground surface go through an annual cycle. In the winter, frosts loosen the ground, puff up the soil surface and so reduce the exposure of the erosion pins. As the summer proceeds, the ground surface dries up, becomes more compacted and reaches its minimum altitude in September. In the vegetated area, the cycle is complicated by the seasonal increase and decrease in organic activity as well as leaf fall. The amplitude of this seasonal cycle is hard to assess; research elsewhere suggests that it could be as great as 5 mm (cf. Haigh and Sansom, 1999).

RESULTS

Raw results from the first six years of observation at Pernik indicate that, overall, rates of ground retreat on the untreated slope (8.1 mm·yr^{-1}) are much higher than those on the slope protected by trees (2.0 mm·yr^{-1}) (Fig. 1). They also suggest that the rates of erosion on the wattle-protected slope are similar to those on the untreated profile (7.3) mm·yr^{-1}).

However, although data from slope-foot deposition are excluded from this study, because runoff from the slope falls directly into a slope-foot trench drain, results from the forested and unvegetated profile include data from gently sloping, crest sites where erosion is less. When the erosion data are mapped across the slope units from crest to lower concavity, and the terrace located in its midslope (slope: 20°) context (Table 2), the terrace erosion scores are shown to be high for the slope zone. The reason is simple. During the course of this study, the contour wattles have undergone rapid breakdown. This has allowed major erosion of the former

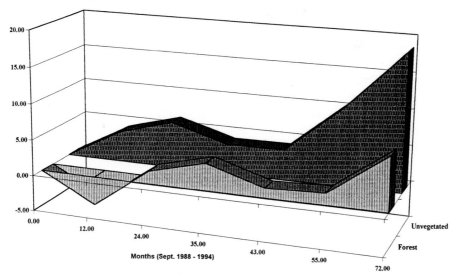

Fig. 2: Cumulative ground retreat 1988–1994; briquette spoil bank, Pernik.

Table 2: Ground retreat 1988–1994 (mm) by slope unit on a coal-briquette spoil bank at Maxim Taban, Pernik

Slope Unit	Slope (°)	Forest	Wattle	Untreated
Crest	11	1.83		35.75
Upper convexity	21	–4.03		47.00
Midslope	20	23.50	43.86	25.97
Lower concavity	22	32.00		97.08

deposition zones that form above each wattle. In sum, the rate of erosion on the contour wattle-protected slope is high and climbing rapidly as wattle breakdown proceeds. However, the wattle-protected slope continues to have conspicuously fewer and more shallow gully channels than the unprotected slope.

Comparison of the erosion scores of forested and unvegetated slope profiles shows major differences on the upper slope units. Here, at the slope-crest (slope: 11°) and upper convexity (slope: 21°), often the sites of greatest ground loss on coal-spoil heaps (Haigh, 1988), the unvegetated slope suffers major ground loss while the forested profile, which is shaded by *Pinus nigra* and *Betula pendula,* has low or negative ground retreat. The difference seems to be largely the result of organic accumulation and, perhaps, loosening of the soil by soil organisms. Farther downslope, under the *Betula pendula,* erosion rates rise, especially in recent years during which these trees have been heavily lopped for firewood and fodder.

The highest erosion scores are recorded on the steeper (22°) parts of the lower slope convexity. These scores are greater on the untreated profile, where there is a much greater slope length available to develop erosive runoff (cf. Haigh, 1988).

Table 3 demonstrates the influence of forest on the chemical properties of the briquette spoils. The most striking impact is on soil pH. The soil under the forest is much less acid than that on untreated sites.

Further changes in the Pernik briquette spoils include evidence of the accumulation of organic carbon and the leaching of calcium down the profile.

DISCUSSION

The moral of the tale, in brief, is that forested slopes suffer much less erosion than unvegetated slopes and that densely forested slopes may show ground advance rather than soil loss. Slopes protected by mechanical means alone do not encourage soil growth and, if the mechanical structures are allowed to break down, they may suffer as much or more

Table 3: Chemical Characteristics of Briquette Spoils, Pernik

Test Plot	Depth (cm)	pH	pH (KCl)	C (%)	N (%)	P (%)	K (%)	Ca (%)
Untreated	00 < 05	2.44	2.20	6.21	0.40	0.05	0.95	0.51
	05 < 10	2.55	2.30	5.81	0.40	0.05	0.94	0.41
	10 < 50	2.43	2.20	8.34	0.42	0.05	0.84	0.44
Forest (av)	00 < 05	4.03	3.22	7.54	0.40	0.06	0.92	0.32
	05 < 10	3.20	2.70	8.02	0.40	0.05	1.00	0.61
	10 < 50	2.95	2.51	7.58	0.32	0.19	0.91	1.94
Robinia pseudoacacia	00 < 05	5.08	4.06	7.35	0.42	0.07	0.67	0.32
	05 < 10	3.97	3.13	9.35	0.51	0.06	0.97	0.22
	10 < 50	3.28	2.78	8.28	0.36	0.49	0.86	0.81
Betula pendula	00 < 05	3.59	2.77	8.60	0.46	0.05	1.03	0.23
	05 < 10	2.82	2.34	8.49	0.37	0.04	0.98	0.69
	10 < 50	2.70	2.17	7.64	0.36	0.03	0.95	1.40
Pinus nigra	00 < 05	3.41	2.83	6.67	0.33	0.05	1.07	0.40
	05 < 10	2.80	2.64	6.21	0.31	0.05	1.04	0.91
	10 < 50	2.86	2.57	6.81	0.25	0.05	0.92	3.60

interrill erosion than untreated sites. However, contour wattles suppress the evolution of gully channels almost as effectively as trees.

Research elsewhere has shown that vegetated surface mine spoils have hydrological characteristics that differ radically from those which are unvegetated. Results from field tests show that grassed mine spoils have a far greater capacity to absorb both rainfall and runoff than unvegetated spoils (Haigh and Sansom, 1999). Vegetated soils have a larger organic content, more water-stable soil aggregates and hence a more free-draining architecture. It is argued that the main stabilising agents of soil aggregates are the secretions of soil organisms. Soils with higher proportions of such aggregates are less liable to compaction. They retain a more open texture, a lower density and are more free-draining. In the Pernik case, preliminary findings suggest that the soil dry bulk densities in the surface layers under the forest may be as low as 0.97 $g \cdot cm^3$ while those at depth and on the untreated slopes, lie in the range 1.35–1.42 $g \cdot cm^3$.

Since vegetation is often associated with acceleration of weathering and increased release of humic acids, it may seem unusual that, in this case, pH is moderated (cf. Ohte et al., 1998). In fact, laboratory studies in the USA have also found that soil organic matter has the capacity to reduce the pH and also to reduce the release of some metal ions from acid pyritic coal spoils (Gentry et al., 1994). Curiously results from Pernik's neighbouring Al. Milenov reclaimed surface coal mine show the reverse pattern of moderation. In this case the pH of neutral/alkaline spoils (pH 7–8) has moderated to pH 5–6.

In two other studies, the impacts of the two tree species on initial pedogenesis were examined on ordinary opencast coal-mine spoil banks near Pernik (Filtcheva et al., 1998; Noustrova et al., 1998). Three 30-year-old test plots were examined. Two were planted for 25 years with: 1) black pine (*Pinus nigra* Arn.) and 2) Black locust (*Robinia pseudoacacia* L.). A third was left largely unvegetated. Litter accumulation under the pine was 7.68 $t \cdot ha^{-1}$ (pH 4.86) compared to 5.53 $t \cdot ha^{-1}$ (pH 7.07) under black locust. Beneath, an organomineral zone developing at the surface of the mine spoils measured 16.18 $t \cdot ha^{-1}$ (pH 6.76) under black pine and 8.21 $t \cdot ha^{-1}$ (pH 7.25) under black locust. The more rapid transformation of black locust litter created more of the mobile organic substances that migrated farther into the mineral profile. Total organic carbon and humic acid levels were greatly enhanced under the trees. Although the microcoenosis was much depleted compared to natural forests, black pine supported a greater microflora, both in the litter and the organomineral layer beneath the litter, while a greater microflora was found in the 0-10-cm layer of the spoils under black locust. The results confirm that the creation of mature soils by forest fallowing is a long slow process. However, forest biological recultivation establishes the preconditions for self-sustaining natural soil development (Noustorova et al., 1998; Filtcheva et al., 1998).

The impact of the forest on the hydrology of the briquette spoils is witnessed by the fact that the trees survive where ground vegetation has failed to become established. During summer, ground surface temperatures on this dark-coloured (2.5YR.3/2–2.5/0), south-facing, slope commonly reach 50°C and may climb as high as 70°C. Simultaneously, during the main part of the growing season, April–September, there is virtually no plant-available moisture in the spoils at 0–20 cm depth. Meanwhile at 40–50 cm the spoils may be moist and at 60–80 cm, there may be free moisture (Gentcheva-Kostadinova, 1994). Despite a total porosity of 64%, the poor soil structure does not encourage capillary rise. However, the deeper tree roots and stronger transpiration system are able to exploit the 40 $mm \cdot m^2$ available moisture in reserves towards 100 cm depth. Elsewhere, where the trees have improved the soil microclimate and their litter has improved the moisture-holding capacity, patches of ground vegetation have begun to develop.

It has been argued, sometimes, that the role of trees in reducing erosion is largely due to their litter or undergrowth. It has even been suggested that tall trees can accelerate rainfall erosion by creating larger droplets on their leaves. The results at Pernik demonstrate that forest reduces erosion even in the absence of ground vegetation or a significant litter layer.

CONCLUSION

Forestation with deep-rooting tree species begins the process of re-establishing the biological factors of natural soil formation. It moderates the soil pH and increases the organic content of the soil to greater depths than revegetation with grass. Even in the absence of ground vegetation and litter, it decreases the rate at which ground is lost—by a factor of 4 (48.5 vs 12.1 mm between 1988–1994) on the 16–17° slope briquette spoil test site at Pernik—and allows ground gain on the forested upper convexity. Mechanical protection alone is far less effective in preventing interrill erosion but nearly as effective in preventing gully incison. In sum, forest fallowing surface coal-mine-disturbed land with appropriate tree species, helps improve soil quality, reduces erosion and may promote soil development, even on steeply sloping sites.

Acknowledgements

The authors thank the British Council, Sofia, for their support of this project.

References

Filtcheva, E., Gentcheva-Kostadinova, Sv., Noustorova, M. and Haigh, M.J. 1998. Forestation as an ecological approach for improving surface coal-mine spoils I: Impact on organic matter accumulation. *J. Balkan Ecology* 1: 47–55.

Gentcheva-Kostadinova, Sv. 1994. Classification and special characteristics of anthropogenic soils. (in Bulgarian). D.Sc. Agric. thesis, Univ. Forest Engineering, Sofia, 350 pp. (unpubl.).

Gentcheva-Kostadinova, Sv. and Haigh, M.J. 1988. Land reclamation and aforestation research on the coal-mine disturbed lands of Bulgaria. *Land Use Policy* 5 (1): 94–102.

Gentcheva-Kostadinova, Sv., Zheleva, E., Petrova, R. and Haigh, M.J. 1994. Soil constraints affecting the forest-biological recultivation of coal-mine spoil banks in Bulgaria. *Int. J. Surface Mining, Reclamation and Environment* 8 (2): 47–54.

Gentry, C.E., Walton, G.S., Davidson, W.H. and Wade, G.L. 1994. Influences of humic and fulvic acids and organic matter on leachate chemistry from acid mine spoils. In: *Reclamation and Revegetation*, pp. 166–174. Bureau of Mines Spec. Publ. 06C–94 (3): 353 pp.

Haigh, M.J. 1977. Use of erosion pins in the study of slope evolution. *British Geomorph. Res. Group, Tech. Bull.* 18: 31–49.

Haigh, M.J. 1988. Slope evolution on coal-mine disturbed land In: A.S. Balasubramaniam, S. Chandra, D.T. Bergado and Prinya Nutalaya (eds.). *Environmental Geotechnics and Problematic Soils and Rocks*, pp. 3–13. A.A. Balkema, Rotterdam.

Haigh, M.J. 1993. Surface mining and the environment in Europe. *Int. J. Surface Mining and Reclamation* 7 (3): 91–104.

Haigh, M.J. and Sansom, B. 1999. Soil compaction runoff and erosion on reclaimed coal-lands (UK.) *Int. J. Surface Mining, Reclamation and Environment* 13: (in press).

Ivanov, D.I. et al. 1973. *80 Gorodini Durdjavna Min. 'Georgi Dimitrov'*, Pernik, Durdjavno Izdatelstvo Tehnika, Sofia: 272 pp.

Malakov, P. 1993. Country Report: Bulgarian Delegation. *UN Econ. Comm. Europe, Commit. Energy, Working Party on Coal, Workshop on Environmental Regulations in Opencast Mining under Market Conditions (Most, Czech Republic) 1993.* (See also: UNECE Secretariat 1994: Outcome of Workshop on Development of Environmental Regulations in Opencast Coal Mining under Market Conditions (Most, Czech Republic, 1993). ENERGY/WP 1/R31: p. 14.

Noustorova, M., Gencheva-Kostadinova, Sv., Filtcheva, E. and Haigh, M.J. 1998. Forestation as an ecological approach for improving surface coal-mine spoils: II: Impact on the microbial action. *J. Balkan Ecology* 1: 56–60.

Ohte, N., Tokuchi, N. and Asano Y. (1998). Geographical variation in acid buffering processes in forest catchments. pp. 69–84. In: M.J. Haigh, J. Krecek, G.S. Rajwar and M.P. Kilmartin (eds.). *Headwaters: Hydrology and Soil Conservation*, A.A. Balkema, Rotterdam, 460 pp.

Onchev, N. 1988. State and problems of the efficient use of eroded lands in Bulgaria. pp. 295–301. In: *UNESCO Int. Symp. Water Erosion* (Varna), Proc., Bulgarian Nat. Comm. Int. Hydrological Programme and MAB, Sofia, 376 pp.

Onchev, N. and Kolchakov, I. 1988. Predicting erosion degradation of some soil types in Bulgaria. pp. 309–314. In: *UNESCO Int. Symp. Water Erosion* (Varna), *Proc.* Bulgarian Nat. Commit. Int. Hydrological Programme and MAB, Sofia, 376 pp.

Onchev, N. and Minev, V. 1988. Methodological foundations of building up a computerised system for monitoring the soil erosion and managing its control. pp. 283–290. In: *UNESCO Int. Symp, Water Erosion (Varn); Proc.*, Bulgarian Nat. Comm. Int. Hydrological Programme and MAB, Sofia, 376 pp.

Zheleva, E. and Haigh, M.J. 2000. Case study: vegetation, erosion and soil development on lignite spoil banks: Maritsa-Iztok, Bulgaria. *Land Reconstruction and Management* 1: 275–280.

12

Forest Recultivation of Coal-Mined Land: Problems and Prospects

Allison Flege

Abstract

Although, in many places, the devastating results of surface mining may provide similar landscapes, site conditions of each are unique. Satisfactory results have been devised for individual sites using a variety of techniques and specific combinations of physical treatments, soil amendments, vegetation and after-care. An extensive literature covers a range of problems and practices. Any attempt to establish vegetation must take into account the nature of the particular site—the physical and chemical characteristics of the created minesoil, the proposed after-use, the availability of nutrients, and the establishment of a functioning biotic community. Although technology exists to reclaim almost any land mined for coal, successful reclamation depends on matching the proposed after use to the economic potential of the site and the concerns of the owner. Ultimately, the success of revegetation depends on political and social, as well as physical, processes.

INTRODUCTION

In most coal-mining regions of the world, some deliberate effort to 'reclaim', or 'restore' or 'rehabilitate' the mine site is required by law. The stated objectives are many and varied—re-establishing a topography and productivity to support post-mining use; maintaining the visual character and usefulness of the surrounding land; ensuring that important pre-mining uses of the land are not lost to society as a result of mining, and creating a new and better landform. However, the immediate concern

Land Reconstruction and Management, Vol. 1, 2000, pp 291–338
ISSN 1389-2541
ISBN 90 5410 793 6
A.A. Balkema, Rotterdam, The Netherlands

and primary goals of reclamation remain slope stability, control of erosion and sedimentation, and off-site effects of runoff and toxic drainage. The solution invariably involves revegetation (Toy and Hadley, 1987). What that vegetation will be depends on the anticipated land-use and harsh realities of the site.

Coal spoils are hostile environments for vegetation. Mining inevitably degrades the land, destroying the natural soil environment, obliterating existing communities of animals, plants and the microorganisms of the soil, and completely disrupting drainage and hydrology. Revegetation, whether deliberate or the result of natural colonisation, provides a resource base for initial soil-forming processes and soil renewal on land disrupted by mining for coal.

RECLAMATION AND VEGETATION

Vegetation exerts major influences on mine-soil properties. Plants modify hydrology and provide mechanical reinforcement. Their roots take up large amounts of water and release it to the atmosphere through evapotranspiration. Foliage, stems and branches intercept rainfall and prevent soil detachment by rainsplash. Root systems restrain soil particles, above-ground portions retard runoff velocity and trap sediment, while plant residues contribute to soil porosity and permeability, delaying and reducing surface runoff (Gray, 1994). In addition, plants mine nutrients from the soil and return them in leaf litter (Gray, 1994). Through photosynthesis, plants supply the energy for biologic processes that break down and recycle organic matter and minerals and thus drive the soil system.

Any dense plant cover provides protection for the restructured landscape and prevents erosion of the newly formed surface, limiting the contamination of water with sediment and the pollution of air with dust. However, a fast-growing herbaceous cover does not necessarily provide long-term stability, pollution control, or visual appeal (Toy and Hadley 1987). In recent years, it has become increasingly clear that revegetation for long-term protection requires looking beyond erosion control and a quick cover of grass—beyond what Hansen (1976) termed the 'green grass syndrome'—towards establishing soil-plant communities able to protect the quality of the environment and the utility of the land over the long term (Fischer, 1986; Jenny, 1980; Krause, 1973).

Ecosystem Approach to Reclamation

A plant community is not simply an assemblage of plants, but an integrated, interactive system (Cairns, 1995; Fischer, 1986). Normally a functioning system possesses stability, so that there is a balance between the input and output of energy and matter. Because a system has inherent structural characteristics, any action taken to modify the system works either with or against these natural patterns. The key to successful reclamation is to regenerate an ecosystem that can be tapped for human needs without compromising the system's ability to recover (Ashby and Kolar, 1981).

Mining for coal, especially surface mining, disrupts virtually every aspect of the environment except climate. It is offen suggested that mine sites can be rehabilitated most quickly, and at least cost, by working with nature and taking advantage of natural processes of ecosystem development (Allen and Friese, 1990; Bradshaw, 1994; Daniels and Zipper, 1995; Gentcheva-Kosadinova and Haigh, 1988; Griffiths, 1992; Tueller, 1990). It is axiomatic that the more effectively the system is restored, the higher the return in continuing productivity (Ashby and Kolar, 1981).

Trees for Long-term Stability

Almost all land stripped for coal can be rehabilitated to support some form of vegetation, at least in the short term. However, not all land is sufficiently valuable to be reclaimed to the highest level possible (Binns, 1983; Leopold and Wali, 1992; Moffat and McNeill, 1995). Such sites are often best reclaimed through forestry (Wilson, 1983). 'Reclamation to forestry can be very much cheaper than reclamation to agriculture' (Teasdale, 1983: 5).

In addition, trees planted for 'biological recultivation' or 'forest fallow' on mined soil can restore the soil-plant system at significantly lower cost than conventional amelioration and maintenance procedures (Haigh, 1992; Gentcheva-Kostadinova and Haigh, 1988). Established trees actively help the land to recover from mining through natural soil-building processes (Haigh, 1992; Gentcheva-Kostadinova et al. 1994; Scullion, 1992; Richardson et al., 1987). However, the process takes time; trees grow slowly on such sites and they are more expensive, initially, than herbaceous ground-cover mixes. Yet, once trees are established, maintenance requirements can become minimal (Richardson et al., 1987).

However, the slow growth of trees on most former coal-mine sites, caused by low fertility and poor soil structure, means that economic targets must be set realistically low. If timber harvest is an objective, attention is generally paid to maintaining the desired species, reducing competition from forbs and grasses, and encouraging maximum productivity. By contrast, after initial establishment, non-commercial reforestation plantings are often left to develop through the forest's natural regenerative capacity and may ultimately achieve the same ecological end.

Self-sustaining forest ecosystems are richly diverse in plant species and life forms—trees, shrubs and herbs (Allen, 1990). They harbour a variety of animals, which provide fertilising dung, distribute seeds and act as vectors for essential soil bacteria, fungi and other micro-organisms

(Parmenter and MacMahon, 1990; Ponder, 1980; Tate, 1985a). Planting trees on strip-mined land provides an initial tree cover and may provide the potential to maximise future forest development (Ashby et al., 1980b). The species selected for reclamation should provide the self-regenerating basis of forest development and thus invite colonisation by additional trees species and other plant forms (Ashby and Kolar, 1984).

Mature, stable forest is the result of a century or more of slow development (Gleason and Cronquist, 1964). Much research is developed towards methods that will encourage more rapid establishment (Coppin and Richards, 1990; Schiechtl, 1980). Establishing forest trees on coal-mined land, whether abandoned or newly reconstructed, involves coping with many site-specific problems directly related to the locality, extent of the area, the created landform, its surface characteristics and the proposed after-use (Chadwick et al., 1987; May, 1986).

Rehabilitating Coal-mined Land with Trees

Before the development of today's huge earth-moving machines, tree planting provided a means of restoring the productivity of strip-mined land at reasonable cost. With the advent of recontouring capabilities and huge rubber-tyred loading and grading tractors able to traverse almost any terrain, extensive new mining-grading-topsoiling practices became the norm (and in some places a requirement). These modern reclamation technologies, with high fertiliser rates and smooth graded surfaces, give the advantage to grasses and agricultural use (Torbert et al., 1995).

The effect of these developments was a drastic reduction in tree planting and a dramatic increase in seeded herbaceous forage (Smith, 1980). However, although sowing grass mixtures may cost less than tree planting, it involves considerable cost in site preparation. Trees, on the other hand, often can be planted directly into mine spoil, greatly reducing the costs of soil handling, which represents most of the cost of coal-mine reclamation (Abbott and Bacon, 1977; Gentcheva Kostadinova et al., 1994; Jochimsen, 1986; Samuel, 1991; Smith, 1980). Planting trees at the outset, with minimal grading and soil handling, can provide a relatively inexpensive cover for mine wastes (Abbott and Bacon, 1977; Smith, 1980).

Restarting Biochemical Processes

The unavoidable disruption of soil and rock profiles during mining suspends the biochemical processes essential to the operation of the soil system and the growth of plants (Juwarkar et al., 1994). Organic matter accumulation, soil particle aggregation and mineralisation of nitrogen and carbon cease to operate following mining and may take several decades to re-establish. Without organic colloids and clay particles with charged sites, which can participate in ion exchange, nutrients in solution are readily leached from the root zone (Tan, 1993; Paul and Clark, 1996). Furthermore, in the absence of buffering effects of soil organic matter, extremes of acidity or alkalinity can tie up nutrients in forms unavailable to plants or release them in quantities that can be toxic for plants and micro-organisms (Brady and Weil, 1996; Katzur and Haubold-Rosar, 1996). Essential soil organisms are influenced by acidity, as well as by physical factors, including temperature, moisture availability, aeration and bulk density (Chadwick, 1982).

Active populations of soil organisms are needed in developing mine soils to break down organic matter, initiate the cycling of plant nutrients and form soil horizons through the downward leaching of the products of microbial decomposition. Various soil organisms release minerals into the soil solution, process organic nutrients to mineral form and modify adverse mine-soil properties (Brady and Weil, 1996; Cundell 1977; Leirós et al., 1996; May, 1986; SCS, 1978). As Daniels and Zipper (1995) point out, provided the initial mine-soil conditions are adequate, the success of revegetation in the long term depends on two major factors: (1) the accumulation of organic matter and nitrogen and (2) the establishment of an organic phosphorous pool and avoidance of phosphorous fixation. Both of these require the introduction of microbial communities and their continued functioning over time.

CONSTRAINTS IMPOSED BY THE SUBSTRATE

The nature of the substrate on reclamation sites, however, presents serious obstacles to establishing vegetation that include both physical and chemical constraints (Chadwick, 1982. Limstrom 1960). Spoil is not soil. A naturally formed, undisturbed soil is an organised body developed from rock near the earth's surface, which interacts with atmospheric, biologic and organic constituents and is modified over time (Bradshaw and Chadwick, 1980; Jeffrey, 1987; Jenny, 1941, 1980; Limstrom, 1960). Soil is the product of interactions of climate, organisms and topography working on mineral bedrock material. It differs chemically, biologically, in form and in structure from the geologic material from which it is derived. Soil is dynamic, containing both inorganic and organic fractions, living and dead.

A characteristic of all soils is an internal organisation not encountered in spoils. This organisation includes soil horizons that form from additions such as organic matter, from losses from the soil due to leaching, from transfers within the soil and from mineral and organic matter transformation. In contrast, the rooting medium on spoil banks and stripped land is a disorganised collection of overturned and mixed mineral overburden derived from any or all strata—from the surface to the mined coal seam—and which retains the physical and chemical characteristics of its geologic origin (Bell et al., 1994; Packer, 1974; Senkayi ɹd Dixon, 1988; Smith and Sobek, 1978). It lacks both organic matter and ɹoil organisms (Juwarkar et ai., 1994; SCS, 1978). Spoils can show a fair degree of uniformity but great diversity even within small distances is more common. Conditions may vary greatly from one site to another, as well as within the same site (Limstrom, 1960).

Limits to Plant Establishment

Physical properties of mine soils, such as compaction, poor texture and inadequate structure, discourage rooting and create ongoing problems of moisture and temperature stress for vegetation (Gentcheva-Kostadinova and Haigh, 1988; Chadwick, 1982; McSweeney and Jansen, 1984; Richards et al., 1993). The newly contoured landscape hides both cuts and fills and "one need not be a soil specialist to know that the amount of site compaction, soil air space and surface water in-soak rate may not be the same on cuts as it is on fills" (Medvick, 1980: 87). For trees planted under these conditions, identical mine soils planted similarly can produce both success and failure.

Some factors such as stoniness, have limited effect on patterns of tree roots, which are readily able to grow around large particles and boulders and to penetrate fractured and weathered rock (Ashby et al., 1984; Flege, 1996; Samuel, 1991).

Soil chemical properties of concern in vegetating mine soil include soil reaction (acidity or alkalinity), mineral reactivity (ion exchange capacity), and the concentration of elements in the soil (Bradshaw and Chadwick, 1980; Chadwick, 1982; Katzur and Haubold-Rosar, 1996). Plants growing in mine spoils face problems due to both the deficiencies of essential mineral nutrients and the presence of minerals in toxic concentrations (Berg and Vogel, 1980).

Site factors limiting plant establishment and survival on reclamation sites vary with climate and geology. Among the problems encountered most frequently on mined sites are high bulk density, high temperature on the dark-coloured spoil surface, moisture stress, acidity, limited supplies of nutrients and toxic concentrations of minerals, often magnesium, sodium and metals (Baig, 1992; Barnhisel 1988, Barnhisel and Massey 1992, Bell 1979). In western Canada, Baig (1992) reports that moisture stress and nutrient deficiency cause the most significant problems. In the eastern United States, Schoenholtz and colleagues (1992: 1177) found plant-available moisture to be the 'most critical factor during the vegetation establishment phase'. Helliwell (1994) points out that roots require oxygen, moisture and nutrients from the substrate, in that order, and cannot grow in soils that are highly compacted, extremely dry, or poorly aerated.

CONCENTRATION OF MINERALS IN THE SOIL

Plants require a number of mineral nutrients, which must be obtained from the soil solution in a form the roots can absorb. Until an efficient plant-soil system is activated, nutrients, especially the macronutrients, are generally deficient in mine soils. A significant factor limiting plant growth on coal-mined land is a deficiency of plant-available nitrogen and phosphorus (Bloomfield et al., 1982; Palmer, 1990; Simcock, 1993; among many).

Nitrogen

Freshly exposed mine spoil has no reserves of nitrogen. Plants need a constant supply of nitrogen, and a functioning nitrogen economy is critical for long-term revegetation success (Bradshaw and Chadwick 1980; Russell and Roberts, 1986). In the absence of organic matter and the nitrogen-processing organisms that it supports, plants suffer from inadequate supplies of usable nitrogen (Jefferies et al., 1981; Plass, 1979; Vogel, 1981). In northern England, Palmer (1990) concludes that there is a 'trigger level' of nitrogen before plants will invade abandoned mine refuse. Nitrogen deficiencies are generally corrected initially by applying fertiliser to the reclamation site. However, although nitrogen fertilisers have a place in site amelioration, their effects are ephemeral (Russell and Roberts, 1986). Long-term nitrogen cycling depends on nitrogen-fixing plants and their fungal symbionts (Beck et al., 1990; Bradshaw and Chadwick, 1980; Russell and Roberts, 1986). By providing a continuous supply of nitrogen, nitrogen-fixing species influence successional progress towards ecological equilibrium (Palmer and Chadwick, 1985; Palmer, 1990). Nitrogen-fixing plants do, however, need an adequate supply of phosphorus and this must be supplied.

Phosphorus

The availability of phosphorus to plants depends on retention reactions within the soil solids. Phosphorus ions adsorbed to soil particles are essentially unavailable to plants, which take up nutrients only in solution. The pH of the soil and the presence of aluminium, iron or manganese control these fixation reactions. As organic matter accumulates and the soil system develops, organic residues release compounds that combine with sites that normally adsorb phosphorus, making more phosphorus available to plants (Brown et al., 1985). Phosphorus is mineralised and released to the soil solution by micro-organisms, especially mycorrhizal fungi, and to a lesser degree by plant roots (Brady and Weil, 1996).

Phosphorus is thought to be the dominant element controlling carbon and nitrogen immobilisation in biological systems (Paul and Clark, 1996).

Soil organisms take up a large amount of phosphorus, immobilising it temporarily but preventing its long-term fixation in minerals. Thus microcrobial biomass is a reservoir of organically-bound phosphorus that is slowly released into the soil solution for plant uptake. The accumulation of carbon, nitrogen, sulphur and phosphorus in soil organic matter depends on how much phosphorus is in the parent material (Paul and Clark, 1996).

Potassium

Plants require potassium in quantities similar to their nitrogen needs. Potassium content varies in mine soils and each site must be studied to determine whether and how much potassium may be needed. Although it is abundant in soil in mineral form (especially as feldspar and mica), only 1 to 2% is readily available and soluble. As a soil develops, a reserve of slowly released potassium can form which resupplies the solution as potassium ions are absorbed by roots. Applications of lime may increase potassium fixation.

Nitrogen and phosphorus are usually applied at the same time, together with potassium if it is needed (Jefferies et al., 1981; Plass, 1976; Vogel, 1981). The need for additional potassium depends on the content of the spoil, which will vary with the overburden material, as well as with the requirements of the species to be planted (Chadwick et al., 1987; Dragovich and Patterson, 1995; Hower et al., 1992).

NEGATIVE EFFECTS OF FERTILISER

Beyond an initial application of nutrients to initiate biologic cycles, fertiliser application can have negative effects on tree establishment. Fertilisers, especially nitrogen, enhance herbaceous productivity at the expense of woody species. Application of fertilisers can also depress soil microbial activity, disrupting the balance of microbial populations (Domsch, 1986; Söderström et al., 1983; Smyk et al., 1986). In plots fertilised with nitrogen and phosphorus, Robertson and Wittwer (1983) reported a reduction in survival of cottonwood (*Populus deltoides*) and sycamore (*Platanus occidentalis*) compared with unfertilised controls. Losses were less with slow-release tablets than with immediate-release fertiliser (Robertson and Wittwer, 1985). This study also suggest a link between phosphorus applied to the root zone and seedling mortality. Inappropriate fertilisation has also been shown to reduce seed germination, due to the interaction of high levels of salts in fertilisers and high levels of toxic ions, not least manganese, found in the growing medium (Brown et al., 1985).

SOIL REACTION

Many of the problems encountered in establishing trees on land mined for coal can be shown to be related, directly or indirectly, to the degree of acidity or alkalinity in the soil (Veith et al., 1985). Elements essential to plants may be available in adequate, deficient, or toxic concentrations, depending on soil reaction, and acidity potentially detrimental to plants can occur in spoils in most coal regions (Veith et al., 1985).

High acid potential results when iron-bearing minerals such as pyrite in the overburden are exposed to atmospheric oxygen at the surface, where the microbially enhanced oxidation of pyrite produces acid, large quantities of sulphate and reduced iron (Pietsch, 1996). Wherever coal is mined there exists the possibility of pyritic spoil (Barnhisel and Massey, 1969; Richards et al., 1993; Leopold and Wali, 1992). The establishment of any vegetation is difficult on acid mine soils and few plants will grow well if the pH drops below about 4.0 (Chadwick et al., 1987). Below pH 3.5 acidity becomes essentially toxic for most plants and few species can survive (Ferchau, 1988). Toxicity may result from the mobilisation of high levels of aluminium and manganese, rather than the acid itself.

Less often, mine-soil conditions for plant growth can be distinctly alkaline, usually as a result of the presence of hydroxides of magnesium, sodium, calcium, or potassium (Leirós et al., 1996; cf. Saiz de Omeñaca et al., 1994). In humid climates, difficulty in establishing plants on saline soils is generally temporary, since the salts are relatively easily leached. In arid climates, where evaporation at times exceeds precipitation, high levels of salts accumulate and can prevent revegetation entirely (Bradshaw and Chadwick, 1980; Chadwick et al., 1987). Salts, sulphates, and other toxic chemicals may be leached more effectively through the use of techniques such as furrow grading, which provide for increased moisture infiltration (Riley, 1973).

Soil pH influences the composition of the vegetation that invades and survives on abandoned mined land (Johnson and Skousen, 1995). However, although the pH of the soil produces various physiological effects on plants, these effects cannot always be isolated from other changes in the system that are also pH dependent. Furthermore, the roots themselves can bring about significant changes in the pH of their mine-soil environment (Miles, 1985; Moore, 1974; Paul and Clark, 1996). Once established, roots create a zone of intense microbial activity. Turnover of roots, especially ephemeral fine roots, produces organic matter to sustain microbial nutrient cycling. Cation exchange is enhanced and the pH near the root can be significantly lowered, affecting microbial species and their functions (Paul and Clark, 1996).

Problem of Acid Mine Drainage

The effects of coal-mine acidity frequently extend well beyond the mined site. When extremely acid drainage, with high concentrations of iron, aluminium and sulphates, seeps into streams and lakes, it eliminates plant and animal life (Caruccio et al., 1988; Feiss, 1965; Leopold and Wali, 1992; Richards et al., 1993). The residual drainage is extremely acid, with pH values between 1.9 and 3.1, and contains large quantities of iron (Pietsch, 1996).

The control and treatment of drainage from pyritic spoil is crucial in rehabilitating the site (Jeffrey, 1987). Treatment of acid mine drainage by liming is frequently disappointing, as the capacity of mine soil to produce acid is high and the process essentially irreversible. Some success has been achieved by deep burial of pyritic overburden during the mining operation or, where strata have not been segregated during mining, covering pyritic spoil with a deep layer of less toxic overburden and soil (Jeffrey, 1987; Pulford, 1991; Strzyczcz 1996, 1993). Acid mine drainage and acid substrates remain virtually intractable problems in many places, despite the accumulation of considerable experience over many years in several countries. Pulford (1991) reviews traditional treatments and examines the choice of ameliorants in relation to the nature of the problem and the anticipated land-use, including creating artificial wetlands, to prevent acid mine drainage from entering streams and rivers.

Effects of Slope Position and Aspect

The position of trees on a slope and the direction the slope faces influence the survival and growth of trees (Leopold and Wali, 1992; Limstrom, 1948; Vogel et al., 1984). Conditions tend to be more favourable for tree survival and growth on lower slopes, where soil moisture, runoff-transported soil and nutrients accumulate, and on gentler slopes, where infiltration potential is greater (Funk, 1973; Gray and Leiser, 1982; Jewell, 1978; Toy and Hadley, 1987; Vinczeffy, 1995). Gentcheva Kostadinova and Haigh (1988) suggest that slope angle has significant effect on growth of planted tree seedlings on coal spoils in Bulgaria. Similar findings have been reported for grasses and other non-woody species. The apparent advantages of lower positions can be nullified, however, if toxic materials from the upper slope are being eroded or leached downslope. In Kansas (USA) Geyer (1973) found that trees planted on upper rather than lower slopes of spoil banks were much taller. Certain species show a preference for slope position related to low-moisture tolerance (Rogers, 1992).

The direction towards which a slope faces, the slope aspect, affects the amount of sunlight it receives and thus both air and soil temperatures. South-facing slopes in the Northern Hemisphere and north-facing slopes in the Southern Hemisphere are generally not only warmer but drier than the opposite slopes, leading to important variations in growing conditions. Indeed, South-facing slopes are considered to create greater physiological stresses for plants even in the latitudes of Northern Europe.

While slope and aspect affect tree growth, trees themselves alter the microclimate and growing conditions creating microenvironments quite different from the average. Near-surface radiation, wind, temperature and particularly soil moisture vary greatly with the type of vegetation (Schiechtl, 1980). Comprehensive reviews of the interaction between trees and the soil environment are found in Greenway (1987) and in Coppin and Richards (1990).

Importance of Roots

Plants and their roots contribute both chemically and physically to mine-soil interactions. Once established, roots exude materials that create a narrow zone of intense microbial activity. Turnover of roots, especially ephemeral fine roots, produces organic matter to sustain microbial nutrient cycling. Cation exchange is enhanced and the pH near the root can be significantly lowered, thus affecting microbial species and their functions (Paul and Clark, 1996). Roots also contribute mechanical strength to unconsolidated surface material (Gray and Leiser, 1982; Gray and Megahan, 1981; Riestenberg and Sovonick-Dunford, 1983). A mat of interwined roots lends cohesion. Shear strength is increased by the tensile strength of the roots and their properties of friction and adhesion (Coppin and Richards, 1990; Gray and Leiser, 1982; Greenway, 1987).

Trees and Moisture

Trees extract soil moisture to meet their biological requirements. Surface material responds directly to hydrologic effects of trees as they intercept rainfall, reduce raindrop impact and extract soil moisture through transpiration. Soil strength is increased and surface erosion reduced (Coppin and Richards, 1990; Greenway, 1987; Gray and Leiser, 1982). The rate of moisture consumption depends on the species of tree and the climate. It varies with the weather and site conditions, such as slope, aspect, soil type and moisture availability (Greenway, 1987). Biddle (1983) found that the water extracted by poplar trees from clay soils in England reduced pore-water pressures in the soil well beyond the root zone. Biddle's findings confirm studies by Brenner (1984) who modelled vegetated and cutover slopes, as well as field measurements by Richards et al. (1983) in Australia comparing soil suction under Acacia trees with that under pasture. Hsuang (1983) points to higher rates of evapotranspiration in deciduous compared with coniferous trees. Tree canopy interception rates of 10% to 25% of precipitation, ranging up to 100% when rainfall is light, are reported (Greenway, 1987). Actual rates of interception are unpredictable and highly variable.

REVEGETATION STRATEGIES

Left to itself, and given sufficient time, most land disturbed by coal extraction can be expected to develop vegetation and soil cover (Daily, 1995; Richardson, 1984; Ross et al. 1995). This natural recovery can prevent further deterioration and conserve the productive potential of land (Daily, 1995; cf. Struthers and Vimmerstedt, 1965).

In general terms, in the absence of aridity, most sites that support grasses are plantable to trees. Trees are planted not only for forests but as shelter-belts for agriculture and to improve amenity values on non-agricultural land (Wilson, 1983). However, desirable trees do not always plant themselves where they are wanted (Medvick, 1980). Reclamation experience suggests that human intervention can be effective, if not essential. It can establish a path and rate of succession that achieves substantive improvements over a time scale of practical use to society.

Trees and other plants exist in a location either because they have been planted or because their popagules have been carried to the site by wind, water, or animals (Bradshaw and Chadwick, 1980; Gleason and Cronquist, 1964). In reclaiming mined land, humans become the agents of dispersal by seeding or planting species that otherwise might not be able to reach the site or require a very long time to colonise it (Bradshaw and Chadwick, 1980; Richardson et al., 1987).

Basic soil parameters such as clay content, soil depth and per cent organic matter content are practical indicators of those trees and shrubs best suited to a particular site, as well as what, if any, amelioration may be required (Scullion and Malinovszky, 1995). It is important to remember that there are no universal prescriptions; all sites are different and need to be evaluated individually (Binns, 1983).

Getting the Trees to Grow

Although virtually any site can be manipulated to culture some sort of plant life, species planted for reclamation will not persist on a site where their seeds cannot germinate or seedlings are unable to survive (Allen, 1990). Plants vary greatly in what they require in order to grow, to reproduce and to distribute their seeds. If selected plants are unable to grow on the site, it is important to ask (1) why they will not grow, (2) what is limiting growth, (3) whether the soil can be amended to support growth, and finally (4) which of a number of organic soil amendments to apply (Aldon and Springfield, 1975). There are several remedies. However, there is no universally applicable formula for site amelioration. Each mine soil must be considered separately (Welsh and Hutnik, 1973) because each site will have special attributes (Binns, 1983; Chadwick et al., 1987; Chadwick, 1982).

Supplying Organic Matter

Although good initial plant cover may be achieved using chemical fertilisers, ecosystems do not develop in the absence of organic matter to support microbial activity (Marx, 1980; Tate, 1985). Initial vegetation often deteriorates before it can begin to improve spoil (Bradshaw and Chadwick, 1980; Juwarkar et al., 1994).

Bulky organic materials added at the surface or mixed with the spoil not only add nutrients, but also decrease bulk density and improve porosity and water-holding capacity (Bradshaw and Chadwick, 1980; Toy and Hadley, 1987). These materials also create a habitat for soil orgnisms and improve conditions for extensive root development (Juwarkar et al., 1994). They reduce erosion and provide carbon and mineral nutrients that encourage the development of beneficial soil micro-organisms (Chadwick et al., 1987; Del Tredici, 1992; Limstrom and Merz, 1949; Toy and Hadley, 1987).

Various substances are applied to mine soils to improve physical and chemical conditions for vegetation. Organic materials such as straw, wood chips, sewage sludge, cattle slurry, chiguano (chickenhouse litter) and sugar mill residues are used to amend mine soils (Atkinson et al., 1991; Dewangan and Mishra, 1994; Leirós et al., 1996; McCormick and Borden, 1973; Webber et al., 1994). Applying sewage sludge and industrial wastes to land has raised concern, however, that high levels of inorganic contaminants and heavy metals in some municipal sludges are serious threats to water systems and food chains, and may affect the microbial population adversely (Joost et al. 1987, Richards et al., 1993; Webber et al., 1994).

Ideally an organic amendment to the soil will inoculate it with beneficial micro-organisms. It will also be slow to decompose and often will be locally available in bulk supply (Chadwick et al., 1987). However, surface mulches can be difficult or impossible to use on exposed, windy slopes, albeit methods of anchoring or stabilising the material with various nettings or chemical substances have been tried with some success (Brown et al., 1981; Chadwick et al., 1987; Kay, 1978; Plass, 1978). Avoiding fine-textured mulches and using partially decomposed, aged materials can minimise problems of high carbon-to-nitrogen ratios. A light topdressing of a balanced (N-P-K) fertiliser can minimise the immobilisation of soil nitrogen by decomposer organisms and speed decomposition as well (Berg 1973, 1980; Chadwick et al., 1987; Del Tredici 1992). However, care is needed when mixing fertilisers into the material, to avoid concentrations toxic for young seedlings (Chadwick et al., 1987).

The purpose of the amendment, the site characteristics, the availability of suitable materials and the cost involved will determine the choice of material. No single material will suit all spoils and all purposes, and combinations of two or more methods of substrate improvement are frequently employed (Leopold and Wali 1992). Summaries of the characteristics and relative merits of various organic soil-amending materials are included in Chadwick et al. (1987), Chadwick (1982), Coppin and Richards (1990); Gray and Leiser (1982), Kay (1978); Toy and Hadley (1987) among many.

Encouraging diversity

There is more to a forest than trees. A natural forest is a dynamic system operating at near-equilibrium. It usually comprises a variety of plant forms (shrubs, vines, grasses and forbs, as well as trees), and planting comparable multilayered vegetation has been suggested for coal-mine reclamation (Ashby and Kolar, 1981; Jha and Singh, 1995). Many researchers consider a variety of plant forms and vegetation types essential to a revegetation scheme (Allen, 1990; Binns, 1983; Chadwick, 1982; Parmenter and MacMahon, 1990; Tueller, 1990). However, others suggest that planting as few as two or three carefully selected site-tolerant tree species can provide early canopy cover, while relying on natural plant invasion to create a more diverse woody and herbaceous vegetation over time (Jencks et al., 1982). A mix of trees increases the probability of successful vegetation establishment (Gray and Leiser, 1982) and will be less susceptible to damage from insect and diseases, which plague monocultures (Hoffard and Anderson, 1982). Furthermore, a diverse planting, including more than one plant form, encourages successional processes, leading to more desirable communities (Ashby et al., 1980a). In either case, mixed leaf litter provides better protection for the soil surface, improving soil quality as well as ecological balance (Kellogg, 1936; Neumann, 1973).

The question of competition

Interactions among neighbouring plants can be beneficial or competitive. For example, trees planted on mined land provide shade and control surface temperatures, modify microclimates to the advantage of developing seedling or later successional species (Nair, 1990). However, plants compete for space, light and carbon dioxide above ground and for water, nutrients and root space below (Galston et al., 1980). All vegetation, whether volunteer or seeded, will be part of the competition with planted trees (Etherington, 1982; Limstrom and Merz, 1949).

Providing a rapidly developing dense ground cover is often considered essential for reclamation, since trees and shurbs take several years to provide the same level of effective surface protection (Jochimsen, 1986 SCS, 1978). Where rapid development of herbaceous ground cover is required, either for erosion control or simply to satisfy legal requirements or improve appearance, numerous investigators have reported adverse effects on tree growth and survival from competition for moisture and nutrients (Anderson et al., 1989; Medvick, 1980; Vogel and Berg, 1973). Reports are contradictory, however, regarding the extent and seriousness of the problem (Plass, 1968).

On reclaimed coal land, erosion-control plantings using heavily seeded, highly aggressive seed mixes and optimum fertiliser applications put tree seedlings at a competitive disadvantage, regardless of whether seeding precedes planting or seeding and planting are done together (Torbert et al., 1995). Often, quick-forming 'reclamation' ground cover smothers any planted tree or shrub seedlings with a mat of aggressive grasses and legumes as soon as or soon after they are planted (Kelly, 1980). Turf grasses are heavy feeders and several investigators have reported that these grasses rapidly deplete soil moisture, so that tree growth is depressed and roots cannot reach deep moisture (Berg, 1973; Bramble et al., 1996; Samuel, 1991; Vogel and Berg, 1973; Watson, 1988). Stem dieback and even the failure of entire tree plantings have been reported (Medvick, 1980; Russell 1972; Samuel, 1991). Weeds also constitute a significant drain on resources, especially soil moisture (Richards et al., 1993). Simcock (1996) found that in New Zealand, with its unique flora and fauna, spreading topsoil and seeding grasses and legumes smothered native seedlings and prevented natural colonisation.

At a time when grading and topsoiling were not general practices, Struthers and Vimmerstedt (1965) found that grasses did not interfere with tree seedlings planted on new rock spoil and 'no such competition' occurred. Experience from the past, however, may not be applicable today, because of changes in mining operations, equipment and reclamation techniques, as well as regulatory practices (Ashby et al., 1980a). More recently, Jochimsen (1986) reported that planting young trees in unvegetated spoil appeared to inhibit the development an herbaceous layer. However, 15-year experiments in Bulgaria, with dense plantings of woody species only, have proven successful in stabilising spoil banks and encouraging herbaceous colonisation (Gentcheva-Kostadinova et al. 1994). These apparently conflicting conclusions are probably both valid in their particular circumstances.

Since grading, topsoiling and fertilisers tend to favour herbaceous cover, engineering practices and mine-soil construction methods may contribute an advantage to the seeded ground-cover species over planted trees (Chaney et al., 1995). Grasses are strong competitors and, when given the advantage of fertiliser, generally outcompete woody species (Chadwick, 1982; Daniels and Zipper, 1995; Schoenholtz et al., 1992). The problem is particularly difficult when grasses are planted for erosion control but the end-use is to be forest (Daniels and Zipper, 1995). Control of grasses and weeds can be critically important not only at planting but until tree roots can extend deep enough to ensure consistent access to moisture (Chaney et al., 1995; Richards et al., 1993; Samuel, 1991). Mowing and cultivating are not favoured because they tend to damage trees while having little effect on perennial weeds, especially grasses (Richards et al., 1993).

Surface Manipulation—Topsoiling and Ripping

Although the storage and respreading of topsoil previously removed as a first step in mining is generally considered an engineering technique, carefully removed and stored topsoil becomes a mine-soil amendment when returned to the reconstructed site (Bradshaw and Chadwick, 1980; Chadwick, 1982; Norland and Veith, 1991). However, both the cost of stripping and storing topsoil and the additional cost of spreading make topsoiling extremely expensive. There is also significant increase in bulk density due to compaction during replacement (Plass, 1968; Vogel and Berg, 1973). Grading to a smooth surface has been shown to inhibit tree growth, due to excessive compaction incurred during the grading and smoothing (Geyer, 1973; Jobling, 1983; Perkins and Vann, 1995; Torbert and Burger 1990. Vogel and Gray, 1987).

There is growing evidence that topsoiling is not needed for tree establishment and may be detrimental to planted trees, due to compaction from heavy equipment (Ashby and Kolar, 1981; Chaney et al., 1995; Gentcheva Kostadinova et al. 1994, Samuel, 1991; Smolik, 1988; Teasdale, 1983). Ashby and his coworkers (1984) found that trees grew taller and had better roots on ungraded, rocky spoils than on graded surfaces. Ungraded mine sites provide furrows that are ideal microsites for tree and shrub establishment (Bromley, 1983; Byrnes and Miller, 1973; Geyer, 1973; Riley, 1973). Spreading topsoil over the spoil surface is unnecessary, but small quantities of topsoil are effective when applied as a mulch or used to backfill planting pockets. These additions of topsoil can provide essential micro-organisms and a seedbank of native plants to enhance diversity (Chadwick, 1982; Dewangan and Mishra, 1994; Samuel, 1991). Long-term growth and survival data are limited for trees planted on topsoiled sites and the influence of topsoiling on establishment and the growth of trees is somewhat unclear (Chaney et al., 1995).

The engineering answer to compaction resulting from grading and smoothing operations is 'ripping'. Ripping, deep cultivation of the compacted surface, loosens the surface material to increase infiltration and allow roots to penetrate more deeply (Chadwick et al., 1987; Coppin and Richards, 1990; Simcock, 1993). However, not grading in the first place may be better where trees are to be planted (Ashby et al., 1984). Torbert and Burger (1990) compared two species of conifers and four broad-leaves trees grown on both smooth-graded and semi-rough, uncompacted surfaces on a reclaimed surface-mine site in the central Appalachian coalfields of the eastern United States. They found survival rates of 70%

on the rough-graded surface but only 42% on the smooth-graded, compacted surface. Compaction due to final grading also reduced the growth rate of the surviving trees. Ripping in the direction of the slope (downslope ripping) is considered a means of directing drainage, as well as loosening the mine-soil surface (Medvick, 1980).

Use of Herbicides

It is often considered important to the survival of tree seedlings to eradicate grasses and weeds, either by hand, which is expensive, or by using chemical controls (Chadwick, 1973; Russell, 1971; Jobling, 1983; Samuel, 1991; Wright and Daniels, 1995). Newly contoured sites are often seeded with grass mixtures a year in advance of tree planting, in order to check erosion and improve visual appearance (Chadwick, 1973; Jobling, 1983; Samuel, 1991; Vogel and Berg, 1973). Where this is done, herbaceous cover can compete with tree seedlings for available moisture and nutrients, reducing both seedling survival rates and growth (Ringe and Graves, 1985). Herbicides are a commonly used tool for controlling competition from herbaceous ground cover during the early establishment of tree seedlings. Most often this is accomplished by spraying contact herbicides rather than residual materials that could reach and damage trees or leach into water bodies (Richards et al., 1993). The advisability and necessity of this practice has been questioned. Scant attention has been paid to the impact of herbicides on soil microflora, symbiotic nitrogen fixers, mycorrhizae, or root pathogens (Rovira et al., 1990). The use of herbicides remains a convenient and frequently employed practice (Ringe and Graves, 1985; Rovira et al., 1990).

Alternatives to Herbicides

Several investigators suggest that using less-aggressive ground-cover species and reducing the seeding rate to a minimum can significantly reduce competition with planted trees and the need for control (Chadwick, 1982; Chaney et al., 1995; Lyle, 1976; Torbert et al., 1995; Vogel and Berg, 1973). Where they can be applied, mulches also reduce competition by suppressing herbaceous cover, as well as conserving moisture and providing organic matter (Del Tredici, 1992). However, the rapid development of a tree canopy provides the most permanent solution to weed competition (Richards et al., 1993).

SELECTING TREES AND SHRUBS FOR THE SITE

The success of a revegetation effort depends on the physical and chemical properties of the spoil and on how the site is graded and prepared. It also depends on the characteristics of the selected plant species and how they are managed (Daniels and Zipper, 1995; Gentcheva-Kostadinova et al., 1994; Moffat and McNeill 1995). The objective of the post-mining land-use is a primary consideration in choosing trees for reclamation (Badrinath et al., 1994; Leopold and Wali, 1992; Samuel, 1991; Wright and Daniels, 1995). Thus, political and economic decisions about land-use constrain the choice of species. The choice is further restricted by the characteristics of the climate, topography and substrate (Samuel 1991). Even more limiting is the question: what will grow on the site? When the objective of revegetation is simply to overcome adverse conditions, Knabe (W. Knabe, discussant in Schlätzer, 1973) suggested testing a large number of tree species, especially those grown successfully elsewhere, and all available seed sources.

Adaptability and Tolerance

Trees and shrubs appropriate for reclamation planting will be undemanding in their basic needs (Binns, 1993; Piha et al., 1995). They must be adaptable to diverse site conditions, establish readily, and require little money and effort to maintain (Binns, 1983; Howell, et al., 1991; May 1986; Piha et al., 1995; Ruffner and Steiner, 1973; Wright and Daniels, 1994). Further, they must be able to survive and reproduce on mined land (Dewangan and Mishra, 1994; SCS, 1978). Although trees on strip-mined land do not need to contend with sod, old roots, or brush and stumps, they are expected to establish themselves on stony sites that are windswept and exposed and may be alternately wet and dry within a brief time. Therefore, the first question to be answered is whether a particular tree or shrub can grow and reproduce in the conditions provided at the site.

Certain inherent characteristics govern a tree's potential to thrive on difficult sites and to resist competition (Grime, 1979; Schiechtl, 1980). In order to grow on the mine site, a tree must be able to tolerate raw mineral soils and disrupted drainage conditions. Adaptability is increased in plants that are able to fix atmospheric nitrogen and in those able to tolerate heavy-textured soils and alternating moist and dry conditions (Grime, 1979). Heat and sun tolerance also lend a competitive edge (Kozlowski et al., 1991).

Relatively few species prove capable of adapting to the combination of extreme and unpredictable environments on reclaimed land (Grime, 1979). Lists of species recommended for specific conditions and purposes for various climates and geographic locations are included in Bagley 1980 ab, Chadwick (1982), Chadwick et al. (1987), Coppin and Richards (1990), Hughes and Styles (1989), SCS (1978) and Thornberg (1982).

Diversity and Forest Development

The tolerance of certain species to harsh and degraded site conditions is well known. These pioneer species are included as a high percentage of reclamation plantings, in order to take advantage of their adaptability and rapid establishment. Hardy, climatically adapted species with low nutrient requirements and those tolerant of highly acid conditions and toxic substances in a particular spoil best support the low-cost, low-input approach to vegetation (Piha et al., 1995). Within these limits, trees and shrubs can be chosen that are likely to increase diversity and minimise competition, thus promoting forest development (Allen, 1990; Badrinath et al., 1994). However, of itself, diversity may not indicate system stability (Allen, 1990); Moffat and McNeill 1995, Tilman, 1996).

Including a mix of species enhances the likelihood of identifying those able to survive and reproduce on the evolving mine-soil substrate. Mixing shrubs and trees reduces the impact of winds and the likelihood of windthrow (Smolik, 1988). Uncommon species with important functions such as nitrogen fixation can be included in the species mix, together with key plants, such as shrubs known to improve microclimates for later plant establishment (Allen, 1990; Binns, 1983). Trees and shrubs that produce litter favourable for soil development contribute to the evolving substrate (Cromack et al., 1979; Hodgson and Buckley, 1975). Unfortunately, preference is often given to a few trees and shrubs with proven pioneer characteristics and to species with readily obtained seeds or those easily started in nurseries (Ashby and Kolar, 1984). The less hospitable the site conditions, the fewer the species likely to become established. Thus, diversity on reclaimed strip-mined land rarely will equal that of unmined land (Allen, 1990).

The trees and shrubs planted should be compatible with the planned permanent cover but could include a preliminary nurse crop to build the soil (Ashby et al., 1980a; Badrinath et al., 1994; Kelly, 1980). A small proportion of more demanding climax shrub and tree species should be included as well. If able to survive and reproduce, these climax species will eventually become dominant as site conditions improve (Allen, 1990; Knabe, 1973; Olschowy, 1973; Pandey and Singh, 1984; Wright and Daniels, 1995). The ratio of pioneers to climax trees and shrub species will depend upon the site and planting objective (Wright and Daniels, 1994).

Nitrogen-fixing Trees and Shrubs

Although generally considered weeds, woody legumes, such as wattles (*Acacia* spp.) and black locust (*Robinia pseudoacacia*), and other nitrogen-fixing pioneer trees, such as alder (*Alnus* spp.), are well adapted to harsh conditions on recontoured mine soil and can be extremely useful in establishing forest cover, especially in areas where recultivation is the sole consideration and forest is the objective (Chadwick, 1982; Daniels and Zipper, 1995; Dewangan and Mishra, 1994; Hughes and Styles, 1989; Piha et al., 1995). These specialist species help initiate nutrient cycling and improve conditions for soil biological systems as well as for more desirable tree species (Chadwick, 1982; Dewangan and Mishra, 1994; Hughes and

Styles, 1989; SCS, 1978). In places where forests are primary sources of fuel, such trees are particularly useful for local people, since they are not sought by other sectors of the economy (Badrinath et al., 1994; Hughes and Styles, 1989).

IDENTIFYING TREES FOR REVEGETATION

Pre-mining vegetation is often useful as an indicator of trees that could do well on the reclaimed site following mining, especially in harsh environments where abiotic factors—heat, cold, salinity, or extreme moisture deficit—are controlling factors (Fuller, 1986; Potter et al., 1951). However, in more favourable climates, extant plants may represent climax vegetation that has evolved over long periods of time under substantially different conditions (Fuller, 1986). Species presently growing in the area may not tolerate the newly upturned mineral substrate, even when amended, and the site may require an interim pioneer vegetation to initiate succession (Fuller, 1986; Hatton and West, 1987). Studies of natural invasion may identify species suitable for the most difficult conditions (Bell, 1979).

Root Characteristics

Plants best able to tolerate conditions on newly contoured spoils will be those having highly adaptable root systems (Hodgson and Buckley, 1975; Kozlowski et al., 1991). All root-system architecture is plastic and responsive to soil, site and weather conditions, but as with other growth characteristics, the degree of plasticity is inherent and varies among species (Jeffrey, 1987). Pioneer species generally show a strong tendency to adapt root-system morphology to the soil environment and to change radically in response to changes in substrate conditions (Fitter and Bradshaw 1987; Keresztesi 1968, Keresztesi 1980). Trees likely to produce deep root systems flexibly adaptable to site conditions are good choices for the harsh and variable conditions on reclamation sites (Hodgson and Buckley, 1975).

Native vs. Exotic

Trees native to an area are generally preferred to introduced species. They will be suited to the climate and can be expected to have superior potential for success in establishing a diverse plant community in harmony with the surroundings (Davidson, 1980; Fuller, 1986; Narten et al., 1983). When exotic species are introduced, the experience gained from the original habitat cannot predict plant performance, even where climate and site characteristics are quite similar (Chadwick et al., 1987). Therefore, to avoid unpleasant surprises it is essential, as Schlätzer (1973: 62) suggested, always to 'ask the plant itself first' through large-scale testing and evaluation. Unfortunately, native trees suitable for reclamation often are not available in the nursery trade, since there is little interest in them; thus introduced species are selected by default (Butterfield and Fisher, 1994). Furthermore, seedlings of some native species may be difficult to reproduce under nursery conditions because of recalcitrant traits such as seed-coat or embryo dormancy (Baig, 1992).

In addition to native trees, familiar introductions that have become naturalised in an area over time are often used in mine reclamation, where they frequently outperform native species (Fuller, 1986). Examples are trees such as the black locust from the eastern United States, planted frequently throughout Europe and in Asian forestry, eucalyptus planted on mine spoils in Africa, and Northern Hemisphere pines, alders and poplars in New Zealand (Chadwick et al., 1987; Keresztesi, 1980; Simcock, 1993). Although exotics and naturalised species are both introductions, much more is known about naturalised plants, and their performance is more predictable. Fuller (1986) and Ruffner and Steiner (1973) describe the work of plant development centres in the United States where prospective revegetation species are identified, seed is gathered and increased, and the plants are tested on-site and evaluated in field-size trials. The programme could serve as a model for such investigations. (see species list on pages 332–337).

Stock Options

The usual method of establishing trees and shrubs on mined land is the often difficult and expensive process of hand-transplanting young trees (Zarger et al., 1969). Transplants have been extremely successful on disturbed sites and in revegetating sites where direct seeding has failed (Gray and Leiser, 1982). In the context of mine-soil revegetation, tree and shrub planting offers advantages over direct seeding. Because transplanting avoids the hazards of the critical initial germination and establishment stages, survival rates are higher and less follow-up care is needed (Bradshaw and Chadwick, 1980; Hektner et al., 1981). Planting is generally more economical than seeding, despite the higher cost of material and labour (Hektner et al., 1981). High quality stock is crucial in establishing seedlings and in their recovery after planting (Coppin and Richards, 1990; Jobling and Stevens, 1980; Leaf et al., 1978; Limstrom 1960).

Bare-rooted material

Trees and shrubs to be transplanted are usually bare-rooted nursery stock, although unrooted cuttings of easily rooted species (such as willow or poplar) and container-grown plants are also used where conditions are suitable (Chadwick, 1982; Moffat and McNeill, 1995). Locally grown stock can be expected to have the best survival rate (Abbott and Bacon, 1977).

Root pruning at the time plants are lifted in the nursery (usually from undercutting) can damage roots and depress root growth of woody plants and hence should be kept to a minimum (Hellum, 1978). The choice of well-balanced transplants of appropriate size is essential to success in establishing trees. A balance must be struck between making certain that transplant stock is large enough to survive and achieve rapid growth and the additional cost and difficulty in handling larger stock (Moffat and McNeill, 1995). Year-old (1-1) nursery-grown transplants, 30 to 60 cm, can be expected to transplant and establish more quickly and grow better than larger stock. Even smaller material—20 to 30 cm—may be best for conifers, which are somewhat top-heavy and subject to 'socketing' in windy circumstances (Wright and Daniels, 1995).

Direct seeding

Some tree species can be established on some sites by direct seeding (Macyk 1985, Plass, 1976; SCS, 1978). However, the results show more failures than successes (Davidson, 1980; Zarger et al., 1969). Where climate and site conditions are suitable, seeding woody species on the recontoured surface can be an attractive low-cost alternative to more costly transplants. This is especially true on sites with high erosion potential, where tree planting can be expensive and hazardous (Davidson, 1980; Plass 1976; Zarger et al., 1969).

In tropical areas, trees are often seeded rather than planted. Seeding is done during the wet season, usually after the first rains, to give the seedlings time to develop roots sufficient for survival in the dry season. In places with two rainy seasons, local conditions dictate preferential planting time (Woods, 1996). However, seeding is unlikely to succeed in a climate where the growing season is short and the seedlings are likely to be stressed by temperature or inadequate moisture or sunlight (Ferchau, 1988). Better growth and survival have been found with seeded conifers than with hardwoods, with certain exceptions, including black locust and black walnut (Davidson, 1980).

Direct seeding is not a simple method and well-documented procedures must be followed (Davidson, 1980). Special attention, for example, must be paid to the problems of competition with herbaceous species. Watson (1994), for example, advises seeding beneath the protection of annual arable crop. Furthermore, the low cost of seeding may be offset by the higher cost of surface treatment in preparing the site, as well as increased follow-up care. The level of engineering and cultivation required for seeding trees adds greatly to the costs of revegetation (Bagley, 1980 a, b). The choice of trees and shrubs for direct seeding is limited by availability of seed for the more aggressive reclamation species, as well as by the success rate for germination and survival of a given species (Butterfield and Fisher, 1994; Piha et al., 1995). However, where trees can be established successfully by direct seeding, transplanting may not make economic sense (Balmer, 1976). Regardless of the method of revegetation chosen, it is vital to determine, as early as possible in the planning process, whether trees will be planted, or seeded on the site (Balmer, 1976).

Container plants

A reliable but expensive means of providing a good start for the planted trees lies in the use of container-grown plants (Balmer, 1976; Moffat and McNeill 1995). Container-grown stock provides some advantages over bare-root transplants (Balmer, 1976; Ross et al., 1995; Van Eerden and Kinghorn, 1978). Container plants provide an undisturbed root system and thus suffer less planting shock. They offer the freedom to plant over a longer planting season and planting can be done by less-experienced hands. Some of the initial cost of container plants may be offset by increased survival rates. However, the high cost of both material and transport make container-grown stock unattractive for most revegetation work except in unusual situations. In comparing effects of various nursery practices, Chaney and colleagues (1995) reported that after 12 years growth, the tree production method no longer influenced survival and productivity of black walnut (*Juglans nigra*) on a revegetated mine soil in the midwestern United States.

Starter pods

Root-training containers, such as planting books, plastic or paper tubes, and planting blocks of peat or other materials, have gained wide acceptance in reclamation work. They are available in many shapes and sizes, but are generally of two types—those meant to be removed before planting and those which are supposedly biodegradable and can be planted with the seedling (Gray and Leiser, 1982). Containers of rigid materials have several advantages. They can be shaped to direct roots downward and discourage root circling, which is thought to limit tree stability. Also because the container is removed, roots are in direct contact with the surrounding substrate after planting. However, the removal of both rigid and biodegradable containers before planting may be advisable, since biodegradable containers break down slowly without sufficient moisture and a dry container wall will be a barrier restricting root growth (Gray and Leiser, 1982).

Handle with Care

How planting stock is handled during lifting, storage, transport and planting will affect both vigour and root development (Coppin and Richards, 1990; Leaf et al., 1978; Limstrom, 1948, 1960). Planting stock needs protection from drying winds and temperature extremes. Every effort must be taken to prevent roots from drying out between the time they are dug from the nursery bed and then planted (Chadwick, 1982; Bagley, 1980a; Jobling, 1983; Samuel, 1991). Therefore, the trees should be planted as soon as possible after they reach the site (Bagley, 1980; Moffat and McNeill, 1995).

Establishing Bare-root Transplants

Richardson et al. (1987: 139) points out that "normal agricultural and forestry techniques are not suitable either in the site preparation, sowing and plantings, or in aftercare" on mine-soils. Proper planting methods are essential to tree survival and growth and cannot be taken for granted if failure is to be avoided. Supervision is needed throughout the planting operation. It is rarely possible to isolate a single cause for poor tree performance, but choosing appropriate plants and using care in planting greatly improve prospects for successful revegetation (Bagley 1980; Binns, 1983; Richardson et al., 1987).

The technique used in planting is a crucial factor in root formation and tree survival (Tinus, 1978). Field investigations confirm that root deformation reduces root growth throughout the life of a tree. The damage to the root system caused by careless planting of tree seedlings cannot be reversed (Grene, 1978). Proper planting requires that roots not be jammed or shoved into planting holes (Abbott and Bacon, 1977; Bagley, 1980; Hellum, 1978). Root form in transplanted seedlings never resembles that of seeded-in-place trees but care in planting can minimise negative effects of transplantation (Long, 1978; Segaran et al., 1978; Tinus 1978). The effects of plant production and planting methods on the form of planted trees are treated at length in Van Eerden and Kinghorn (1978).

To notch or not to notch

Poor root development can lead to the death of trees and other vegetation in the droughty conditions common to coal spoils (Baig, 1992; Fitter and Bradshaw, 1974). Of particular concern on the dense substrates encountered in mine reclamation is the additional compaction contributed by the reaming action of planting spades, mattocks, wedges, or punches used in notch planting. The resultant increase in mechanical resistance and loss of aeration may severely limit root penetration (Appleroth, 1978; Balmer, 1976; Grene, 1978; Schultz and Thompson, 1990). Notch planting usually asks too much of the plants, many of which do not survive, so that extensive follow-up and replanting are required (Bagley, 1980; Richardson et al., 1987).

Pocket planting

More labour-intensive methods using augered or hand-dug planting holes can generally be expected to provide a margin of success for young trees (Balmer, 1976; Richardson et al., 1987). In the harsh conditions of mine revegetation sites, plants need all the help they can get, and pocket planting can minimise the damaging effects of the substrate by providing an improved growth medium for seedling trees during their early establishment (Balmer, 1976; Hodgson and Buckley, 1975).

The use of enriched soil in planting holes provides an immediately available, nutrient-rich growth medium to encourage the development of vigorous root systems able to penetrate well beyond the planting pocket (Balmer, 1976; Hodgson and Buckley 1975; Richardson et al., 1987). Although it has been suggested that roots will remain in the planting pocket, unable to penetrate the surrounding substrate, studies in the eastern United States and Wales showed these concerns to be unfounded (Flege, 1993, 1996; Wu et al., 1995).

Planting should be done at the time most advantageous to the plant, avoiding periods that are too wet or too dry. In temperate climates, tree planting is recommended between late autumn and early spring (Moffat and McNeill, 1995; Samuel, 1991). Where moisture and temperature conditions permit, better survival rates can be expected with spring-planted trees, which are not subject to root damage from frost heaving. However, the effect of the season of planting varies among species (Davis, 1973). Spring planting is best done as early as possible and should not be extended beyond the season (Abbott and Bacon, 1977; Davis, 1973; Wright and Daniels, 1995).

In general, careful planting calls for planting trees at the same depth as they were grown in the nursery. However, except on poorly drained sites, Stroempl (1990) recommends deeper planting to encourage root formation. His comparative studies showed that deeper planting can encourage root formation and increase the survival of newly planted bare-root trees. Coal spoils dry out quickly after rain and Stroempl found that often new tree seedlings die not from drought but because high surface temperatures cause failure at the root collar, the sensitive transition zone where root and shoot meet and where, in some species, either roots or shoots can form.

MEASURING SUCCESS IN REVEGETATION

In the past, reclamation success was equated with extensive surface cover or high productivity (the annual increment of new growth), leading to the assumption that if it's green, it's finished (Toy and Hadley, 1987).

However, early success of revegetation has not proven a good guide to long-term results (Binns, 1983).

If 'diverse' and 'self-sustaining' are operative terms, the measure of reclamation success becomes evidence of a robust, self-replicating, interactive soil-plant system (Daniels and Zipper 1995, Darmer, 1992; Ross et al., 1995; Whitford and Elkins, 1986). It is not always a simple matter, however, to achieve these apparently straightforward objectives. Establishing realistic and objective criteria to ascertain whether vegetation will become 'permanently' established and continue to satisfy the objectives of reclamation is a tall order indeed (Fischer, 1986).

Indicators of Success

Frequently used measures of vegetation development include direct counts of the number of species present (species diversity) and the number of plants of one species relative to the number of individual plants present (species richness) (Chambers et al., 1990; Chambers and Wade, 1990). These are then compared with nearby reference plots or a similar standard (Fischer, 1986). However, Carter (1994) and others found these variables to be poor predictors of long-term success. Measurements of diversity and density of a plant community reflect only the vegetation present on a site at the time it is sampled, not whether it is capable of reproducing itself (Allen, 1990), and species diversity can be changeable over space and time (Pandey and Singh, 1984). Diversity indices and their applications are reviewed by Chambers (1983).

To date, no single indicator has been found that can promise success in mined-land reforestation. We can look for signs of developing plant-soil-atmosphere systems. We can examine the trees—do they appear to be healthy? We can measure their growth. We can ask whether soil particles are aggregating and a soil horizon forming, and if the developing soil shows positive signs of improvement, such as lower bulk density and increased nutrient content. We can determine whether soil organisms are colonising the site—from the micro-organisms needed to break down organic material and recycle minerals to the burrowing earthworms and other creatures which stir, mix and bind the soil particles (Scullion, 1992). We can look for mycorrhizal fungi in the soil and for nodules on symbiotic roots. We can measure the carbon-to-nitrogten ratios to monitor some of these processes (Cromack, 1981). These provide clues to the state of a developing vegetation system. The question remains—how can we know that what we measure, howsoever accurately and efficiently, is a valid measure of success?

A commonly used measure of successful rehabilitation is the quality of the vegetation cover. This directly affects erosion and sediment rates and also improves the appearance, and hence the public's perception of, the repaired landscape (Dragovich and Patterson, 1995). However, the condition of vegetation at the time of measurement does not reveal the trend toward natural succession and a stable system nor the likelihood of its persistence. The stability of the system cannot be determined simply by the achievement of a certain level of cover and productivity (Fischer, 1986). These measurements reveal neither the state of the soil environment nor the health of microbial communities, without which any initial development is likely to be short-lived (Curry, 1975; Fischer, 1986; Shubert and Starks, 1985; Tueller 1990). Thus, how to measure the success of a functioning ecosystem remains a problem (Ashby and Kolar, 1984).

In future, investigations into landscape ecology and habitat analysis may provide the basis for determining success (Fischer, 1986; Olschowy, 1973; Tueller, 1990). Clues to long-term recovery are likely to come from soil microbial populations, which are rapid and sensitive indicators of the stability and resilience of the plant-soil system (Shubert and Starks, 1985; Tate, 1985). Soil biologic factors are preferred to analyses of above-ground vegetation, such as diversity, productivity and cover, since reclaimed sites are rarely planted with the same species in exactly the same proportions and under precisely the same conditions (Shubert and Starks, 1985). Algae, in particular, may have value in investigating and managing mined-land reclamation. By identifying limiting factors in substrate material and monitoring changes, algal communities indicate the progress of reclamation (Shubert and Starks, 1985). Soil enzymes are also suggested as a tool to diagnose and monitor reclamation by providing information on the effects of spoil materials on plant community development. Another suggested indirect indicator of system stability is the census of animals frequenting the site (Parmenter and MacMahon, 1990).

Priorities and Prospects

The goal of revegetation has been defined as plant cover that fulfils the purpose intended, is ecologically sound, is reasonable in cost and visually pleasing (Morrison, 1987; Olschowy, 1973). A review of coal-mine revegetation literature reveals a wide range of reclamation objectives:

- to meet regulatory obligations and deter erosion (Daniels and Zipper, 1985);
- to serve the objectives of the landowner (Moody and Kimbrell, 1980);
- to make the land as good as, or better than, it was before mining (Curry, 1975);
- to improve visual quality while assuring economic return (Abbott and Bacon, 1977);
- to conserve important land-use opportunities present in the pre-mining landscape (NRC, 1981);
- to seize the opportunity to maximise post-mining land-use potential (Bradshaw, 1995 Griffiths, 1992; Murphy and Bace, 1976, Reiss, 1973);
- to build soil (Struthers and Vimmerstedt, 1965);

- to reconstruct a diverse and self-sustaining ecosystem (Chambers and Wade 1990; Daniels and Zipper, 1995; Leopold and Wali, 1992; Toy and Hadley, 1987; Whitford and Elkins, 1986);
- to modify the spoil so that it approaches productive soil in character (Leirós et al., 1996).

These objectives are not necessarily mutually exclusive. Revegetation can have multiple purposes encompassing both short-term and long-term aims.

Unfortunately, both economic and regulatory pressures direct research efforts towards plants that give quick cover, control erosion and protect water resources, rather than considering what will grow best or be the most useful in establishing a permanent vegetation system (Murphy and Bace, 1976). Mining companies seek a quick fix to fulfil their obligations and landowners expect a rapid return to economically productive use of the land after mining (Whitford and Elkins, 1986). Landowners often see pasture as a shortcut to economic return but initial success is often short-lived, if fields are not carefully managed. Without continued applications of fertilisers and control of grazing, the land deteriorates and may never fully recover. (Ashby and Kolar, 1981). Thus, as more longer-term studies become available, it becomes increasingly apparent that the success of revegetation depends on more than rapidly developed grass and herbaceous cover (Chambers et al., 1990; Fischer, 1986; Haigh, 1992; Toy and Hadley, 1987; Whitford and Elkins, 1986). Working with nature, planting forest trees and allowing them to return life and fertility to the post-mining landscape may be the best, most cost-effective means of restoring life after coal mining (Chadwick 1982).

Life tends to improve conditions that support life; revegetation becomes truly self-sustaining only when planted trees are augmented by invading indigenous species, not only of trees but of herbs and shrubs as well. Long-term success in revegetation is most likely to be achieved by a holistic approach to evaluation, so that success is then defined as a functioning, self-sustaining natural ecosystem with evidence of evolving successional processes (Allen, 1990; Griffiths, 1992).

The technology exists to grow plants on practically any site. Most sites are biologically restorable (Dick and Thirgood, 1975; Leopold and Wali, 1992; Wali, 1992). However, the ability to particular plants on a site does not mean that this is a reasonable, economic, or logical course to follow. The value of the land may not equal the cost of reclamation and furthermore, the long-term outcome may be unacceptable. From a scientific point of view, it is possible now to know with fair certainty, before mining commences, which areas can be successfully rehabilitated and which cannot (Wali, Pers. comm). Multivariate analysis, geographic information systems (GIS) and remote sensing technology provide powerful tools for predicting the outcome of various revegetation strategies on any site (Tueller, 1990;) With recent progress in understanding landscape ecology, analyses of data from pre-mining sampling of overburden, together with climatic variables, can predict likely rates of physical and chemical weathering, as well as which species are likely to colonise a site and their probable persistence over time (Haering et al., 1993). Whether informed decisions will be made based on this knowledge is not so certain.

SUMMARY

Surface-mining reclamation involves the rehabilitation of a system that has been drastically disturbed. The role of vegetation is of paramount importance in this process. Revegetation serves a variety of purposes, from short-term erosion control to sustained productivity of the land. Trees and shrubs can be a cost effective and environmentally sound answer to revegetating strip-mined land by working with nature, harnessing the natural regenerative power of forests of renew the degraded environment. Attention to the choice of trees and shrubs and care in handling and planting them can determine whether the planting will succeed or fail. However, the outcome of reclamation rests on economic, political and social decision-making, as well as on the vegetation.

If reclamation efforts are to be worthwhile, it is important that decision-makers recognise that the needs of reclamation place limits on mining practices and, furthermore, that site factors will limit the possibilities for revegetation (Dick and Thirgood, 1975). The ultimate goal is the restoration of a functioning natural system—a goal most readily achieved in many instances by planting forest trees and shrubs. If the biological component of reclamation does not succeed, the ensuing failure involves not only the small portion of costs associated with revegetation but also the much greater investment (perhaps 50 times as much) in engineering the landform (Daniels and Zipper, 1994; Samuel 1991; Tueller, 1990). The high cost of failure and the importance of biological recovery underscore the need for careful planning and implementation of revegetation operations from the outset. Trees take time to do their work, perhaps 50 years or more. The prospects for successful reclamation ultimately depend upon the commitment—social, political and economic—to repair the damage done in extracting the earth's underground riches.

References

Abbott, D. and Bacon, G.B. 1977. Reclamation of coal-mine wastes in New Brunswick. *Can. Inst. Mining Bull.*, 70: 112–119.

Aldon E.F. and Springfield, H.W. 1975. Problems and techniques in revegetating coal-mine spoils in New Mexico. In: M.K. Wali (ed.). *Practices and Problems of Land Reclamation in Western North America*, pp. 122–132. Univ. North Dakota Press, Grand Forks, ND.

Allen, E.B. 1990. Evaluating community-level processes to determine reclamation success. In: J.C. Chambers and G.L. Wade (eds.). *Evaluating Reclamation Success: the Ecological Consideration*. US For. Serv., NE For. Exper. Station, Broomall, PA. Gen. Tech. Rept. NE-164.

Allen, M.F. and Friese, C.F. 1990. Mycorrhizae and reclamation success: importance and measurement. In: J.C. Chambers and G.L. Wade (eds.). *Evaluating Reclamation Success: the Ecological Consideration*. US For Serv., NE For. Exper. Station, Broomall, PA, Gen. Tech. Rept. NE-164.

Andersen, C.P., Bussler, B.H. Chaney, W.R., Pope, P.E. and Burnes, W.R. 1989. Concurrent establishment of ground cover and hardwood trees on reclaimed mined land and unmined reference sites. *Forest Ecol. Management* 28: 1–99.

Appleroth, S. 1978. A work-study man's view on the root form of planted trees. In: E. Van Eerden and J.M. Kinghorn (eds.). *Root Form of Planted Trees*. BC. Ministry of Forests/Can. For. Serv., Victoria, BC, Canada, Joint Rept. Nr. 8.

Arrhenius, W.R.G. 1922. Some factors affecting the hydrogen-ion concentration of the soil and its relation to plant distribution. *Trinity Coll., Dublin, Botan. School, Notes* 3: 133–177.

Ashby, W.C. and Kolar, C.A. 1981. Productivity with trees and crops on surface-mined lands. *Trans. Soc. Mining Eng., AIME* 274: 1813–1817.

Ashby, W.C. and Kolar, C.A. 1984. Ecological forces affecting reclamation with trees. pp. 395–409. In: *Proc. Conf. Better Reclamation with Trees*. Amer. Soc. Surface Mining and Reclamation. Princeton, WV,

Ashby, W.C., Kolar, C.A. and Rogers, N.F. 1980a. Results of 30-year-old plantations on surface mines in the central states. In: *Trees for Reclamation*. US For. Serv., Broomall, PA, Gen. Tech. Rept. NE-61.

Ashby, W.C, Rogers, N.F. and Kolar, C.A. 1980b. Forest tree invasion and diversity on stripmines. pp. 178–281. In: H.E. Garrett and G.S. Cox (eds.). *Proc. Central Hardwood Forest Conf., III*, Columbia, MO.

Ashby, W.C., Vogel, W.G., Kolar, C.A. and Philo, G.R. 1984. Productivity of stony soils on strip mines. In: D.M. Kral (ed.). *Erosion and Productivity of Soils Containing Rock Fragments*. Soil Sci. Soc. Amer., Madison, WI, Spec. Publ. 13.

Atkinson, S.L. Buckley, G.P. and Lopez-Real, J.M. 1991. Evaluation of composted sewage sludge/straw for the revegetation of derelict land. In: M.C.R. Davies (ed.). *Land Reclamation: an End to Dereliction?* pp. 329–335. Coll. Cardiff, Univ. Wales (UK).

Badrinath, S.D., Chakraborty, A. and Khan, S. 1994. Forest ecosystem and mining activity. pp. 159–172. In: *Proc. Int. Symp. Impact of Mining on Environment, Nagpur, India*, Prints, New Delhi,

Bagley, F.L. 1980a. Tree planting: strip-mined area in Maryland, Part I. pp. 27–28. In: *Trees for Reclamation*. US For. Serv., Broomall, PA, Gen. Tech. Rept. NE-61.

Bagley, F.L. 1980b. Tree planting: strip-mined area in Maryland, Part II. In: *Trees for Reclamation*. US For. Serv., Broomall, PA, Gen. Tech. Rept. NE-61 pp. 29–36.

Baig, M.N. 1992. Natural revegetation of coal-mine spoils in the Rocky Mountains of Alberta and its significance for species selection in land restoration. *Mountain Research and Development* 3: 285–300.

Balmer, W.E. 1976. Use of container grown seedlings on disturbed surface areas, pp. 43–47. In: *Proc. Conf. Forestation Disturbed Surface Areas*. US For. Serv., SE Region, Birmingham, AL.

Barnhisel, R.I. 1988. Correction of physical limitation to reclamation. pp. 191–211. In: L.R. Hossner (ed.). *Reclamation of Surface-mined Lands*, CRC Press, Inc., Boca Raton, FL.

Barnhisel, R.I. and Massey, H.F. 1992. Chemical, mineralogical and physical properties of eastern Kentucky acid-froming coal spoil materials. *Soil Science* 108: 367–372.

Beck, R.L., Fehr, L.S. and McDonald, R.R. 1990. *Effects of Controlled Overburden Placement on Topsoil Substitute Quality and Bond Release: Phase III.* Office of Surface Mining, NTIS, Wash., DC.

Bell, T.J. 1979. Factors which affect the establishment of natural vegetation on a coal strip mine spoil bank in southeastern Ohio. Unpublished Ph.D. dissertation. Ohio State University.

Bell, J.C., Cunningham, R.L. and Anthony, C.T. 1994. Morphological characteristics of reconstructed prime farmland soils in western Pennsylvania. *J. Environmental Quality* 23: 515–520.

Berg, W.A. 1973. Evaluation of P and K soil fertility tests on coal-mine spoils. pp. 93–131. In: R.J. Hutnik and G. Davis (eds.). *Ecology and Reclamation of Devastated Land,* vol. 1, Gordon and Breach, New York, NY.

Berg, W.A. 1980. Nitrogen and phosphorus fertilization of mined lands. pp. 20/1-20/8. In: *Proc. Symp. Adequate Reclamation of Mined Lands, Billings, Montana.* Soil Cons. Soc. Amer. Madison, WI,

Berg, W.A. and Vogel, W.G. 1980. Toxicity of acid coal-mine spoils to plants pp. 57–68. In: R.J. Hutnik and G. Davis (eds.). *Ecology and Reclamation of Devastated Land,* vol. 2, Gordon and Breach, New York, NY.

Biddle, P.G. 1983. Patterns of soil drying and moisture deficit in the vicinity of trees on clay soils. *Geotechnique,* 33 (2), 107–126.

Binns, W.O. 1983. Treatment of surface workings. In: Reclamation of Mineral Working to Forestry. *Forestry Comm., Edinburgh, Scotland, R. & D Paper* 132, 9–16.

Bloomfield, H.E., Handley, J.F. and Bradshaw, A.D. 1982. Nutrient deficiencies and aftercare of reclaimed derelict land. *J. Appl. Ecol.,* 19: 151–158.

Bradshaw, A.D. 1995. Alternative endpoints for reclamation. pp. 165–185. In: J. Cairns, Jr. (ed). *Rehabilitating Damaged Ecosystems,* Lewis Publ., Boca Raton, FL.

Bradshaw, A.D. and Chadwick, M.J. 1980. *The Ecology and Reclamation of Derelict and Degraded Land.* Univ. Calif. Press, Berkeley, CA.

Brady, N.C. and Weil, R.R. 1996. *The Nature and Properties of Soils.* Prentice Hall, Upper Saddle River, NJ.

Bramble, W.D., Byrnes, W.R., Hutnik, R.J. and Liscinsky, S.A. 1996. Interference factors responsible for resistance of forb-grass cover types to tree invasion on an electric utility right-of-way. *J. Arboriculture* 22: 99–105.

Brenner, F.J. 1984. Restoration of natural ecosystems on surface coal-mine lands in the northeastern United States. pp. 195–210. In: T.N. Veziroglu (ed.). *The Biosphere: Problems and Solutions,* Elsevier, Amsterdam.

Bromley, R.F. 1983. Practical works involved in the restoration of the land form for planting. London, HMSO, *Forestry Commission Research and Development Paper 132,* 6–8.

Brown, J.E., Maddox, J.B. and Bartley Jr., G.N. 1981. Acrylic emulsion as a straw binder in reclaiming coal surface-mined land. *Tennesse Valley Authorities, Div. Land-Forest Resources, Norris, TN, Tech. Note* B43.

Brown, J.E., Farmer Jr., R.E. and Splittstoesser, W.E. 1985. The establishment of mixed plant communities on surface-mined land for timber production, timber and wildlife only, and a native mixed forest. pp. 431–435. In: *Proc. 1985 Symp. Surface Mining, Hydrology Sedimentology, and Reclamation.* Coll. Eng., Univ. Kentucky, Lexington, KY.

Brown, R.W. 1973. Transpiration of native and introduced grasses on a high-elevation harsh site. pp. 467–481. In: R.J. Hutnik and G. Davis (eds.). *Ecology and Reclamation of Devastated Land,* Vol. 1, Gordon and Breach, New York, NY.

Butterfield, R.P. and Fisher, R.F. 1994. Untapped potential: native species for reforestation. *J. Forestry* 92: 37–40.

Byrnes, W.R. and Miller, J.H. 1973. Natural revegetation and cast overburden properties of surface-mined coal lands in southern Indiana. pp. 285–306. In: R.J. Hutnik and G. Davis (eds.). *Ecology and Reclamation of Devastated Land,* vol. 1, Gordon and Breach, New York, NY.

Cairns, John Jr. 1995. Restoration ecology: protecting our national and global life support systems. pp. 1–12. In: *Rehabilitating Damaged Ecosystems*. Lewis Publ., Boca Raton, FL,

Carter, R.A. 1994. Reclamation targets are coming into focus. *Coal* 99: 38–44.

Caruccio, F.T., Hossner, L.R. and Geidel, G. 1988. Pyritic materials: acid drainage, soil acidity, and liming, pp. 159–189. In: L.R. Hossner (ed.). *Reclamation of Surface-mined Lands*, CRC Press, Inc., Boca Raton, FL.

Chadwick, M.J. 1983. Fertilizer and herbaceous cover influence establishment of directseeded black locust on coal-mine spoils. In: R.J. Hutnik and G. Davis (eds.). *Ecology and Reclamation of Devastated Land*, vol. 2, pp. 175–188. Gordon and Breach, New York, NY.

Chadwick, M.J. 1975. The cycling of materials in disturbed environments. pp. 3–16. In: M.J. Chadwick and G.T. Goodman (eds.). *Ecology of Resource Degradation and Renewal*, John Wiley & Sons, New York, NY.

Chadwick, M.J. 1982. *An Assessment of the Potential of Derelict and Industrial Wasteland for the Growth of Energy Crops*. Univ. York, York (UK).

Chadwick, M.H. Highton, N.H. and Lindman, N. 1987. *Environmental Impacts of Coal Mining and Utilization*. Pergamon Press, Oxford (UK).

Chambers, J.C. (1983). Measuring species diversity on revegetated surface mines: an evaluation of techniques *Research Paper Int.* 332: 16.

Chambers, J.C. and Wade, G.L. (eds.). 1990. *Evaluating Reclamation Success: the Ecological Consideration*. US For. Serv, NE For. Exper. Station, Radnor, PA, Gen. Tech. Rept. NE-164.

Chambers, J.C., MacMahon, J.A. and Wade, G.L. 1990. Differences in successional processes among biomes: importance of obtaining and evaluating reclamation success. In: *Evaluating Reclamation Success: the Ecological Consideration*. US For. Serv., NE For. Exper. Station Broomall, PA, Gen, Tech. Rept. NE-164, pp. 59–72.

Chaney, W.R., Pope, P.E. and Byrnes, W.R. 1995. Tree survival and growth on land reclaimed in accord with Public Law 95–87. *J. Environmental Quality* 24: 630–634.

Coppin, N.J., and Richards, I.G. 1990. *Use of Vegetation in Civil Engineering*. Butterworths London.

Cromack, K. Jr. 1981. *Below-ground Processes in Forest Succession*. Springer-Verlag, New York, NY.

Cromack, K. Jr., Delwiche, C.C. and McNabb, D.H. 1979. Prospects and problems of nitrogen management using symbiotic nitrogen fixers pp. 210–233. In: J.C. Gordon, C.I. Wheeler and D.A. Perry (eds.). *Symbiotic Nitrogen Fixation in the Management of Temperate Forests*. Oregon State, Univ. Press, Corvallis, OR.

Cundell, A.M. 1977. The role of micro organisms in the revegetation of strip-mined land in the western United States. *J. Range Management* 30 (4): 299–305.

Curry, R.R. 1975. Biogeochemical limitation on western reclamation. pp. 18–47. In: *Practices and Problems of Land Reclamation in Western North America*. Univ. North Dakota Press, Grand Forks, ND,

Daily, G.C. 1995. Restoring value to the world's degraded lands. *Science* 269: 350–354.

Daniels, W.L. and Amos, D.F. 1984. Generating productive topsoil substitutes from hard rock overburden in the Southern Appalachians. pp. 37–57. In *Proc, 1984 National Meeting*, American Society for Surface Mining and Reclamation. Owensboro, KY, 10–13 July, 1984.

Daniels, W.L. and Zipper, C.E. 1995. Improving coal surface mine reclamation in the central Appalachian region. pp. 187–217. In: J. Cairns, Jr. (ed.), *Rehabilitating Damaged Ecosystems*, Lewis Publ., Boca Raton, FL.

Darmer, G. 1992. *Landscape and Surface Mining: Ecological Guidelines for Reclamation (tr.* Landschaft und Tegebau. Oekologische Leitbilder für die Rekultivierung (1979). Van Nostrand Reinhold, New York, NY.

Davidson, W.H. 1980. Direct seeding for forestation. pp. 61–62. In: *Trees for Reclamation*. US For. Serv., Broomall, PA, Gen. Tech. Rept. NE-61.

Davis, G. 1973. Comparison of fall and spring planting on strip-mine spoils in the bituminous regions of Pennsylvania. pp. 525–538. In: R.J. Hutnik and G. Davis (eds.). *Ecology and Reclamation of Devastated Land*, vol. 1, Gordon and Breach, New York, NY.

Del Tredici, P. 1992. Make mine mulch. *Arnoldia* 52 (3): 30–32.

Dept. of the Interior, USA. 1992. *Surface Coal Mining Reclamation: 15 Years of Progress, 1977–1992* (part 1). Office of Surface Mining, Wash., DC.

Dewangan, M.P. and Mishra, G.N. 1994. Development and afforestation of mine spoil sumps in manganese ore, Nagpur. pp. 349–354. In: *Proc. Int. Symp. Impact of Mining on Environment, Nagpur, India.* Prints, New Delhi.

Dick, J.H. and Thirgood, J.V. 1975. *Development of Land Reclamation in British Columbia: Practices and Problems of Land Reclamation in Western North America.* Univ. North Dakota, Grand Forks, ND,

Domsch, K.H. 1986. Influence of management of microbial communities on soil. pp. 355–367. In: V. Jensen, A. Kjoller and L.H. Sorensen (eds.). *Microbial Communities in Soil*, Elsevier Publ., Barking, Essex (UK).

Dragovich, D. and Patterson, J. 1995. Condition of rehabilitated coal-mines in the Hunter Valley, Australia., *Land Degradation & Rehabilitation* 6: 29–39.

Etherington , J.R. 1982. *Environment and Plant Ecology* (2nd ed.). John Wiley & Sons, Chichester (UK).

Feiss, J.W. 1965. Coal-mine spoil reclamation: scientific planning for regional beauty and prosperity. pp. 12–23. In: G. Davis, W.W. Ward and R.E. McDermott (eds.). *Proc. Coal-Mine Spoils Reclamation Symp.*, Penn. State Univ. School of Forest Resources, State College, PA.

Fenton, M.R. 1973. Landscape design principles for strip-mine restoration. pp. 485–595. In: R.J. Hutnick and G. Davis (eds.). *Ecology and Reclamation of Devastated Land, vol. 2,* Gordon and Breach, New York, NY.

Ferchau, H.A. 1988. Rehabilitating ecosystems at high altitudes. pp. 193–209. In: J. Cairns, Jr. (ed.). *Rehabilitating Damaged Ecosystems*, CRC Press, Boca Raton, FL.

Fischer, N.T. 1986. Vegetation ecology, sample adequacy, and the determination of reclamation success. pp. 182–215. In: C.C. Reith and L.D. Potter (eds.). *Principles and Methods of Reclamation Science: with Case Studies from the Arid Southwest*, Univ. New Mexico Press, Albuquerque, NM.

Fitter, A.H. and Bradshaw, A.D. 1974. Root penetration of *Lolium perenne* on colliery shale in response to reclamation treatments. *J. Appl. Ecol.*, 19: 609–616.

Flege, A.S. 1993. An examination of black locust (*Robinia pseudoacacia*) as a suitable species for stabilization of clay-mantled hillslopes at Cincinnati, Ohio. MA thesis, Dept. Geog., Univ. Cincinnati, Cincinnati,, OH.

Flege, A.S. 1996. Unpublished research.

Fuller, W.W. 1986. The selection of species for revegetation. pp. 109–133. In: *Principles & Methods of Reclamation Science.* Univ. New Mexico Press, Albuquerque, NM,

Funk, D.T. 1973. Growth and development of alder plantings on Ohio strip-mine banks. pp. 483–492. In: R.J. Hutnik and G. Davis (eds.). *Ecology and Reclamation of Devastated Land, vol. 2*, Gordon and Breach, New York, NY.

Galston, A.W., Davies, P.J. and Satter, R.L. 1980. *The Life of the Green Plant.* Prentice-Hall, Englewood Cliffs, NJ.

Gentcheva-Kostadinova, Sv. and Haigh, M.J. 1988. Land reclamation and afforestation research on the coal-mine-disturbed lands of Bulgaria. *Land Use Policy*, 5(1). 94–102.

Gentcheva-Kostadinova, Sv., Zheleva, E., Pectrova, R. and Haigh, M.J. 1994: Soil constraints affecting the forest-biological recultivation of coal-mine spoil banks in Bulgaria. *International Journal of Surface Mining, Reclamation and Environment* 8 (2): 47–54.

Geyer, W.A. 1973. Tree species performance on Kansas coal spoil. pp. 81–90. In: R.J. Hutnik and G. Davis (eds.)(. *Ecology and Reclamation of Devastated Land*, vol. 1, Gordon and Breach, New York, NY.

Gleason, H.A. and Cronquist, A. 1964. *A Natural Geography of Plants.* Columbia Univ. Press, New York, NY.

Gray, D.H. 1995. Influence of vegetation on the stability of slopes. pp. 1–24. In: D.H. Barker (ed.), *Vegetation and Slopes: Stabilization, Protection and Ecology*, Inst. Civil Eng., London.

Gray, D.H. and Leiser, A.T. 1982. *Biotechnical Slope Protection and Erosion Control.* Van Nostrand Reinhold Co., New York, NY.

Gray, D.H. and Megahan, W.F. 1980. Forest vegetation removal and slope stability in the Idaho Batholith. *United States Department of Agriculture, Forest Service Research Note* INT-271: 23pp.

Greenway, D.R. 1987. Vegetation and slope stability. pp. 187–230. In: M.G. Anderson and K.S. Richards (eds.). *Slope Stability—Geotechnical Engineering and Geomorphology*, John Wiley & Sons, Chichester (UK).

Grene, S. 1978. Root deformations reduce root growth and stability. pp. 150–155. In: *Proc., Root Form of Planted Trees Symp.* Ministry of Forests/Can. For. Serv. Jt. Rept. Nr. 8.

Griffiths, D.G. 1992. Land reclamation in Wales: aims and strategy of the Welsh Development Agency. *Land Degradation & Rehabilitation* 3: 157–159.

Grime, J.P. 1979. *Plant Strategies and Vegetation Processes.* John Wiley & Sons, Chichester (UK).

Hackett, C. 1968. Quantitative aspects of the growth of cereal root systems. pp. 134–135. In: W.J. Whittington (ed.). *Root Growth.* Plenum Press, New York, NY.

Haering, K., Daniels, W.L. and Roberts, J.A. 1993. Changes in mine soil properties resulting from overburden weathering. *J. Environmental Quality* 22: 194–200.

Häge, K., Drebenstedt, C. and Angelov, E. 1996. Landscaping and ecology in the lignite mining areas of Maritza-East, Bulgaria. *Water, Air, and Soil Pollution* 91: 135–144.

Haigh, M.J. 1992. Degradation of 'reclaimed' lands previously disturbed by coal mining in Wales: causes and remedies. *Land Degradation and Rehabilitation* 3: 169–180.

Hansen, R.P. 1976. Statutory and regulatory aspects of mined land reclamation. In: pp. 1–7. K.C. Vories (ed.). *Reclamation of Western Surface Mined Lands*, Ecology Consultants, Inc.,

Harwood, G.D. and Thames, J.L. 1988. Design and planning considerations in surface-mined land. pp. 137–158. In: L.R. Hossner (ed.). *Reclamation of Surface-mined Lands*, CRC Pres, Inc., Boca Raton, FL.

Hatton, T.J. and West, N.E. 1987. Early seral trends in plant community diversity on a recontoured surface mine. *Vegetatio* 73: 21–29.

Haynes, R.J. 1985. Natural vegetation development on a 43-year-old surface-mined site in Perry County, Illinois. pp. 437–466. In: *Proc. 1985 Symp. Surface Mining, Hydrology, Sedimentology, and Reclamation.* Coll. Eng., Univ. Kentucky, Lexington, KY.

Hektner, M.M., Reed, L.J., Popenoe, J.H., Mastroguiseppe, R.J., Vezie, D.J. and Sugihara, N.B. 1981. A review of the revegetation treatments used in Redwood National Park: 1977 to present. pp. 70–71. In: *Proc. Symp. Watershed Rehabilitation in Redwood National Park and Other Coastal Areas.* Nat'l Park Serv., Arcata Ca.,

Helliwell, D.R. 1995. Rooting habits and moisture requirements of trees and other vegetation. pp. 260–263. In: D. Barker (ed.). *Vegetation and Slopes: Stabilisation, Protection and Ecology,* Inst. Civil Eng., London.

Hellum, A.K. 1978. The growth of planted spruce in Alberta. pp. 191–196. In: E. Van Eerden and J.M. Kinghorn (eds.). *Root Form of Planted Trees*, BC Ministry of Forests, Victoria, BC, Canada.

Hodgson, D.R. and Buckley, G.P. 1975. A practical approach towards the establishment of trees and shrubs on pulverized fuel ash. pp. 305–330. In: M.J. Chadwick and G.T. Goodman (eds.). *The Ecology of Resource Degradation and Renewal*, John Wiley & Sons, New York, NY.

Hoffard, W.H. and Anderson, R.L. 1982. *A Guide to Common Insects, Diseases, and Other Problems of Black Locust.* US For. Serv., Atlanta, GA.

Howell, J., Clark, J., Lawrence, C. and Sunwar, I. 1991. *Vegetation Structures for Stabilising Highway Slopes:* a Manual for Nepal. UK/Nepal Dept. of Roads, Kathmandu.

Hower, J.M., Barnhisel, R.I. and Hopkins, T.C. 1992. Physical and chemical properties of potential topsoil substitutes in Kentucky oil shale mining reclamation. *J. Environmental Quality* 21: 502–508.

Hsuang, Y.H. 1983. *Stability Analysis of Earth Slopes.* Van Nostrand Reinhold Co., New York.

Hughes, C.E. and Styles, B.T. 1989. The benefits and risks of woody legume introductions. pp. 505–531. In: C.H. Stirton and J.L. Zarucchi (eds.). *Advances in Legume Biology*, Missouri Botanical Garden, St. Louis, MO. Monographs in Systematic Botany 29.

Jefferies, R.A., Bradshaw, A.D. and Putwain, P.D. 1981. Growth, nitrogen accumulation and nitrogen transfer by legume species established on mine spoils. *J. Appl. Ecol.*, 18: 945–956.

Jeffrey, D.W. 1987. *Soil Plant Relationships: an Ecological Approach*. Croom Helm Ltd., London.

Jencks, E.M., Tyron, E.H. and Contri, M. 1982. Accumulation of nitrogen in mine spoils seeded to black locust. *Soil Sci. Soc. Amer. J.*, 46: 1290–1293.

Jenny, H. 1941. *Factors of Soil Formation*. McGraw-Hill, New York, NY.

Jenny, H. 1980. *The Soil Resource: Origin and Behavior*. Springer-Verlag, New York, NY.

Jewell, K.E. 1978. Soil forming factors and yellow-poplar seedling growth on eastern Ohio minesoils. Ph.D. diss., Ohio State Univ., Columbus, Ohio.

Jha, A.K. and Singh, J.S. 1995. Rehabilitation of mine spoils with particular reference to multipurpose trees. 237–249. In: P. Singh, P.S. Pathak and M.M. Roy (eds.). *Agro forestry Systems for Sustainable Land Use*, SP Sci. Publ., Lebanon, NH.

Jobling, J. 1983. *Treatment of deep-mine colliery spoil. In: Reclamation of Mineral Workings to Forestry*. For. Comm. R & D Paper 13. HMSO, London.

Jobling, J. and Stevens, F.R.W. 1980. *Establishment of Trees on Regraded Colliery Spoil Heaps*. For. Comm. Occasional Paper 7. HMSO, London.

Jochimsen, M. 1986. Begrünungversuche auf Bergematerial der Halde Evald/Herten. *Ökologie* 14: 223–228.

Johnson, C.D. and Skousen, J.G. 1995. Minesoil properties of 15 abandoned mine land sites in West Virginia. *J. Environmental Quality* 24: 635–643.

Joost, R.E., Olsen, F.J. and Jones, J.H. 1987. Revegetation and minesoil development of coal refuse amended with sewage sludge and limestone. *Journal of Environmental Quality*, 16 (1), 65–68.

Juwarkar, A.S., Thawale, P.R., Mowade, S., Srivastava, S., Deshbh, P.B. and Juwarkar, A. 1994. Reclamation of coal-mine spoil dump through integrated biotechnological approach pp. 339–347. In: *Proc. Int.. Symp. Impact of Mining on Environment, Nagpur, India*. vol. 2, Prints, New Delhi.

Katzur J. and Haubold-Rosar, M. 1996. Amelioration and reforestation of sulfurous mine soils in Lusatia (Eastern Germany). *Water, Air & Soil Pollution* 91 (1–2): 17–32.

Kay, B.L. 1978. *Mulches for Erosion Control and Plant Establishment on Disturbed Sites*. Agronomy Prog. Rept. Nr. 87. Univ. Calif. Agric. Exper. Station, Davis, CA.

Kellogg, L.F. 1936. *Growth of Black Locust*. For. Serv., Columbus, OH. CSFS Station Note No. 6.

Kelly, A.W. Jr. 1980. The role of West Virginia's Division of Forestry. pp. 63–64. In: *Trees for Reclamation*. US For. Serv., Broomall, PA, Gen. Tech. Rept. NE-61.

Keresztesi, B. 1968. Morphological characteristics of the Robinia root system on different sites of the Great Hungarian Plain. pp. 86–95. In: USSR Academy of Sciences (ed.), *Methods of Productivity Studies in Root Systems and Rhizosphere Organisms, International Symposium, Moscow 28 August–12 September 1968*. Leningrad: Nauka.

Keresztesi, B. 1980. *The Black Locust*. Akadémiai Kiado, Budapest, Hungary.

Killham, K. 1994. *Soil Ecology*. Cambridge Univ. Press, Cambridge (UK).

Knabe, W. 1973b. Investigations of soils and tree growth on five deep-mine refuse piles in the hard-coal region of the Ruhr. pp. 273–394. In: R.J. Hutnik and G. Davis (eds.). *Ecology and Reclamation of Devastated Land*, vol. 1, Gordon and Breach, New York, NY.

Kozlowski, T.T., Kramer, P.S. and Pallardy, S.G. 1991. *Physiological Ecology of Woody Plants*. Acad. Press, San Diego, CA.

Krause, R.R. 1973. Predicting mined-land soil. In: R.J. Hutnik and G. Davis (eds.). *Ecology and Reclamation of Devastated Land*, vol. 1, pp. 121–131. Gordon and Breach, New York, NY.

Kurz, H. 1923. Hydrogen-ion concentration in relation to ecological factors. *Botanical Gazette* 76: 1–29.

Leaf, A.L., Rathakette, P. and Solan, F.M. 1978. Nursery seedling quality in relation to plantation performance. In: E. Van Eerden and J.M. Kinghorn (eds.). *Root Form of Planted Trees*. BC Ministry of Forest/Can. For Serv., Victoria, BC, Canada, Joint Rept. Nr. 8, pp. 45–50.

Leirós, M.C., Gil-Sotres, F., Tresar-Cepeda, M.C., Saá, A. and Seoane, S. 1996. Soil recovery at the Mierama opencast lignite mine in northwest Spain: a comparison of the effectiveness of cattle slurry and inorganic fertilizer. *Water, Air, and Soil Pollution* 91 (1–2): 109–124.

Leopold, D.J. and Wali, M.K. 1992. The rehabilitation of forest ecosystems in the eastern United States and Canada. pp. 187–231. In: M.K. Wali (ed.). *Ecosystem Rehabilitation, Vol. 2: Ecosystem Analysis and Synthesis*, SPB Acad. Publ., the Hague.

Limstrom, G.A. 1948. Extent Character and Forestation Possibilities of Land Stripped for Coal in the Central States. US For. Serv. Tech. paper CS-109.

Limstrom, G.A. 1960. *Forestation of Strip-mined Land in the Central States.* United States Department of Agriculture, Forest Service, Washington, DC.

Limstrom, G.A. 1964. Revegetation of Ohio's strip-mined land. *Ohio J. Sci.* 64: 15–16.

Limstrom, G.A. and Merz, R.W. 1949. *Rehabilitation of Lands Stripped for Coal in Ohio.* Central States For. Exper. Station, US Dept. Agric., Wash., DC, Tech. paper 113.

Long, J.N. 1978. Root system form and its relationship to growth in young planted conifers. In: E. Van Eeorden and J.M. Kinghorn (eds.). *Root Form of Planted Trees. BC Ministry of Forests/Can. For Serv., Victoria, BC, Canada, Joint Rept.* No. 8: 222–234.

Lyle, E.S. 1976. Grass, legume and tree establishment on Alabama coal surface mines. pp. 12–19. In: *Proc., Conf. Forestation of Disturbed Surface Areas.* US For. Serv., Wash., DC.

MacMahon, J.A. 1981. Successional processes: comparisons among biomes with special reference to probable roles of and influences on animals. pp. 277–304. In: D.C. West, H.H. Shugart and D.B. Bodkin (eds.). *Forest Succession: Concepts and Application*, Springer-Verlag, New York, NY.

Macyk, T.M. 1985. Reclamation research in the foothills/mountain regions of Alberta: a case study. pp. 13–17. In: *Proc. Second Ann. Meeting Amer. Soc. Surface Mining and Reclamation.* Amer. Soc. Surface Mining and Reclamation, Princeton, WV.

Maiti, D.H., Maiti, S.K. and Banerjee, S.P. 1994. Revegetation of coal-mine spoils by application of domestic raw sewage. pp. 369–381. In: *Proc. Int. Symp. Impact of Mining on Environment, Nagpur, India.* Prints, New Delhi.

Marx, D.H. 1980. Role of mycorrhizae in forestation of surface mines. pp. 109–116. In: *Trees for Reclamation.* US For. Serv., Broomall, PA, Gen. Tech. Rept. NE-61.

May, J. T. 1986. Complexities of reclamation. pp. 3–11. In: K.A. Utz (ed.). *Proc., Cong. Forestation of Disturbed Surface Areas*, US For. Serv., Wash., DC.

McCormick, L.H. and Borden, F.Y. 1973. Percolate from spoils treated with sewage effluent and sludge. pp. 239–247. In: R.J. Hutnik and G. Davis (eds.). *Ecology and Reclamation of Devastated Land*, vol. 2, Gordon and Breach, New York, NY.

McSweeney, K. and Jansen, I.J. 1984. Soil structure and associated rooting behaviour in minesoils. *Soil Sci. Soc Amer. J.*, 48: 607–612.

Medvick, C. 1980. Tree planting experience in the Eastern Interior Coal Province. pp. 65–80. In: *Trees for Reclamation.* US For. Serv., Broomall, PA, Gen. Tech. Rept. NE-61,

Miles, J. 1985. The pedogenic effects of different species and vegetation types and the implications of succession. *J. Soil Science* 36: 571–584.

Moffat, A.J., Bending, N.A.D., and Roberts, C.J. 1991. The use of sewage sludge as a fertiliser in the afforestation of opencast coal spoils in South Wales. pp. 391–392. In: M.C.R. Davies (ed.), *Land reclamation: an End to Dereliction?* Univ. Wales, Cardiff.

Moffat, A. J. and McNeill, J. 1995. *Reclaiming Disturbed Land for Forestry.* For. Bull. 110. HMSO, London.

Moffat, A.S. 1996. Biodiversity is a boon to ecosystems, not species. *Science*, 271, (March): 1497.

Moody, C.W. and Kimbrell, D.T. 1980. Trees for reclamation in the eastern United States. pp. 7–8. In: *Trees for Reclamation.* US For. Serv., Broomall, PA, Gen. Tech. Rept. NE-61.

Moore, D.P. 1974. Physiological effects of pH on roots. pp. 5–16. In: *The Plant Root and Its Environment.* Virginia Polytech. Inst., Blacksburg, VA.

Morrison, D. 1987. Landscape restoration in response to previous disturbance. pp. 159–172. In: M. Turner (ed.). *Landscape Heterogeneity and Disturbance.* Springer-Verlag, New York, NY.

Murphy, H.E. and Brace, A.C. Jr. 1976. Potentials in the use of spoil banks. pp. 66–70. In: K.A. Utz (ed.). *Proc., Conf. Forestation of Disturbed Surface Areas,* US For. Serv., Wash., DC.

Nair, P.K.R. 1990. *The Prospects for Agroforestry in the Tropics.* Int. Bank Reconstruction and Development, Wash., DC, World Bank Tech. Paper 131.

Narten, P.F., Litner, S.F., Allingham, J.W., Foster, L., Larsen, D.M. and McWreath, H.C. 1983. *Reclamation of Mined Lands in the Western Coal Region.* US Geol. Survey, Wash., DC, Circular 872.

Neumann, U. 1973. Succession of soil fauna in afforested spoil banks of the brown-coal mining district of Cologne. pp. 335–348. In Hutnik, R.J., and G. Davis eds., *Ecology and Reclamation of Devastated Land.* Gordon and Breach, New York.

Norland, M.R. and Veith, D.L. 1991. *Soil Characterization and Soil Amendment Use on Coal Surface Mined Lands: an Annotated Bibliography.* Dept. of the Interior, Wash., DC, Information Circular 9285.

NRC. 1981. *Surface Mining. Soil, Coal, and Society.* Nat Res. Council, Wash., DC.

Olah, P. 1990. Analysis of tree establishment on stripmined land in southeastern Ohio. MA thesis, Ohio State Univ., Columbus, OH.

Olschowy, G. 1973. Landscape planning on an ecological basis. pp. 477–484. In: R.J. Hutnik and G. Davis (eds.). *Ecology and Reclamation of Devastated Land,* vol. 2, Gordon and Breach, New York, NY.

Packer, P.E. 1974. *Rehabilitation Potentials and Limitations of Surface-mined Land in the Northern Great Plains.* US For. Serv., Ogden, Utah, Gen. Tech. Rept, INT-14.

Palmer, J.P. 1990. Nutrient cycling: the key to reclamation success? pp. 27–36. In: J.C. Chambers and G.L. Wade (eds.). *Evaluating Reclamation Success: the Ecological Consideration.* US For. Serv., NE For. Exper. Station, Broomall, PA, Gen. Tech. Rept. NE-164.

Palmer, J.P. and Chadwick, M.J. 1985. Factors affecting the accumulation of nitrogen in colliery spoil. *J. Appl. Ecol.,* 22: 249–257.

Pandey, A.N. and Singh, J.S. 1984. Mechanism of ecosystems recovery; a case study from Kumaun Himalaya. *Reclamation and Revegetation Research* 3: 272–292.

Parmenter, R.R. and McMahon, J.A.. 1990. Faunal community development on disturbed lands: an indicator of reclamation success. pp. 73–89. In: J.C. Chambers and G.L. Wade. (eds.). *Evaluating Reclamation Success: the Ecological Consideration.* US For. Serv., NE For. Exper. Station, Broomall, PA, Gen. Tech. Rept. NE-164.

Pasch, R.N. 1984. An ageing study on two Wyoming spoils. pp. 55–74. In: M.G. Berg (ed.). *Proc., Conf. Overburden Requirements for Successful Revegetation,* U.S.G.P.O. Wash. DC. Denver, Co.

Paul, E.A. and Clark, F.E. 1996. *Soil Microbiology and Biochemistry.* Acad. Press, San Diego, CA.

Pearcy, J.N. 1991. Performance of selected *Alnus glutinosa* genotypes in two mine resoiling materials: glasshouse study of early growth characteristics and their heritability. MA thesis, Ohio State Univ., Columbus, OH.

Perkins, P.V. and Vann, A.R. 1995. The early performance of deciduous trees grown in acidic minespoil ameliorated with pulverized fuel ash. *Land Degradation and Rehabilitation* 6: 57–67.

Pietsch, W.H.O. 1996. Recolonization and development of vegetation on mine spoils following brown coal mining in Lusatia. *Water, Air, and Soil Pollution* 1: 17–22.

Piha, M.I., Vallack, H.W., Reeler, B.M. and Michael, N. 1995. A low input approach to vegetation establishment on mine and coal ash wastes in semi-arid regions: tin mine tailings in Zimbabwe. *J. Appl. Ecol.,* 32: 372–381.

Plass, W.T. 1968. *Tree Survival and Growth on Fescue-covered Spoil Banks.* US For. Serv., Broomall, PA, USFS Res. Note NE-90.

Plass, W.T. 1976. Direct seeding of trees and shrubs on surface-mined lands in West Virginia. pp. 32–46. In: K.A. Utz (ed.). *Proc., Conf. Forestation of Disturbed Surface Areas*, US For. Serv., Wash., DC.

Plass, W.T. 1978. Use of mulch and chemical stabilizers for land reclamation in the eastern United States. pp. 329–337. In: F.W. Schaller and P. Sutton (eds.). *Reclamation of Drastically Disturbed Lands*, Amer. Soc. Agron., Madison, WI.

Plass, W.T. 1979. The establishment and maintenance of vegetation on minesoils in the eastern United States. pp. 431–437. In Wali, M.K., (ed.), *Ecology and Coal Resource Development*, vol. 1. New York: Pergamon.

Ponder, F. Jr. 1980. *Rabbits and Grasshoppers: Vectors for Endomycorrhizal Fungi on New Coal Mine Spoil*. US For. Serv., North Central For. Exper. Station, St. Paul, MINN, NCFES Res. Note NC-250.

Potter, H.S., Weitzman, S. and Trimble, G.R. 1951. *Reforestation of Strip-mined Lands in West Virginia*. US For. Serv., Broomall, PA, NE Exper. Station Paper 43.

Pulford, I.D. 1991. A review of methods to control acid generation in pyritic coal-mine waste. pp. 269–278. In: M.C.R. Davies (ed.). *Lands Reclamation: an End to Dereliction?*, Coll. Cardiff, Univ. Wales (UK).

Reiss, I.H. 1973. Strip-mine reclamation: challenges, planning and concepts. *Mining Cong. J.*, 59: 41–45.

Reith, C.C. and Potter, L.D. 1986. *Principles and Methods of Reclamation Science: with Case Studies from the Arid Southwest*. Albuquerque, NM., Univ. of New Mexico Press.

Richards, G.I. Palmer, J.P. and Barratt, P.A. 1993. *The Reclamation of Former Coal-Mines and Steelworks*. Elsevier, Amsterdam.

Richardson, J.A. 1984. An early reclamation of colliery waste heaps re-examined. *Scottish Forestry* 38: 115–121.

Richardson, J.R., Burn, I. and Craig, G. 1987. Long-term prospects for farm grassland and woodland on reclaimed coal spoil sites. *Landscape and Urban Planning* 14: 131–142.

Riestenberg, M.M. and Sovonick-Dunford, S. 1983. The role of woody vegetation in stabilizing slopes in the Cincinnati area. *Bull. Geol. Soc. Amer.*, 94: 506–518.

Riley, C.V. 1973. Chemical alterations of strip-mine spoil by furrow grading: revegetation success. pp. 315–329. In: R.J. Hutnik and G. Davis (eds.). *Ecology and Reclamation of Devastated Land*, vol. 2, Gordon and Breach, New York, NY.

Ringe, J.M. and Graves, D.R. 1985. Economic considerations in establishing European alder in herbaceous cover on surface-mined land. pp. 417–420. In: *Proc. 1985 Symp Surface Mining, Hydrology, Sedimentology, and Reclamation*. Coll. Eng., Univ. Kentucky, Lexington, KY.

Robertson, S.D. and Wittwer, R.F. 1983. The effects of slit-applied fertilizer treatments on growth and survival of sycamore and cottonwood planted on minespoil. pp. 483–488. In: *Proc. 1985 Symp. Surface Mining, Hydrology, Sedimentology, and Reclamation*. Coll. Eng., Univ. Kentucky, Lexington, KY.

Rogers, P.N. 1992. Investigation of overburden dumps at Stockton coal-mines carried out in December, 1991. Stockton Coal-Mines, Westport, New Zealand (unpubl. rept.).

Ross, C. and Williams, P. 1995/1996. *Pakihi Restoration Trial, Giles Creek Coal-Mine Site, Reefton*. Landcare Research, Christchurch, New Zealand.

Ross, C.W., Davis, M. and Langer, L. 1995. Rainforest restoration after gold and coal mining in Westland, New Zealand. pp. 331–344. In: *Proc. 20th Ann. Environ. Workshop: Managing Environmental Impacts—Policy and Practice*. Mineral Council Australia, Dickson, Australia.

Rovira, A.D., Elliott, L.F. and Cook, R.J. 1990. The impact of cropping systems on rhizosphere organisms affecting plant health. pp. 289–345. In: J.M. Lynch (ed.). *The Rhizosphere*, John Wiley & Sons, New York, NY.

Ruffner, J.D. and Steiner, W.W. 1973. Evaluation of plants for use on critical sites. pp. 3–12. In: R.J. Hutnik and G. Davis (eds.). *Ecology and Reclamation of Devastated Land*, vol. 2: Gordon and Breach, New York, NY.

Russell, M.J. and Roberts, B.R. 1986. Revegetation of coal-mine spoil using pasture on the Darling Downs of Queensland, Australia. *Reclamation and Revegetation Research* 5: 509–519.

Russell, E.W. 1973. *Soil Conditions and Plant Growth* (10th ed.). Longman, London.

Saiz de Omeñaca, J., Solar, M., Gómez, D. Marcos, J., Saiz de Omeñaca, M.G. and Saiz de Omeñaca, J.A. 1994. Opencast mine Reclamation in Spain: a review with particular reference to the Montehano mine (Cantabria, northern Spain). In *Proc. Int. Symp. Impact of Mining on Environment, Nagpur, India.* Preprints, New Delhi, pp. 355–368.

Samuel, P. 1991. Revegetation of reclaimed land. In: M.C.R. Davies (ed.), *Land Reclamation: an End to Dereliction?*, pp. 366–376. Coll. Cardiff, Univ. Wales (UK)

Schiechtl, H.M. 1980. *Bioengineering for Land Reclamation and Conservation* (tr. Sicherungsarbeiten in Landschaftsbau, 1973). Univ. Alberta Press, Edmonton, Alberta, Canada.

Schlätzer, G. 1973. Some experiences with various species in Danish reclamation work. pp. 33–64. In: R.J. Hutnik and G. Davis (eds.). *Ecology and Reclamation of Devastated Land,* vol. 2, Gordon and Breach, New York, NY.

Schoenholtz, S.H. and J.A. Burger 1984. Influence of cultural treatments on survival and growth of pines on strip-mined sites. *Reclamation, Revegetation, Restoration* 3: 223–237.

Schoenholtz, S.H., Burger, U.A. and Kreh, R.E. 1992. Fertilizer and organic amendment effects on mine soil properties and revegetation success. *Soil Sci. Soc. Amer. J.,* 56: 1177–1184.

Schultz, R.C. and Thompson, J.R. 1990. Nursery practices that improve hardwood seeding root morphology. *Tree Planters' Notes* 14 (3): 21–31.

SCS. 1978. *Plant Performance on Surface Coal Mine Spoil in the Eastern United States.* Soil Cons. Serv., USDA, Wash., DC, SCS-TP-155, p. 76.

Scullion, J. 1992. Re-establishing life in restored topsoils. *Land Degradation and Rehabilitation* 3(3), 161–168.

Scullion, J. and Malinovszky, K.M. 1995. Soil factors affecting tree growth on former opencast coal land. *Land Degradation & Rehabilitation* 6: 239–249.

Segaran, S., Dojack, J.C. and Rathwell, R.K. 1978. Assessment of root deformities of jack pine (*Pinus banksiana* Lamb) planted in southeastern Manitoba. pp. 197–200. In: E. Van Eerden and J.M. Kinghorn (eds.). *Root Form of Planted Trees.* Ministry of Forests/Can. For. Serv., Victoria, BC, Canada, Joint Rept. Nr. 8.

Senkayi, A.L. and Dixon, J.B. 1988. Mineralogical consideration in reclamation of surface-mined lands. pp. 105–124. In: L.R. Hossner (ed.). *Reclamation of Surface-mined Lands,* CRC Press, Inc., Boca Raton, FL.

Shubert, L.E. and Starks, T.L. 1985. Diagnostic aspects of algal ecology in disturbed lands. pp. 83–106. In: R.L. Tate III and D.A. Klein (eds.). *Soil Reclamation Processes: Microbiological Analyses and Applications,* Marcel Dekker, New York, NY.

Simcock, R. 1993. Reclamation of Aggregate Mines in the Manawatu Rangetikei and Horrowhenua Districts of New Zealand. Ph.D. thesis, Massey Univ., New Zaland.

Simcock, R. 1996. Pers. comm. Landscare Research, Christchurch, New Zealand.

Smith, R.M. and Sobek, A.A. 1978. Physical and chemical properties of overburdens, spoils, wastes, and new soils. pp. 149–172. In: F.W. Shaller and P. Sutton (eds.). *Reclamation of Drastically Disturbed Lands,* Amer. Soc. Agron., Madison, WI.

Smith, W.D. 1980. Has anyone noticed that trees are not being planted any longer? pp. 53–55. In: *Trees for Reclamation.* US For. Serv., Broomall, PA, USFS Gen. Tech. Rept. NE-61.

Smolik, D. 1988. Practical results of the reclamation of technogenous landscape in the Ostrava-Karviná coal district. *Studia Geologica* 1: 68–78.

Smyk, B.C., Rózycki, and Barabasz, W. 1986. The effect of fertilization with N and NPK on the microbiocenoses and the ecological stability of selected mountain grassy ecosystems. pp. 229–243. In: Jensen and A.V. Kjoller (eds.). *Microbial Communities in Soil,* Elsevier, Barking, Essex (UK).

Söderström, B.B., Båath, E. and Lundgren, B. 1983. Decreases in soil microbial activity and biomasses owing to nitrogen amendments. *Can. J. Microbiol.,* 298: 1500–1506.

Stroempl, G. 1990. Deeper planting of seedlings and transplants increases plantation survival. *Tree Planters' Notes* 41 (2): 17–21.

Struthers, P.H. 1965. Rapid spoil weathering and soil genesis. pp. 86–90. In: G. Davis, W.W. Ward and R.E. McDermott (eds.). *Coal Mine Spoil Reclamation: Scientific Planning for Regional Beauty and Prosperity*, Pennsylvania State Univ., School of Forest Res., State Coll., PA.

Struthers P.H. and Vimmerstedt, J.P. 1965. Stripmine reclamation. *Ohio Report* 50 (1): 8–9.

Strzyszcz, Z. 1993. Biologische Revkultivierung der Bergehalden des oberschlesischen Steinkohlebeckens. *Mensuration, Photogrammétrie, Génie Rural*, April; 276–279.

Strzyszcz, Z. 1996. Recultivation and landscaping in areas after brown-coal mining in middle-east European countries. *Water, Air and Soil Pollution* 91: 145–157.

Tan, K.H. 1993. *Principles of Soil Chemistry* (2nd ed.). Marcel Dekker, New York, NY.

Tate, R.L. III. 1985. Micro-organisms, ecosystem disturbance and soil-formation processes. pp. 1–33. In: R. Tate and D. Klein (eds.). *Soil Reclamation Processes: Microbiological Analyses and Applications*, Marcel Dekker, New York, NY.

Teasdale, J.B. 1983. Forestry Commission and local authority liaison; the Minerals Act. In: *Reclamation of Mineral Workings to Forestry*. For. Comm. R & D Paper 132. HMSO, London.

Thornberg, A.A. 1982. *Plant materials for use on surface-mined land, in arid and semi-arid regions*. US. Envir. Prot. Agency, Cincinnati, OH. SCS-TP-157, EPA-600/7–79–134.

Torbert, J.L. and J.A. Burger (1990). Tree survival and growth on graded and ungraded minesoil. *Tree Planters' Notes* 41 (2): 3–5.

Tilman, D. 1996. Biodiversity: population versus ecosystem stability. *Ecology* 77 (2): 350.

Tinus, R.W. 1978. Root form. In: *Root Form of Planted Trees*. Ministry of Forests/Canadia Forestry Service, Joint Report 8–1, Victoria, B.C.

Torbert, J.L, Burger, J.A. and Probert, T. 1995. Evaluation of techniques to improve white pine establishment on an Appalachian minesoil. *J. Environmental Quality* 24: 869–873.

Toy, T.J. and Hadley, R.F. 1987. *Geomorphology and Reclamation of Disturbed Lands*. Acad. Press, Orlando, FL.

Tueller, P.T. 1990. Landscape ecology and reclamation success. In: *Evaluating Reclamation Success: the Ecological Consideration*. US For Serv., NE For. Exper. Station, Broomall, PA, Gen. Tech. Rept. NE-164.

Van Eerden, Every and Kinghorn, J.M. 1978. *Root Form of Planted Trees*. BC Ministry of Forests, Victoria, BC, Canada, Joint Rept. 8.

Veith, D.L., Bickel, K.L., Hopper, R.W. and Norland, M.R. 1985. *Literature on the Revegetation of Coal-Mined Lands: an Annotated Bibliography*. US Dept. of the Interior, Wash., DC, Information Circular 9048.

Veziroglu, T.N. 1984. *The Biosphere: Problems and Solutions*. Elsevier, Amsterdam.

Vinczeffy, I. 1995. Possibilities offered by grasslands, pp. 21–45. In: *Agroforestry Systems for Sustainable Land Use*. SP Sci. Publ., Lebanon, NH

Vogel, W.G. 1980. Revegetating surface-mined lands with herbaceous and woody species together. pp. 117–126. In: *Trees for Reclamation*. US For. Serv., Broomall, PA, USFS Gen. Tech, Rept. NE-61.

Vogel, W.G. 1981. *A Guide to Revegetating Coal Minespoils*. US For. Serv., Broomall, PA, Gen. Tech. Rept. NE-68.

Vogel, W.G. and Berg, W.A. 1973. Fertilizer and herbaceous cover influence establishment of direct-seeded black locust on coal-mine spoil. pp. 189–198. In: R.J. Hutnik and G. Davis (eds.). *Ecology and Reclamation of Devastated Land*, vol. 2: Gordon and Breach, New York, NY.

Vogel, W.G. and Gray, B. 1987. Will trees survive on topsoiled surface mines? pp. 301–305. In: *Proc. Symp. Surface Mining, Hydrology, Sedimentology, and Reclamation*. Univ. Kentucky, Lexington, KY.

Vogel, W.G., Richards, T.W. and Graves, D.H. 1984. Survival of northern red oak and white oak seedlings planted in tall fescue and black locust-crown vetch covers. In: *Proc., Conf.*

Better Reclamation with Trees. Amer. Soc. Surface Mining and Reclamation. Princeton, WV, pp. 33–41.

Wali, M.K. 1992. Ecology of the rehabilitation process. pp. 3–26. In: *Ecosystem Rehabilitation: Preamble to Sustainable Development*, vol. 1. *Policy Issues.* Acad, Publ., the Hague.

Watson, G.W. 1988. Organic mulch and grass influence on tree roots. *J. Arboriculture* 17 (8): 200–203.

Watson, J.W. 1994. Temperate taungya woodland establishment by direct seeding of trees under an arable crop. *Quart J. Forestry* 88: 199–204.

Webber, M.D., Pietz, R.I., Granate T.C. and Svoboda, M.L. 1994. Plant uptake of PCBs and other organic contaminants from sludge-treated coal refuse. *Journal of Environmental Quality* 23: 1019–1026.

Welsh, G.W. and Hutnik, R.J. 1973. *Growth of Tree seedlings and Use of Amendments on Bituminous Refuse.* Pennsylvania Dept. Envir. Res., Harrisburg, PA, Spec. Res. Rept. SR-92.

Whitford, W.G. and Elkins, N.Z. 1986. The importance of soil ecology and the ecosystem perspective in surface-mine reclamation. pp. 151–187. In: *Principles and Methods of Reclamation Science.* Univ. New Mexico Press, Albuquerque, NM.

Wilson, A.Y. 1983. Treatment of deep-mine colliery spoil. In: *Reclamation of Mineral Workings to Forestry.* For Comm. R & D paper 132. HMSO, London.

Woods, P. 1996. Pers. comm. ERA Ranger Mine, Jabiru, Northern Territory, Australia.

Wright, P.E. and Daniels, N.P. 1995. Vegetation of reclaimed colliery tips. pp. 287–290. In: D.H. Barker (ed.). *Vegetation and Slopes: Stabilisation, Protection and Ecology.* Inst. Civil Eng., London.

Wu, T.H., Riestenberg, M.M. and Flege, A.S. 1995. Root properties for design of slope stabilisation. pp. 52–59. In: D.H. Barker (ed.). *Vegetation and Slopes: Stabilisation, Protection and Ecology.*

Zarger, T.G., Curry, J.A. and Allen, J. 1969. Seeding of pine on coal spoil banks in the Tennessee Valley. pp. 509–523. In: R.J. Hutnik and G. Davis (eds.). *Ecology and Reclamation of Devastated Land*, vol. 1. Gordon and Breach, New York, NY.

APPENDIX

Some Tree and Shrub Species Used in Revegetating Mined Land

Latin Name	Common Name	UK	W-Europe	E-Europe	Russia	Ukraine	India	Australia	NZ	US/Canada
Acacia spp. [1,3]										
A. argyrophylla	wattle						X	X		X
A. auriculiformis	wattle							X	X	
A. calamifolia[4]	Australian babul						X			
A. catechu	wattle							X		
A. melanoxylon	Tasmanian blackwood						X			
A. nilotica	wattle						X			
A. saligna	Cyprus wattle							X		
A. sophorae[4]	Coastal wattle							X		
A. tortillis	wattle						X			
Acer spp.	maple									
A. campestre	field/hedge maple		X	X						X
A. pseudoplatanus	sycamore maple		X	X						
A. saccharinum	silver maple			X						X
Albizzia spp. [1]	silk tree; 'mimosa'									
A. lebbeck							X			
A. tortilis							X			
Alnus spp. [1,2]	alder	X	X	X						
A. glutinosa	European/black alder	X	X	X						X
A. incana	white/gray alder	X	X	X						X
Alstonia							X			
Azadirachta indica							X			

Contd.

Contd. **Appendix**

Latin Name	Common Name	UK	W-Europe	E-Europe	Russia	Ukraine	India	Australia	NZ	US/Canada
Betula spp.	birch	X	X	X		X				
B. papyrifera[5]	paper birch									X
B. pendula	European birch	X								
B. pubescens	pubescent white birch	X								
B. verrucosa	birch			X						
Casuarina equisetifolia[1]	casuarina						X			
Caragana arborescens [1, 4]	Siberian peashrub		X							X
Celtis spp.	hackberry									X
Chamaecytisus palmensis	tagasaste								X	
Cornus spp.	dogwood									
C. amomum[5]	silky dogwood									X
C. stolonifera	red osier dogwood									X
Cotoneaster spp.	cotoneaster	X								
Crateagus spp.	hawthorn	X	X	X						
C. oxyacantha[8]	hawthorn			X						
Cytisus scoparius[7]	Scotch broom	X								
Dalbergia sissoo[1]	sissoo						X			
Desmodium tortosum							X			
Eleagnus spp.	oleaster; olive									
E. angustifolia	Russian olive		X							
E. umbellata[7]	autumn olive		X							
Encelia farinosce	Buittlebush									X
Enchylaena tomentosa[4]	Rubybush	X						X		
Eucalyptus spp. [3, 4]	eucalyptus						X	X		
E. calophylla								X		
Fraxinus spp.	ash									

Contd.

Contd. **Appendix**

Latin Name	Common Name	UK	W-Europe	E-Europe	Russia	Ukraine	India	Australia	NZ	US/Canada
F. americana	white ash	X								X
F. excelsior	European ash	X	X							X •
F. pennsylvanica	green ash									
Hippophae rhamnoides	sea buckthorn	X	X	X	X					
Juglans nigra[9]	black walnut									X
Juniperus spp.	juniper									
J. scopulorum	Rocky Mtn. juniper									X
J. virginiana	eastern red cedar									X
Larix spp.	larch	X		X						X
L. decidua	European larch	X		X						
Leptospermum scoparium	manuka								X	
Leucaena leucocephala	leucena									
Ligustrum spp.	privet			X						
L. vulgare	privet			X						
Lonicera spp.	honeysuckle									
L. maackii[6]	Amur honeysuckle									X
L. tatarica	honeysuckle/woodbine									X
Maclura pomifera	Osage orange									X
Malaleuca spp.[4]	paperbark							X		
Malus spp.	apple					X				
M. baccata	Siberian crabapple									X
Melia azedarach	Pride-of-India; mahogany tree						X	X		
Nerium oleander	oleander							X		
Nitraria billiardierii	dillon bush							X		
Picea spp.	spruce			X	X					X

Contd.

Contd. **Appendix**

Latin Name	Common Name	UK	W-Europe	E-Europe	Russia	Ukraine	India	Australia	NZ	US/Canada
P. mariana	black spruce									X
P. pungens	Colorado blue spruce									X
Pinus spp.	pines	X	X	X	X	X		X	X	X
P. contorta	lodgepole pine	X	X		X					X
P. banksiana	jack pine									X
P. nigra	Austrian/Corsican pine	X								X
P. peuce	Balkan pine			X						
P. radiata[3]	pitch pine							X	X	
P. resinosa	red pine									X
P. sylvestris	Scots pine	X								X
Pongamia pinnata							X			
Populus spp.	poplar	X	X	X	X	X	X	X		X
P. balsomifera	taccamahac, balsam poplar		X							X
P. canadensis	Carolina poplar		X							X
P. candicans	balm of gilead		X							X
P. nigra italica	Lombardy poplar		X					X		
P. tremula	aspen	X	X	X	X	X	X	X	X	X
Prunus spp.	plum, cherry		X						X	
P. avium	cherry		X							
P. serotina	black cherry		X					X		
Pyrus spp.	pear		X							
Quercus spp.	oak		X	X						
Q. petraea	durmast oak		X	X						X
Q. robur	English/truffle/summer oak	X	X	X						X

Contd.

Contd. **Appendix**

Latin Name	Common Name	UK	W-Europe	E-Europe	Russia	Ukraine	India	Australia	NZ	US/Canada
Q. rubra (borealis)	red oak									X
Rhus spp.	sumac									
R. tribolata	skunkbush sumac		X							X
R. typhina[10]	staghorn sumac		X							
Robinia pseudoacacia[1]	black locust			X		X				X
Rosa spp.	native roses	X	X					X		
R. multiflora	multiflora rose	X	X							
Salix spp.	willow	X	X	X						X
S. alba	white willow	X								X
S. callianta	willow			X						
S. cordata	willow			X						
S. dasyclados	willow			X						
S. purpurea	willow			X						
Sambucus racemosa	European red elder		X							
Sesbania aegyptiaca[1]	sesbania						X			
Shepherdia argentea[4]	silver buffaloberry									X
Sorbus spp.	rowan, mountain ash	X	X	X						
S. aucuparia	rowan mountain ash	X	X	X						
S. x intermedia	rowan, mountain ash		X							
Spirea spp.	spirea		X	X						
S. billiardii	billiard spirea		X							
Tamarix spp.	tamarisk	X						X		
T. asphylla[4]	athel							X		
Taxus spp.	few		X							
T. baccata	English/common yew		X							
Tectona grandis	teak						X			

Contd.

Contd. **Appendix**

Latin Name	Common Name	UK	W-Europe	E-Europe	Russia	Ukraine	India	Australia	NZ	US/Canada
Tilia spp.	lime; linden									
T. platyphyllos	linden		X	X		X				
Ulmus spp.	elm			X		X				
U. effusa	common elm			X						
Ulex europaeus[1]	gorse	X								
Viburnum prunifolium[4]	blackhaw									X

Uses and characteristics
[1]Nitrogen-fixing species
[2]Acid sensitive
[3]Acid tolerant
[4]Arid or semi-arid climate
[5]Moist climate
[6]Invasive pest
[7]Salt tolerant
[8]Fine-textured soil
[9]Allelopathic
[10]Erosion control

13

Case Study: Geoecological Reconstruction of Coal-mine-Disturbed Lands in India's Jharia Coalfield

Rekha Ghosh

Abstract

Jharia, like most Indian coalfields, has been exploited by a combination of opencast and underground mining, much of it undertaken by private enterprises with no proper planning. Until recently, the duty of land reclamation has been neglected. Large tracts of waste dumps and abandoned open-pit quarries exist. Widespread damage has been done to the natural geoecological system, not the least of which has occurred in aquifers and watercourses. This chapter proposes a comprehensive land-use management planning (LUMP) model. Detailed schemes are presented for the concurrent reclamation of new opencast mines, greening of waste-tips, in filling and utilisation of quarries and subsidence hollows, and reconstruction of the hydrogeological system.

INTRODUCTION

Indian coal seams occur as continuous sedimentary units, generally with low dip. The seams outcrop on the surface at their up-dip end and drift to greater and greater depths down-dip. Opencast is the most common coal mining activity. The mines work the seams where they outcrop, proceeding down-dip up to the greatest practicable depth, leaving the remaining reserves to be exploited by underground mining. Thus most of the Indian coalfields combine opencast and underground mining. However, the cumulative effect of the two types of mining over the entire region is damage to the geoecosystem and generation of barren lands.

Land Reconstruction and Management Vol. 1, 2000, pp. 339–368.
ISSN 1389-2541
ISBN 90 5410 793 6
A.A. Balkema, Rotterdam, The Netherlands

These require runoff and erosion control treatment, reconstruction of the geoecosystem, including regeneration of water resources (above as well as below ground) and the regeneration of greenery over barren lands.

In the early days of the coal mining industry in India, mine owners sought to achieve maximum profits at minimum cost. They were not prepared to risk capital in developing an environmentally satisfactory land reclamation system in advance of taking their profits. Even health and safety issues were largely ignored. When mining terminated, due to an accident or non-availability of coal, the land was simply abandoned. This caused great ecological imbalance. The story is much the same for all the old-coal-mining areas in India that experienced unregulated private sector mining.

The situation improved somewhat after nationalisation of this industry in 1973, although in most cases the improvement was not up to the desired level. The most serious problem was that the cumulative damage caused by private mine owners until 1973 was colossal. The newly nationalised mining sector simply was not in a position to spend the huge amount of money required to reclaim the old abandoned mining areas, nor to mitigate their impacts—many of which are very severe.

Recently, the Ministry of Environment and Forests (MOEF) decreed that no mining project shall be allowed to start until it has obtained approval of its Environmental Management Plan (EMP). Land reclamation and land-use management plans form an important part of the EMP.

This chapter attempts to evaluate the potential for environmental regeneration in the Jharia coalfield, one of the oldest and most important coalfields in India. Procedures for geoecological reconstruction of coal-mine-disturbed lands are defined, which could be adapted for similar situations in other areas.

CASE STUDY AREA

The Jharia coalfield is the chief storehouse of prime coking coal in India. It has experienced more than a hundred years of mining. Mining has passed through the private sector, nationalised sector and MOEF control. Opencast mining has affected huge areas and is still active with modern mines exploiting seams up to a maximum depth of about 70 metres. During the 68-year period of its mining history, the area has lost 260 km of its natural surface watercourses and about 33 sq km (28.83%) of its green cover. Despite being a sedimentary terrain, it no longer has a continuous water table. Because of these facts, it has been ranked among the most damaged geoecosystems in the coalfields in India. Hence, it is an ideal place for conducting studies to achieve better geoecological reconstruction of coal-mine-disturbed lands.

The Jharia coalfield is situated in eastern India (Fig. 1) at the heart of the Damodar Valley, mainly along the northern bank of the river. It is located about 260 km north-west of Calcutta, in the Dhanbad district of Bihar.

Geologically, Jharia is part of the EW trend of the Gondwana basin chain in eastern India. Rocks of the Gondwana supergroup make up the coalfield (Fox, 1930, and Fig. 2; Mehta, 1957). Lower Gondwana rocks (Autounian and Thuringian) are well developed; the exposed lithologies are mainly sandstones (and some shales) of the Talchir, Barakar, Barren measures and Raniganj formations, together with intrusives of later age.

The field is almost sickle-shaped, covering an area of about 450 sq km. The strike of sedimentary rocks changes from EW in the western part, through intermediate directions, to NS in the eastern part. The strata have an average dip of 15° southwards in the western part and westwards in the eastern part of the field. The Archaean basement is exposed in the north with gradually younger rocks exposed southwards (Fig. 2). The Great Boundary Fault of the Gondwana basin, having a throw of about 1500 metres runs along the southern margin of the coalfield. South of this, Archaean rocks outcrop once again.

The field is characterised by very gently undulating topography with contours ranging from 140 m to 240 m above m.s.l. Rare hillocks are present with peaks of about 400 m in height. The field lies north of the Damodar River and is traversed by a number of its tributaries. Its general slope is ESE.

TRANSFORMATION OF JHARIA

Planning for geoecological reconstruction requires knowledge of the ecological set-up of the region in pre-mining days. Information about the ecological set-up is reflected in the land-use pattern. Ecological conditions are controlled by many factors. Two of the most important are topography and water availability. Water availability is affected by rainfall characteristics, surface water resources and groundwater resources. These aspects of the Jharia coalfield are discussed below.

Topography

The unplanned mining with no consideration of land reclamation that characterised private sector operations, has greatly disturbed the topography of the Jharia area. Mehta (1957) commented: "The area has undergone such tremendous development and transformation in the last 25 years that the old topographic details are no longer there at many places.' This comment is even more apt today. Even in the early days of nationalised sector mining, no real attempt towards land reclamation was

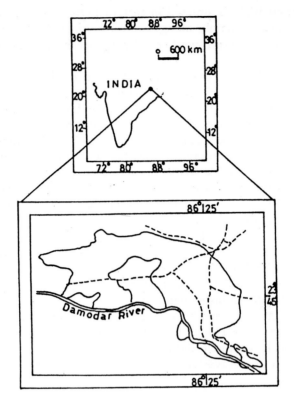

Fig. 1: Location of Jharia coalfield.

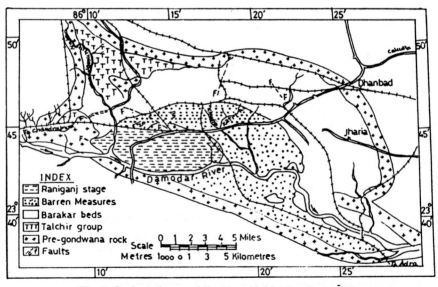

Fig. 2: Geological map of Jharia coalfield (after Fox, 1930).

made. Only very recently, in fact, has backfilling of quarries been made mandatory by MOEF. Until the latter half of the 1980s, quarries and overburden dumps were considered no-one's responsibility.

The trend of topographic damage in the area, may be demonstrated by hypsometry (Strahler, 1952). Figure 3 shows two hypsometric curves—one for 1925 and the second for 1974 (Fig. 3). These curves represent area-altitude ratios. It is evident that the 1925 topography is notably more uniform than the 1974. The obvious causal factor is the development of new quarries and overburden dumps.

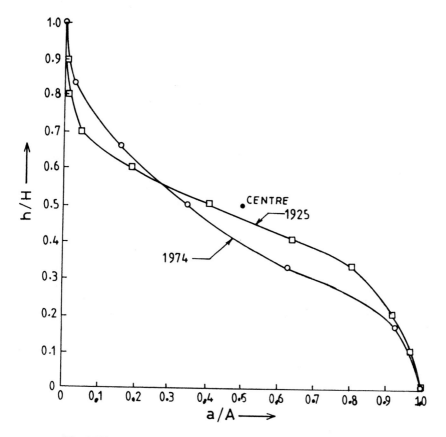

Fig. 3: Hypsometric analysis of Jharia coalfield for 1925 and 1974.

Rainfall

Variations in the amount of rainfall between 1924 and 1983 were examined and the data averaged for every ten years (Table 1). It can be seen that rainfall has decreased over this 60-year period. Undoubtedly, diminishment in precipitation has had some impact on the ecosystem.

Table 1: Monthly rainfall over a 60-year period, averaged for every 10 years (data of D.V.C., Maithon) (mm)

Years	June	July	August	September	October
1924–33	154	437 (591)	329 (920)	182 (1102)	92 (1194)
1934–43	168	333 (501)	357 (858)	230 (1088)	83 (1171)
1944–53	183	358 (541)	319 (860)	215 (1075)	83 (1158)
1954–63	144	302 (446)	260 (706)	253 (959)	103 (1062)
1964–73	166	293 (459)	327 (786)	227 (1013)	70 (1083)
1974–83	180	312 (492)	262 (754)	229 (983)	55 (1038)

N.B.: Cumulative values are those within brackets.

Natural Surface Drainage

Table 2 and Fig. 4 show the changes in the natural drainage system in the area between 1925 and 1993. These indicate that within the interval of 67 years, the region lost 260 km (66%) of its natural surface drainage.

Water Table

When opencast excavations extend below the water table, the open pit receives a continuous inflow of water. Mining under such conditions requires continuous pumping of water out of the excavation site, which lowers the water table. Since this practice has been common in Jharia, obviously, the water table of the region has been seriously disturbed.

Black-and-white aerial photographs taken in 1982 were analysed to obtain some idea of the damage sustained by the water table (Ghosh, 1993). The principal factor in this analysis was the observation that soil containing high moisture absorbs more light. Thus high-moisture zones appear grey-toned in black-and-white photographs (Nefedov and Popova, 1972).

Ghosh (1993) found that as a result of disturbance to the original landscape, natural control of groundwater level has been completely lost. The groundwater level now depends mainly on present-day topography and geomorphic features such as waste-tips, abandoned channels and mining-induced recharge conditions.

Land-use

No technical source of information about the land-use pattern prevailing in pre-mining days exists. The literature (Ruthermond et al., 1980) suggests that in the 1890s, the coalfield was mainly forest and agricultural land. There was no dearth of water and, apparently, no barren land or topographic irregularities other than the small natural hillocks. Intensive mining began in 1925 and this is the first year for which technical mining information is available. Land-use maps and land-use patterns for different timespans since that year have been prepared by Ghosh (1987a, 1987b;

Table 2: Land-use in Jharia coalfield—1925, 1974, 1987 and 1993

Sl. No.	Land-use	Area %				Change % between					
		1925	1974	1987	1993	1925–74	1925–87	1925–93	1974–87	1974–93	1987–93
1.	Villages, settlements and townships	8.6	16.0	32.3	33.10	7.4	23.7	24.50	16.3	17.1	0.8
2.	Land in mining use including open pits	4.7	17.4	12.5	19.42	12.7	7.8	14.72	-4.9	2.0	6.9
3.	Water bodies	7.3	6.7	3.1	2.90	-0.6	-4.1	-4.40	-3.6	-3.8	-0.2
4.	Forests and plantation	4.9	0.7	0.7	2.45	-4.2	-4.2	-2.45	0.0	1.8	1.8
5.	Agriculture and natural vegetation	65.4	56.8	49.4	39.02	-8.6	-16.0	-26.38	-7.4	17.8	-10.4
6.	Fallow land and pasture	9.1	2.4	2.0	3.11	-6.7	-7.1	-5.99	-0.4	0.7	1.1
7.	Natural surface drainage length (km)	395	251	181	135	-36	-54	-66	-28	-46	-25

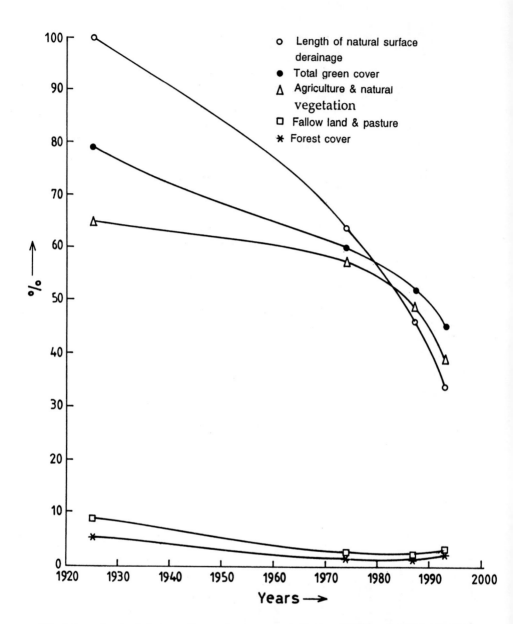

Fig. 4: Loss of natural drainage lines and green cover in Jharia coalfield over a 68-year period.

Ghosh and Ghosh, 1990a). The cumulative and comparative picture is given in Table 2 and Fig. 5.

   The Table is self-explanatory and charts both change in land-use and damage to the ecosystem. The pattern suggests that if mining were to

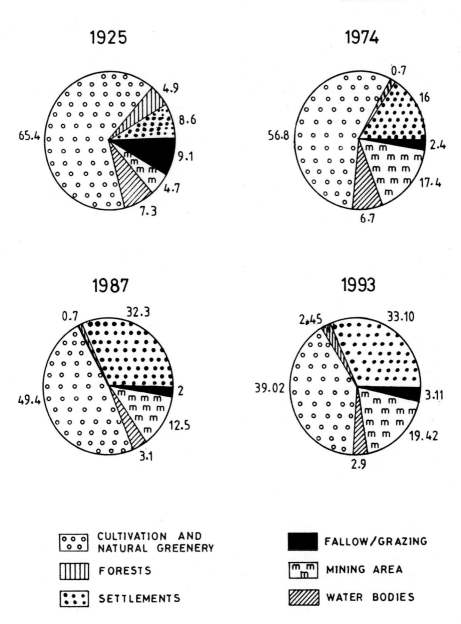

Fig. 5: Change in land-use pattern over a 68-year period in Jharia coalfield

proceed in a similarly uncontrolled manner, the region would not remain habitable for long.

The factors leading to these damages include abandoned quarries, overburden dumps, fire areas and subsided lands, damage to surface water and groundwater resources, and direct and indirect damage to vegetated areas. However, probably the most widespread damage is caused by erosion due to runoff water. This erosion has caused the materials in overburden dumps and other irregularities generated by mining to be distributed over large areas and, further, has led to siltation of water bodies. Extensive erosion on sloping lands has resulted in loss of topsoil, leaving barren lands. The foregoing, together with loss of water bodies due to siltation, plus the damage done to aquifers, are basically responsible for most of the damage sustained by the ecosystem.

Protection from such eventualities in future requires the development of strategies for (1) reclamation of quarries and overburden dumps, both old and new, and (2) control of runoff and erosion on sloping lands. Such measures are important prerequisites for geoecological reconstruction in this coal-mining area.

RECONSTRUCTION STRATEGIES

Opencast coal mining has severely damaged the geoecosystem in the Jharia coalfield. The problem is shared with almost all the coalfields in India that have been exploited through opencast and underground mining. The reconstruction strategies discussed here address the problems caused by opencast mining. However, the long-term consequences of subsidence and mine fires associated with underground mining may also lead to the development of barren lands. Some of the strategies discussed in this chapter also take into consideration the reclamation of those barren lands that have developed as end-effects of underground mining operations.

Land damage problems in such old mining areas are varied because such districts include both old abandoned mines and presently active or newly completed mining sites. These different contexts often demand different reclamation strategies. Strategy selection must obviously be case specific.

The main considerations for the geoecological reconstruction of these coal-mine disturbed lands should include:
• filling quarries and greening (or using up) overburden dumps,
• greening all barren surfaces as a top priority,
• aquifer regeneration.
Some techniques that might achieve this in Indian conditions are detailed below.

Reclamation of Excavations

Opencast mining methods practised in India include area strip mining, contour strip mining, quarrying etc. However, nowhere has there been significant success regarding revegetation. One main reason for this is that the topsoil and subsoil from the pre-mining environment have been neither properly preserved nor properly reinstated. Even when removed for the purpose of restoration, the topsoil was reinstated after such a long period that its fertility had been lost. Most modern land reclamation techniques recognise the importance of replacing the topsoil as quickly as possible (Ramsey, 1986). Experience from various regions of India indicates that in tropical regions, topsoils lose their nutrients when stored beyond 1 to 1.5 years. Contrarily, the work in mines and quarries often continues for 20–30 years or more.

In the reclamation of opencast mining areas 'concurrent reclamation' is without doubt the best approach since it minimises the amount of land disturbed at one time. The official land reclamation policies of many nations advocate this strategy.

The model presented here, 'continuous and concurrent reclamation mining' (Ghosh and Ghosh, 1990b), takes into consideration all the problems mentioned above and offers a feasible solution.

The strategy is based on the 'continuous reclamation mining method' outlined in Coates (1981) and a somewhat similar method used by EXXON, which seeks to undertake surface mining without leaving significant scars on the land. A major advance in the proposed model, over earlier methods used in India, is the special care taken to preserve and utilise the topsoil and subsoil separately and effectively.

As a first step, the total area for opencast mining should be divided into more or less uniform plots so sized that mining in each plot can be finished in six months. Excavation starts with the corner plot, i.e., plot No. 1, with the removal and storage of topsoil and subsoil (separately) on the fringe of this plot. These dumps need to be grassed and protected from erosion. Natural depressions in the area should be preferred dumping sites. Contour bunds may be utilised to reduce rain erosion. One easy, inexpensive method for preserving the fertility of the soil is to layer and mulch the storage dumps with cut twigs and leafy parts of plants and trees. These can be collected during site preparation. The leafy portions and branches of shrubs and trees and also local grasses with their roots and 15 cm of soil can be used to cover the surfaces of the topsoil and subsoil dumps. These mulch covers will not only protect the soil heaps from rain erosion, but the leaves and twigs preserve moisture and enrich the soil as they rot and decompose. The roots of grass and the cut plants will also bind the soil until they decompose. After decomposition, all the plant tissues add nutrients to the underlying soil, while some of the roots and branches may even yield plants. In other words, this natural vegetation while protecting the soil from erosion, concomitantly provides a green manure for the soil to be reinstated (Haigh, 1976).

Under this plan, the fertile soil is stored for just one year and costs are minimal.

The next step involves excavation of the overburden in plot 1. This is dumped at a separate, properly selected, site and likewise preserved by planting rapidly growing grass and shrubs. Studies of slopes on lands disturbed by surface mines suggest that those covered by a turf experience less than a third of the soil loss recorded for non-vegetated parts (Haigh, 1976).

The next phase is excavation and removal of the coal. Towards the end of coal extraction in plot 1, site preparation should begin in the adjacent plot 2. This starts with removal of vegetation and its temporary dumping over the topsoil and subsoil heaps from plot 1. Removal of the topsoil and subsoil from plot 2 follows. These soils are immediately covered with the freshly heaped vegetation mulches priorly placed on the soil heaps of plot 1. By this time, mining operations should be finished in plot 1, so removal of the overburden from plot 2 can commence. This newly removed overburden is used to fill the pit of plot 1. Then concomitant with coal excavation of plot 2, the subsoil, topsoil and decomposed plant tissues from plot 1 are transferred to the surface of the refilled pit.

Mechanical compaction is controversial but many workers prefer that it should accompany each of the aforesaid stages. Other researchers recommend that the soils and subsoils be left loose tipped at the lowest possible density.

Ideally, this cycle of four stages is completed within one year, before the topsoil and subsoil can degrade. Over longer periods, the microecological system of the soil deteriorates. The soil loses its nutrients, structure and fertility. Hence the capacity to engender fresh vegetation is much reduced.

Repetition of this sequence successively in adjacent plots, one after the other, continues over the entire area to be mined. Finally, the overburden from plot 1 is used to fill the quarry in the last plot, which is one reason why special care should be taken to select an appropriate dumping site for overburden from plot 1 and why special attention must be paid to protect this dump from causing sediment and chemical pollution problems in its vicinity. A procedure for dump stabilisation is, thus, a prerequisite (as detailed in the next section). An operational plan for the scheme outlined above is presented in Fig. 6.

This particular scheme for concurrent reclamation has two specific advantages. First, each plot receives its topsoil and subsoil back in its original position before one year has passed and before soil nutrients are lost. In most cases, the new land will revegetate naturally and quickly show natural soil conservation. According to Haigh (1977), a natural vegetation cover is the best way to check accelerated soil erosion on surfaces disturbed by mining activities. Afforestation or arable cultivation may be introduced later, as required. The new areas should now be

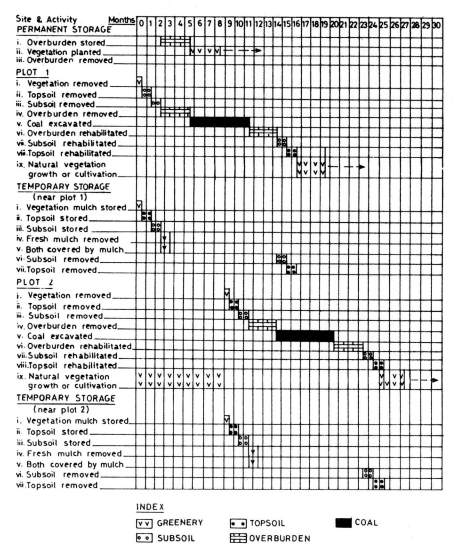

INDEX

| ⟦v v⟧ GREENERY | ⟦• •⟧ TOPSOIL | ▮ COAL |
| ⟦• •⟧ SUBSOIL | ▤ OVERBURDEN | |

Note: The same sequence is to be continued till the last plot is excavated.

Fig. 6: Continuous and concurrent reclamation mining.

covered by a relatively loose soil containing composted mulches and, therefore, will be suitable for cultivation.

Second, in this strategy, only two plots of land are damaged in a single unit of time. The rest of the area remains undisturbed and protected by vegetation.

The vegetation, mainly grass, which regenerates naturally on the rehabilitated topsoil must be protected against grazing during the early phases by fencing.

Once a natural vegetation cover has re-established, conversion to other land-uses may be considered. This type of revegetation, in two phases, is also practised in the outer Himalaya (Shastry et al., 1981). As indicated by Kilmartin and Haigh (1988), the first priority is to clothe the ground with some kind of vegetation cover, and it is easier to construct a stable self-sustaining ecosystem using local species and local natural biological resources than with exotic species (Flege, this volume). The best solution might be to cultivate plants from the list of species that naturally colonise sites in the local environment presently or just prior to mining. It has been observed that, at least in riparian areas, most trees grow better when planted along with leguminous species such as subabul (*Leucaena leucocephala*) and siris (*Albizzia lebbek*) (Singh and Joshi, 1994). Criteria for species selection should include plants most likely to grow on the land concerned and the species best suited to meet the needs of the local community.

Accepting all these factors, it is suggested that, when agricultural recultivation is attempted, the best starting point is the type of cultivation practised in the area in its pre-mining days. If the final goal is forest, then this might best be based in some of the elements of the original flora. Significant components of woodlands in the Jharia coalfield include babul (*Acasianilotica indica*), neem (*Azadirachta indica*), choukundi (*Cassia siamea*), gulmohur (*Gamelina arborea*), karanj (*Pongania pinata*), arjun (*Terminalia arjuna*), sal (*Shorea robusta*) and palash (*Butea monosperma*). Of course, the species would differ in other parts of the country.

Overburden Dump Stabilisation

Keeping in view the standard techniques of overburden dump stabilisation such as greening, terracing etc., a proposal is offered here that might improve the procedure and achieve maximum possible erosion protection.

Selection of the dumping site should be related to the topographical, geological and structural characteristics of the region. Preferably, the dumping ground should be a topographic depression but it must definitely not be a usable water resource. Further, it should have low infiltration potential; otherwise leachates may affect the quality of groundwater.

Runoff and erosion protection may be achieved by compaction of the dump, stabilising its slopes and by allowing minimal contact between dump surfaces and running water. This can be done by constructing an effective drainage network, by building retention walls and ensuring a rapid greening of the dumps (Fig. 7).

In detail, the process involves forming the dump with the lowest possible sideslope angles, building the dumps in small, levelled and

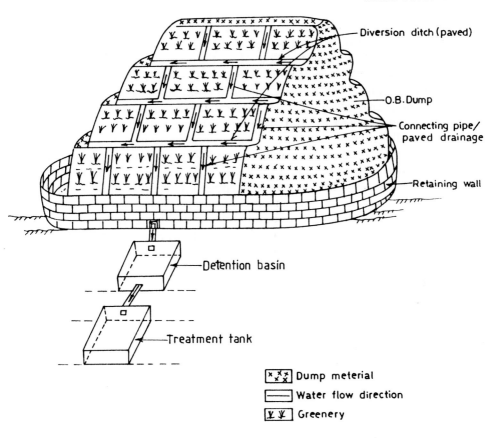

Fig. 7: Dump stabilisation.

compacted layers, thereby carefully creating a stable terraced slope. After dumping is complete, the process of further stabilisation involves the construction of drainage diversion ditches above the slopes and creating pipe or paved drainage to take water down from the upper terrace to the lower terraces. The entire dump should then be covered with green mulch and seeded with rapidly growing grasses, shrubs and deep-rooting trees. In fact, any local species will serve the purpose well. Good results have been obtained, in some cases, in the Jharia coalfield with arhar (*Cajanus cajan*), castor (*Ricinus communis*) and lemongrass (*Cymbopogon flexuosus*). Bradshaw and Chadwick (1980) recommend that before spreading the plant seeds may be mixed with chopped straw, wood pulp and shredded plastic to ensure shade against sun and wind and also to augment moisture-retaining capacity of the surface layers. Latex or some oil-based emulsion, which forms a porous but firm crust, may be added to give further surface stability.

Retaining walls should be constructed around the base of the dump to prevent spread of washed sediment and to trap and direct this runoff to detention and treatment basins. Here the water should rest for some time to allow settlement of solid particles. It should also be tested and treated (if required) to eliminate chemical pollutants that could otherwise disturb the surrounding ecosystem. The detention basin can be dredged as and when required to regenerate its capacity.

Generally, overburden dumps contain carbonaceous shale with some sulphide, which on oxidation generates acidity. Acid drainage may disturb the ecosystem in surrounding areas and reduce the greening of the dump. To neutralise this acidity, lime can be applied at the rate of 30–50 t.ha to a depth of 20–30 cm by deep-rippng the dumps (Valdiya, 1987).

Finally, the dump is greened with suitable legumes or grass. Hydroseeding may be very useful for such purposes (Table 3). The vegetation will act to bind the soil and as a cover, thereby decreasing erosion and dump surface wash.

Table 3: Hydroseeding specifications for establishing grass/legume cover in temperate climates (Bradshaw and Chadwick, 1980)

	Material	Rate of Application
1	Mixture (appropriate) of grass seeds (including nursecrop in exposed situations)	70 kg · ha
2.	Wild white clover or other legume (innoculated and pre-germinated, if necessary	10 kg · ha
3.	Mulch-woodfibre, chopped straw or glasswool	1–2 t · ha
4.	Stabliser-woodfibre, alginate, PVA or latex (if likelihood of severe erosion)	Depending on stabliser
5.	Fertiliser-complete 15:15:15 (slow release N-fertiliser if legume is not included) or dry organic matter	200 kg · ha followed by 300 kg · ha after 8 weeks 500 kg · ha
6.	Lime (if pH requires this and may have to be spread separately)	0–5000 kg · ha

Reclamation of Old Abandoned Quarries

Ghosh and Ghosh (1990c) observed that much of the land in the opencast mining area remains occupied by unfilled and abandoned quarries and by unused overburden dumps, most created by unplanned mining activities in days gone by. Some of these quarries have become filled, naturally, by siltation and some of the very old dumps have regenerated scrub vegetation. However, large areas remain totally unfit for any kind of fruitful land-use.

The need of the day, therefore, is to reclaim this huge tract of damaged land. The present proposal envisages reclamation of abandoned quarries by infilling with wastes from the overburden dumps and reject piles. Some of the old overburden dumps already show green covers provided by scrub vegetation, indicating that the debris has some fertility. The nutrient capacity can be further enhanced by providing a layer of grass/herbaceous mulch or a mulch of tree litter just below the topsoil layer. This exercise will quickly bring the infilled land surface to a standard suitable for agricultural purposes. A few years of careful cultivation will further consolidate the land surface and make it usable for other development projects, construction etc., as desired.

However, it is likely that the materials in the overburden dumps will not be sufficient to fill all the quarries. In this case, some of the quarries can be converted into lakes and water reservoirs for the population living in the vicinity. These could also be designed to serve as sites for the development of fisheries and tourist recreational facilities. Such projects would, additionally, help to stabilise the water table, which has been severely disturbed (Ghosh, 1993). Incidentally, the water supply potential in the Jharia coalfield area is still being wasted due to mining activities. Several million gallons of water are pumped out of the excavation sites every day. Some of this huge volume of water could be stored in the empty quarries.

The remaining unfilled quarries could serve as garbage dumps for urban centres. A manifold increase in such waste is expected in the near future (Ghosh, 1987). However, using wastes from human settlements as infill materials in open quarries carries the risk of contamination of the groundwater by leachates. Hence, extreme care must be taken during planning and site selection. The first precaution would be to ensure that the quarry/quarries to be filled with garbage are situated at a safe distance from any domestic water supply source.

Even given proper site selection, care must be taken to ensure that the leachates do not seep beyond the limits of the infilled quarry. Should such leakages occur, the effluents will pollute the groundwater. An additional precaution, exercised in countries such as the USA (Cargo and Malory, 1974), is the creation of an impermeable barrier, e.g. an impermeable clay lining for any quarry designated for waste disposal. The garbage and its leachates are thereby completely segregated from precipitation, surface flow and groundwater.

Howard and Remson (1978) indicate three classes of disposal sites from the point of view of safe distance from sources of potable water. Class I includes the disposal sites that can accommodate all types of waste except radioactive. In this class, the disposal site should lie on a non-water-bearing rock or on an isolated body of unusable water. Usable surface water bodies should be situated more than 150 metres from such sites. If a disposal site lies near running water, the stream channels would have to be diverted to a safe distance.

Class II disposal sites may accommodate only solid waste and decomposable organic wastes. These disposal sites may be underlain by usable, confined or free groundwater. However, the lowest level of the disposal site should be at least 1 metre above the highest level of groundwater.

Class III disposal sites should contain only non-water-soluble, non-decomposable, inert materials because almost no protection is available for underlying or adjacent bodies.

In the Jharia coalfield, disposal sites of class I should be sought as they can accommodate all kinds of wastes without risk of water pollution. It is generally expected that domestic wastes from urban townships will not contain radioactive materials. Thus it is safe to select unfilled quarries as disposal sites at a distance of 150 metres from usable sources of water. Unfilled quarries at lesser distance from water sources might be used for dumping solid water-insoluble wastes.

As additional insurance against the pollution of groundwater by wastes dumped in unfilled quarries, the hydrogeology of the area should be studied. Common local rock types are sandstones that grade into shale. The coarser sandstones have higher permeability while the finer shales are less permeable. Generally, the coal seams are associated with sandstones interbedded with thin layers of shale or clay. Hence it may be expected that an abandoned quarry will rest on sandstone, which is highly porous and hence not suitable for waste disposal. If, from borehole records, it becomes evident that a shale or clay layer may be reached by proceeding a bit more, the quarry selected for waste disposal should be cut down to the shale level. It is best, by far, to select for waste disposal a quarry that rests on clayey or other such impermeable rocks. Experiments by Raymahashaya (1987) show that clayey soils act as a barrier against the dispersal of toxic metals and detergents.

Indeed, impermeable rocks excavated during mining ought, therefore, to be set aside for use as side and top lining of the waste dumps. Day-to-day procedure should include dumping the day's waste in the abandoned quarry at a landfill site previously provided with an impervious base and then covering it with the impervious materials preserved for this purpose. The next day's haul should likewise be covered after deposition. In other words, covering the garbage with impervious material should be an automatic routine; waste should never be left exposed to the atmosphere. A proper slope must also be maintained on the impervious cover so that only a minimum amount of rain water can percolate through it.

In the long run the disposed waste will decay and compact. Further compaction can be induced mechanically and eventually, when the waste pile reaches ground level, the area may be used for agriculture. The decayed waste materials will convert into organic manure—far preferred to chemical fertilisers. Bownder (1981) states that chemical fertilisers deplete the C:N ratio while organic manures keep this ratio in equilibrium.

Thus this process would enable utilisation of totally useless quarries to solve the problem of urban waste disposal while concomitantly regenerating valuable land resources.

Aquifer Regeneration

Mining activities damage aquifers as well as surface water resources. The two are inseparable in the hydrologic cycle and damage to their natural functioning makes landscape rehabilitation more difficult. Thus aquifer regeneration technology is an important aspect of the geoecological reconstruction of mining areas.

If abandoned quarries are cleaned and dressed properly, these can automatically store rain water to form a surface water resource. Further, they will allow recharge of the ground water regime through their porous base and walls and thereby enhance depleted groundwater reserves. Planning the regeneration of aquifers in quarry-backfilled lands requires a thorough knowledge of the geological and hydrogeological profile of the area.

If the exact level of the base of the natural aquifer is known, the quarry can be backfilled with overburden material to that level. Mechanically compacted, the newly formed base of the partially filled quarry can then be lined with an impermeable material (clay, cement, latex, polyethylene sheeting etc.). This newly developed plane can be adjusted as necessary as shown by a study of the geological sections of the region, to form a continuation of the previous aquifer base in the vicinity. If this adjustment appears impracticable, quite possibly the area of the newly generated aquifer could be fitted to that of the quarry. In this case, the sides of the quarry must be dressed properly and the impermeable lining continued along these sidewalls.

In the next step, a layer of grass-weed mulch is spread on the base lining to form a spongy lining, after which the quarry is filled with overburden material. Care must be taken not to damage the impermeable lining. Filling may continue to a level about 0.5 m below that of the surrounding topography. The final layer should include the grass-weed mulch. Mixed seeds of local hardy species (grass, shrub and tree) can be sprayed on the surface at the onset of the monsoon. These seeds will grow and the grass-weed mulch gradually decompose and support the plants. Spreading some composted cow/poultry dung or domestic garbage after spraying the seeds will assist plant growth.

A pilot experiment in a quarry replica yielded good results with cow dung compost and seeds of arhar (*Cajanus cajan*), castor (*Ricinus communis*), jackfruit (*Artocarpus integrifolia*) and date palm (*Phoenix dactylifera*). The site required watering only once, at the end of the first summer, and the plants have thrived for several years. But it must be mentioned that this result was assisted by providing the site with an artificial impermeable lining that preserved water. Losses to evaporation were reduced by the layer of compost and rotten green mulch. After the monsoon (during which the seeds germinated and grew) the plant leaves provided shade and reduced evaporation. It is expected that, after a few more seasons, the process will regenerate the groundwater aquifer.

The depth of the base of the aquifer may be varied with respect to the type of vegetation to be regenerated and the local geological conditions. The impermeable character of the artificial layer may be lost after some years but, hopefully, by this time the geoecosystem will have been reconstructed.

SUMMARY: GREENING BARREN LAND

Study of the Jharia coalfield identifies six varieties of wasteland (Ghosh, 1991). Most are barren and rocky with negligible soil cover. In addition, their hydrology is poor due to very high infiltration potential in some cases and very high runoff potential in others. Geoecological reconstruction of these areas requires a strategy with the following objectives:
• to improve the runoff-infiltration balance with the ultimate aim of stabilising the groundwater table;
• to generate a sustainable green cover for the land;
• to generate some cover over the land which can substitute for topsoil.
The cumulative effect of these measures should automatically lead to runoff and erosion control.

The greening of barren land, as conducted to date, involves making pits of optimum size, infilling and surfacing them with topsoil (brought from some other place) and then planting trees on top. This produces greenery at one place at the cost of greenery growing potentiality at the place from which the topsoil was brought. This is because topsoil is a natural resource which takes a long time to form by natural processes.

After many years of experimentation in fields having different runoff: infiltration ratios and no topsoil, the following recommendation for a working model for soil rehabilitation/regeneration as well as attainment of an appropriate balance between runoff: infiltration are proposed. The strategy may be presented in the form of a flow chart (Fig. 8). Species listed in the flow chart are local. Full botanical names are given in Table 4.

To make the procedure applicable to other areas, several adjustments would have to be made. These are listed below. Species selection would have to respect the following factors:
• climate and water availability
• soil composition and soil availability
• water infiltration rate/water-holding capacity of the site

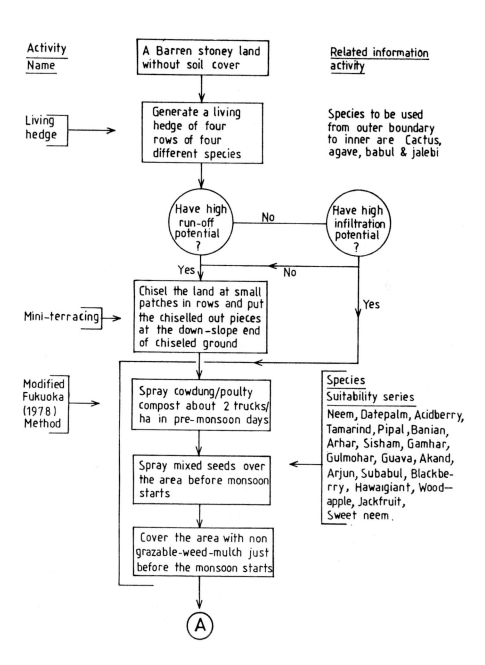

Activity
Name

Related information
activity

A Barren stoney land without soil cover

Living hedge → Generate a living hedge of four rows of four different species

Species to be used from outer boundary to inner are Cactus, agave, babul & jalebi

Have high run-off potential ? — No — Have high infiltration potential ?

Yes

No

Mini-terracing → Chisel the land at small patches in rows and put the chiselled out pieces at the down-slope end of chiseled ground

Yes

Modified Fukuoka (1978) Method → Spray cowdung/poulty compost about 2 trucks/ha in pre-monsoon days

Species
Suitability series
Neem, Datepalm, Acidberry, Tamarind, Pipal, Banian, Arhar, Sisham, Gamhar, Gulmohar, Guava, Akand, Arjun, Subabul, Blackberry, Hawaigiant, Wood-apple, Jackfruit, Sweet neem.

Spray mixed seeds over the area before monsoon starts

Cover the area with non grazable-weed-mulch just before the monsoon starts

(A)

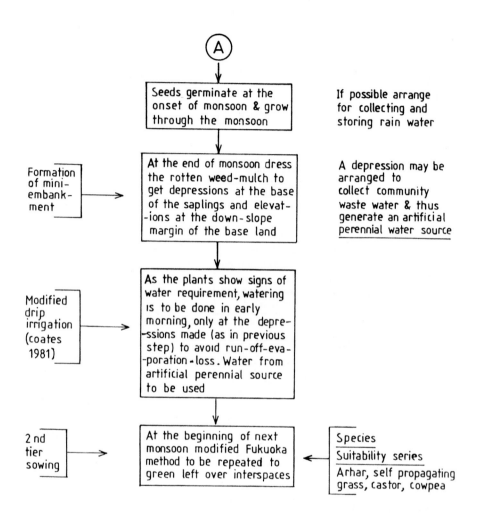

Fig. 8: Flow chart of activities for soil rehabilitation in mining degraded stoney lands.

Table 4: Botanical names of species selected for soil regeneration (Jharia)

Sl. No.	Local Name	Botanical Name
1.	Cactus	*Opuntia* spp.
2.	Agave	*Agave veracruz*
3.	Babul	*Acacia arabica*
4.	Jalebi	*Inga dulcis*
5.	Neem	*Azadirachta indica*
6.	Date palm	*Phoenix dactylifera*
7.	Acid berry	*Zyzyphus mauritiana*
8.	Tamarind	*Tamarindus indica*
9.	Pipal	*Ficus religiosa*
10.	Banian	*Ficus bengalensis*
11.	Arhar	*Cajanus cajan*
12.	Sisham	*Dalbergia sissoo*
13.	Gamhar	*Gamlina arborea*
14.	Gulmohar	*Delonix regia*
15.	Guava	*Psidium guajava*
16.	Akand	*Adhatoda vasica*
17.	Arjun	*Terminalia arjuna*
18.	Subabul	*Leucaena leucocephala*
19.	Blackberry	*Eugenia jambolana*
20.	Huwai giant	*Prosopis julifera*
21.	Woodapple	*Aegle marmeles*
22.	Jackfruit	*Artocarpus heterophylus*
23.	Sweet neem	*Melia azadirachta*
24.	Castor	*Ricinus communis*
25.	Cowpea	*Vigna sinensis*

- requirement of the soil to improve its character and to prevent further erosion
- expectations of local people from the land
- common local species of today and in the past
- grazability
- availability of seeds and saplings.

Keeping the above in view, plans would have to be made for the following:
- Providing a grass cover to prevent further erosion and to initiate ecological improvement in the area.
- Introducing plants that can thrive on poor mined soils.
- Introducing plants that can crack the stone and push roots through it.
- Introducing plants that are hardy, drought-tolerant and preferably of some medicinal/economic value.

- Introducing plants that resist grazing, i.e. thorny species for regenerating living hedges.

Several trials were required in the Jharia coalfield to identify the best management practice with reference to the following points:
- species suitability,
- preparation of plot,
- stages of sowing,
- process of sowing
- aftercare,
- watering.

COST BENEFIT ANALYSIS

The methodologies evolved for the geoecological reconstruction of mining areas must be economically viable to be really useful. Cost benefit analysis for the methodologies discussed in this chapter were carried out for the Jharia coalfield. These indicated that targets could be reached at minimal cost but only with meticulous planning. Details of these analyses are not given here. The controlling factors are too local and specific. However, the key ecological benefits are listed below:
- Weed mulch after rotting together with the compost applied helps in the formation of a very fertile soil cover.
- Whatever greenery grows, even if it dries up or rots, helps soil formation.
- The process, in its entirety, establishes a new runoff percolation balance, which helps to stabilise the water table.

Further points to be noted are:
- The suggested techniques can reconstruct the geoecosystem on a variety of degraded mining lands.
- The techniques require no chemical fertilisers, no collection, no purchasing and no transportation of soil from other places and only limited financial support towards labour and cow dung/poultry/domestic compost.
- The key ingredients are planned activities and people participation.

CONCLUSION

Methodologies are suggested that may help soil formation, water table stabilisation and aquifer reconstruction and thus the rehabilitation of barren wastelands that receive some precipitation during the year. They are commended as useful working models for the regeneration of degraded mining lands.

Protecting Water Resources

The factors affecting damage to water potentialities in Jharia coalfield may be classified into three major groups: (a) unplanned dumping of overburden and unreclaimed excavations, (b) subsidence and spontaneous fires and (c) unplanned discharge of pumped-out mine water. The first is mitigated by the suggested 'continuous and concurrent reclamation mining' strategy (Ghosh and Ghosh, 1990a) and by the technique suggested for the 'reclamation of old abandoned opencast mining areas' (Ghosh and Ghosh, 1990b). The problems which remain are subsidence and spontaneous fires and the unplanned discharge of pumped-out mine water.

Conceptual model

Four methods for increasing aquifer recharge are discussed. The choice of the most suitable method or combination of methods depends on the local situation (Fig. 9; Ghosh and Ghosh 1994).

a) The pumped-out water may be directed to some existing body of water, thereby minimising evaporation and surface runoff and enhancing groundwater recharge.

b) More rapid groundwater recharge can be achieved by 'water spreading' (Karnath, 1987) or through construction of 'recharge pits', or 'recharge wells' (Coates, 1981). However, such recharge methods may contribute to mine flooding and necessitate additional pumping.

(c) The flooding of abandoned underground workings is also an option. This may help prevent spontaneous fire and concomitantly create an underground reservoir of water that may be useful in the dry season. Mine water discharge may also be used for dust suppression. However, the establishment of underground reservoirs is perhaps the most efficient means of aquifer recharge.

d) The supply of groundwater (after treatment, if required) to local settlements is also an option. During dry periods the population in some local settlements has to depend on mine water. However, this water must be analysed chemically and its treatment and use determined according to test results. If it contains pollutants beyond permissible limits, it may be poured first into covered and lined chemical treatment tanks and only then into the community water supply system.

Community settlement sewage may, in turn, be used to water the fields or it may be directed into treatment plants and settling tanks. The treated water can then be used for agricultural purposes or groundwater recharge. The receiving water bodies, as far as practicable, should be covered to reduce evaporation.

As suggested, the filling of abandoned quarries, together with mining, and proper backfillng and reclamation, may eventually give rise to a

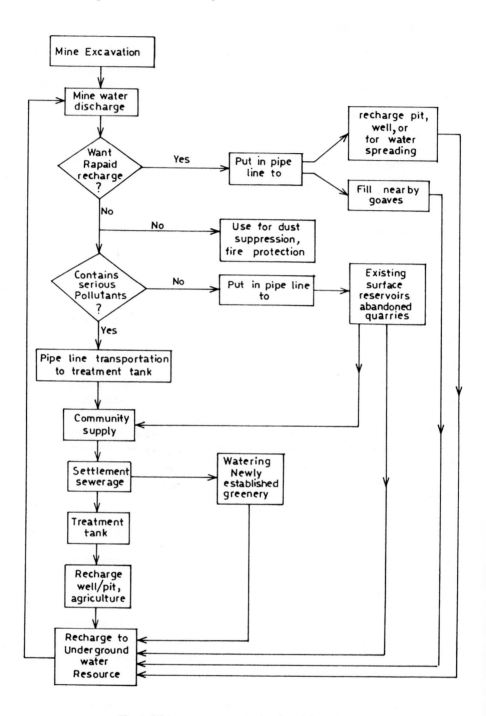

Fig. 9: Water resource protection in mining areas.

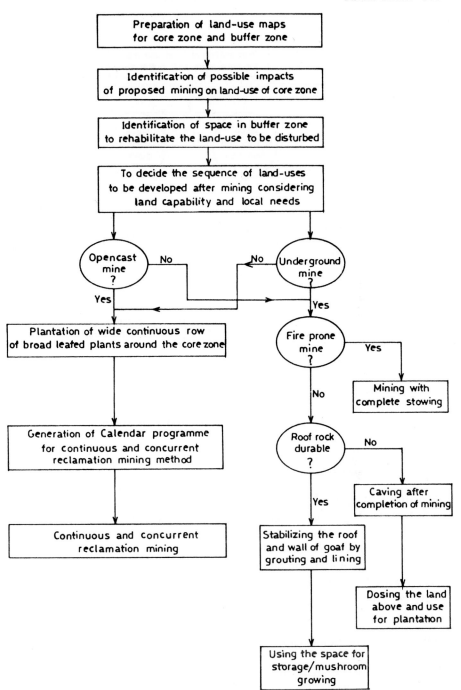

Fig. 10: Activities recommended for deciding the Land-Use Management Plan (LUMP) for geoecological reconstruction in coal-mining areas in India.

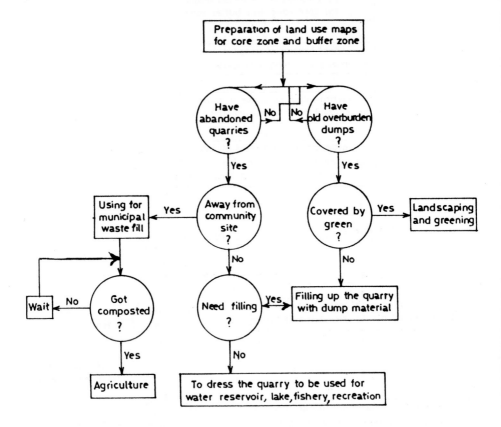

Fig. 11: Activities recommended for deciding the LUMP for geoecological reconstruction in abandoned coal-mining areas in India.

smooth topography. The planned restoration of agriculture/vegetation in such terrain, if done as suggested in this 'conceptual model', will help to stabilise the water table and restore the water resources in the region.

Further, during reclamation, if the mining areas are formed so that the new land slopes at a low angle towards a central zone, the total precipitation receipt will be collected, resulting in minimum runoff and maximum recharge. This will help to ensure the water resource potential long term.

THE ULTIMATE SOLUTION: 'LUMP' MODEL

As already explained, most Indian coalfields employ a combination of opencast and underground mining. The factors damaging the geo-ecosystem are excavation, earth movement, transferring of excavated material to a new place, subsidence and mine fires, damage to aquifers and surface water resources. In concert, these result in damage to the functionality of the geoecosystem at large.

The impact the region faces is cumulative. The reconstruction strategy must also tackle the problem from all possible angles. Individual techniques have already been detailed and are summarised in the flow charts of Figs. 10 and 11 (Ghosh, 1992). The charts aim to help the planner decide which technology should be used in which area. They show the logical sequence of activities recommended to achieve geoecological reconstruction in coal-mining areas of India and thus achieve a sustainable *Land-Use Management Plan* (LUMP).

References

Bownder, B. 1981. In Bownder, B. 1984. *Report on Environmental Management Problems in India.* Dept. Environment, Govt. India, New Delhi and Centre for Energy, Environment and Technology, Adm. Staff, Coll. India, Bella Vista, Hyderabad, 49 pp.

Bradshaw, A.C. and Chadwick, M.J. 1980. *The Restoration of Derelict Land.* Blackwell Oxford, 317 pp.

Cargo, D.N. and Malory, B.F. 1974. *Man and His Geological Environment.* Addison Publ., Wesley, Mass., 548 pp.

Coates, D.R. 1981. *Environmental Geology.* John Wiley & Sons, New York, 701 pp.

Fox, C.S. 1930. The Jharia Coalfield. *Geol. Surv. India, Memoir* 56.

Ghosh, R. 1987a. Land use and land protection in mining areas—a case study in Jharia coalfield. *J. Geol. Soc. India* 29 (2): 250–255.

Ghosh, R. 1987b. Geological considerations for urban development in Jharia coalfield, Eastern India. *Geol. Soc. Hongkong, Bull.* 3: 87–93.

Ghosh, R. 1991. Reclaiming wastelands of Jharia coalfield, Eastern India. *Int. J. Surface Mining and Reclamation* 5 (4): 185–190.

Ghosh, R. 1992. Land use management plan, the case in Jharia coalfield, Eastern India. *J. Tech. Univ. Wroclaw, Poland* 15 (3): 49–58.

Ghosh, R. 1993. Remote sensing for analysis of groundwater availability in an area with long unplanned mining history. *Photonirvachok, J. Indian Soc. Remote Sensing* 21 (3): 119–126.

Ghosh, R. and Ghosh, D.N. 1990a. Land use map of·Jharia coalfield, Eastern India aided by remote sensing. *Photonirvachok, J. Indian Soc. Remote Sensing* 18 (1–2): 23–28.

Ghosh, R. and Ghosh, D.N. 1990b. Land protection and land reclamation in opencast mining areas—a case study in Jharia coalfield, Eastern India. *Indian J. Earth Sci.,* 17 (2): 108–112.

Ghosh, R. and Ghosh, D.N. 1990c. Land reclamation in mining areas—a model of Jharia coalfield, Eastern India. *Proc. Indian Nat. Science Acad.,* 56(2): 145–152.

Ghosh, R. and Ghosh, D.N. 1994. Towards protecting public health in mining areas—a conceptual model. *J. Inst. Public Health Eng., India* 2: 38–47.

Haigh, M.J. 1976. Environmental problems associated with reclamation of old stripmined land. *Oklahoma Geol. Notes* 36: 200–202.

Haigh, M.J. 1977. The retreat of surface mine spoil back slopes. *Prof. Geog.,* 29 (1): 62–65.

Howard, A.D. and Remson, I. 1978. *Geology in Environmental Planning.* McGraw-Hill Book Co., New York, NY, 478 pp.

Karnath, K.R. 1987. *Groundwater Assessment, Development and Management.* Tata-McGraw-Hill Publ. Co., Ltd., New Delhi, 720 pp.

Kilmartin, M.P. and Haigh, M.J. 1988. Land reclamation policies and practices. pp 441–467. In: Joshi, S.C., Bhattacharya, G., Pangtey, Y.P.S., Joshi, D.R. and Dani, D.D. (eds) *Mining and Environment in India.* Himalayan Research Group, Nainital, India.

Mehta, D.R.S. 1957. Revision of the geology of coal resources in Jharia Coalfield. *Memoir Geol. Surv. India,* 84 pt. 2.

Nefedov, K.E. and Popova, T.A. 1972. *Deciphering Groundwater from Aerial Photographs.* Amerind Publ. Co., New Delhi, 191 pp.

Ramsey, W.J.H. 1986. Bulk soil handling for quarry restoration. *Soil Use and Management* 2 (1): 30–39.

Raymahashaya, B.C. 1987. Keynote address. pp. 173–180. In: *Proc. Nat. Symp. Role of Earth Science in Environment.* IIT, Bombay,

Ruthermund, D., Kropp, E. and Dienemann, G. 1980. *Urban Growth and Rural Stagnation.* Manohar Publ., New Delhi, 493 pp.

Shastry, G., Mathur, H.N. and Tejwani, K.G. 1981. Lessons from a landslide reclamation project in the Himalayan foothills. *Oxford Polytechnic Discussion Papers in Geography* 22: 56–77.

Singh, H. and Joshi, P. 1984. *Economic Utilisation of Bouldery Riverbed Land for Fodder, Fibre and Fuel under Rainfed Conditions.* Ann. Rept., Central Soil and Water Conservation Research and Training Inst., Dehra Dun, U.P.

Strahler, A.N. 1952. Hypsometric (area-altitude) analysis of erosional topography. *Bull. Geol. Soc. Amer.* 63 (11): 1117–1142.

Valdiya, K.S. 1987. *Environmental Geology—Indian Context.* Tata-McGraw-Hill Publ. Co. Ltd., New Delhi, 583 pp.

14

Case Study: Ecological and Economic Efficiency of Forest Reclamation of Coal-spoil Banks in the Kolubara Basin, Serbia

Miodrag Zlatić, Renand Ranković and Milvoj Vučković

Abstract

Forest reclamation of 900 ha of coal-spoil banks created by lignite mining in the Kolubara basin, central Serbia, proved economically effective. Producing timber on these embankments generated a higher cost: benefit ratio than local maize cultivation. Furthermore, forests encourage the initial stages of soil formation.

INTRODUCTION

In Serbia, surface coal mining has reduced productive agricultural and forest lands to infertile mineral wasteland. Erosion of these deposits is a source of air, water and soil contamination. Given Serbia's limited resources of productive agricultural land, it is important that these wastelands be rendered harmless to their environs and, if possible, economically productive. The Kolubara basin provides a major focus for Yugoslavian research into the productive recultivation of lignite spoils (Institute of Forestry Belgrade, 1997). This chapter presents an economic evaluation of some of this work.

COAL-MINE SPOIL BANKS, KOLUBARA BASIN: DESCRIPTION

The Kolubara basin, in the hilly region of central Serbia, is an area of mainly agricultural soils and forests. The Kolubara coal-mine spoil banks

Land Reconstruction and Management Vol. 1, 2000, 369–377.
ISSN 1389-2541
ISBN 90 5410 793 6
A.A. Balkema, Rotterdam, The Netherlands

lie in a zone of natural, autochthonous, pseudogley soils. These are heavy soils with low assimilative capacity and a low level of saturation with basic cations in the upper horizons. They contain a very small quantity of humus and reach depths of 30–120 cm. Before contamination, they were used for agricultural production.

The material piled during coal exploitation creates terrain that has a microrelief of depressions and mounds with steep slopes. This is especially expressed on clay, where water remains in depressions and forms ponds. Two types of substrate are common. Clay substrate occurs over small areas, is uniform in structure and contains very little fine sand. Great compaction and stickiness are the limiting factors of this substrate. Contrarily, the sandy substrate is mainly formed of fine sand (more than 80%). Clay and silt contents are very low (less than 20%). This substrate is light, well aerated and exhibits a neutral to acid reaction. The presence of clay has a positive effect as it prevents a sudden loss of water.

Reclamation of the first of the larger mounds in the Kolubara region commenced in the early 1950s; intensive work on their recultivation began in 1973. Afforestation was taken up in the next ten years, covering 500 ha of land; as of 1993, the afforested area covered about 900 ha. Primarily, four tree species were used: larch (*Larix europaea*) (33.3 ha), Douglas fir (*Pseudotsuga menziesii*) (12.7 ha), Weymouth pine (*Pinus strobus*) (21.0 ha) and black pine (*Pinus nigra*) (192.9 ha). Today, 17 years after afforestation over both substrata, the initial phases of soil formation are evident. These are characterised by a thin humus horizon covered with a layer of needles, underlain by an only-slightly-changed parent rock (Institute of Forestry Belgrade, 1998).

METHODS OF RESEARCH

The purpose of the present study was to determine the biological and economic parameters of success in the reclamation of anthropogenically degraded sites. The research plots analysed were covered with deep and heavy clay, constituting the most unfavourable conditions for successful afforestation. Various conifers (larch, pine, Weymouth pine, Douglas fir etc.) and broad-leaved species (black locust, birch, lime, maple and alder) were experimentally established in these plots (seedlings in containers or with bare roots in holes 40 cm × 40 cm × 40 cm together with 3 kg of peat). Seventeen years later, the diameters and heights of all the trees were measured. Finally, three trees from the dominant layer (mean of the 20% thickest trees) were felled on each sample plot for the reconstruction of height and diameter increments.

To evaluate the economic effects, a cost:benefit analysis (B/C and NPV) was done separately for the production of maize (the most frequently sown agricultural crop in the Kolubara basin) and of timber (planted on the spoils). The main elements comprised-cost of maize production, earnings from sale of maize, costs of timber production and income from timber sales. The present values were calculated at a discount rate of 12% (real rate of interest). The cost:benefits of both extensive and intensive production of maize were examined and the costs and earning of the two variants then calculated. The costs of extensive production of maize amounted to 600 DEM·ha^{-1}. The costs of intensive production when all the agro-engineering measures were applied was 800 DEM·ha $^{-1}$. The yield of maize was 3.5 t.ha^{-1} under extensive production against 5.8 t.ha^{-1} under intensive production. The price of maize in the domestic market is 150 DEM·ha^{-1}.

RESEARCH RESULTS

The data presented in Tables 1 and 2 show that all four species of trees thrive better on sand than on clay (Fig. 1a, b). Thanks to a great increment in the first years of development diameter increment, larch develops, significantly faster than the other species. However, the dominance of larch decreases with advancing age. This is most observable on clay, where larch has lately been showing signs of stragnation, expressed through a drastic reduction in height increment and incomplete formation of annual rings—and even the absence of growth rings. This serious loss of vitality makes the success of this species on clay quite uncertain. However, it should be noted that culmination of height and diameter increment in the unfavourable conditions of clay, occurs when the height of the tree is substantial. This phenomenon has also been observed in plantations on natural sites (Vuckovic, 1989).

Cost:benefit analysis shows that, under pre-mine conditions, the extensive production of maize is not profitable. Profitability can only be realised by intensive production (Table 3). However, even in intensive production, profitability is marginal since the ratio B/C is very close to 1. This leads to the conclusion that, in the Kolubara basin, even with high investment in agro-engineering measures, a high level of agricultural profitability cannot be expected.

The profitability of wood production on former coal-mine spoil banks varies with tree species and site quality (Table 4). Practically speaking, however, only on clay is the degree of profitability of investments close to marginal (only the Douglas fir production lies below the margin of profitability). On sandy spoils, a relatively high level of profitability is realised (larch: 2.44 and Weymouth pine: 2.66) (Table 5). This means that after 17 years, the earnings from sales of wood are higher than costs, and satisfying levels of profitability can be achieved. In such assessments and

Table 1: Elements of growth on coal-mine stockpile (per ha)—clay

$d_{1.3}$	Number of trees (N) $pcs \cdot ha^{-1}$				Basal area (G) $m^2 \cdot ha^{-1}$				Timber volume (V) $m^3 \cdot ha^{-1}$				Timber volume increment (Iv) $m^3 \cdot ha^{-1}$			
cm	Larch	W. pine	D. fir	B. pine	Larch	W. pine	D. fir	B. pine	Larch	W. pine	D. fir	B. pine	Larch	W. pine	D. fir	B. pine
5	158	150	248	1285	0.5	0.4	0.7	4.0	3.0	1.0	2.0	7.0	0.3	0.2	0.4	1.0
10	1304	800	836	1728	10.1	6.3	6.6	13.0	62.0	28.0	26.0	36.0	4.9	5.8	4.4	5.2
15	119	450	402	310	1.6	8.0	6.9	5.0	9.0	40.0	30.0	17.0	0.6	6.7	4.3	2.8
20	–	100	–	–	–	2.7	–	–	–	14.0	–	–	–	2.0	–	–
Σ	1581	1500	1486	3322	12.2	17.4	14.2	22.0	74.0	83.0	59.0	60.0	5.8	14.6	9.0	9.0

	Larch (*Larix europaea*)	Weymouth pine (*Pinus strobus*)	Douglas fir (*Pseudotsuga oriziesii*)	Black pine (*Pinus nigra*)
d_g *cm*	10.1	12.2	11.0	8.9
h_L *m*	11.2	10.4	9.5	6.6

Table 2: Elements of growth on coal-mine stockpile (per ha)—sand

d₁.₃	Number of trees (N) pcs·ha⁻¹				Basal area (G) m²·ha⁻¹				Timber volume (V) m³·ha⁻¹				Timber volume increment (Iv) m³·ha⁻¹			
cm	Larch	W. pine	D. fir	B. pine	Larch	W. pine	D. fir	B. pine	Larch	W. pine	D. fir	B. pine	Larch	W. pine	D. fir	B. pine
5	–	75	61	671	–	0.2	0.1	1.5	–	1.4	0.2	2.7	–	0.1	0.0	0.1
10	300	550	394	2414	2.4	4.8	3.5	19.5	15.8	28.5	15.3	74.1	2.2	3.6	3.5	8.8
15	667	600	667	604	11.8	11.1	11.6	9.0	93.1	68.9	55.5	40.2	7.1	10.7	11.9	5.1
20	267	375	182	–	8.4	10.8	5.2	–	71.7	68.6	26.2	–	4.4	10.8	4.9	–
Σ	**1234**	**1600**	**1303**	**3690**	**22.5**	**26.9**	**20.4**	**30.0**	**180.6**	**167.4**	**97.2**	**117.0**	**13.7**	**25.2**	**20.3**	**14.0**

	Larch (*Larid europoea*)	Weymouth pine (*Pinus strobus*)	Douglas fir (*Pseudotsuga menziesii*)	Black Pine (*Pinus nigra*)
dg cm	15.2	14.6	14.2	10.2
hL m	15.7	13.4	11.1	8.0

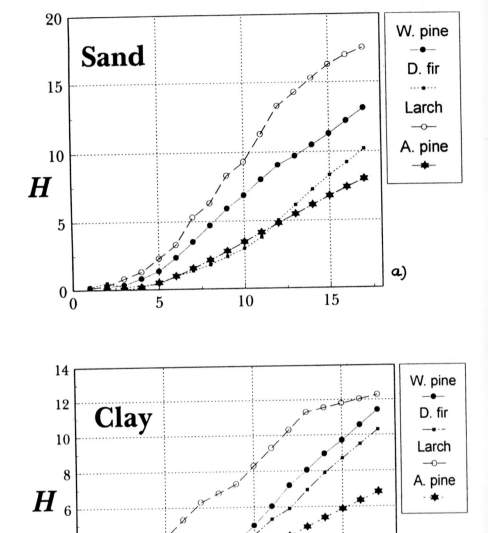

Fig. 1: Growth rates (in metres) of trees on coal-spoil banks in the Kolubara basin, Serbia. a) on sand and b) on clay

Table 3: Maize production (in DEM·ha⁻1, discount rate 12%)

Year	Benefits		Costs	
	Intensive	Extensive	Intensive	Extensive
1.	870.00	525.00	800.00	600.00
2.	852.90	514.70	784.30	588.20
3.	836.20	504.60	768.90	576.70
4.	819.80	494.70	753.90	565.40
5.	803.80	485.00	739.10	554.30
6.	788.00	475.50	724.60	543.40
7.	772.50	466.20	710.40	532.80
8.	757.40	457.00	696.50	522.30
9.	742.50	448.10	682.80	512.10
10.	728.00	439.30	669.40	502.10
11.	713.70	430.70	656.30	492.20
12.	699.70	422.20	643.40	482.60
13.	686.00	414.00	630.80	473.10
14.	672.50	405.80	618.40	463.80
15.	659.40	397.90	606.30	454.70
16.	646.40	390.10	594.40	445.80
17.	633.80	382.40	582.80	437.10
Σ	12682.60	7653.30	11662.20	8746.60

analyses, we should consider the fact that under the prevailing conditions on coal-mine stockpiles, agricultural production can hardly be economically justified—and becomes especially difficult at the time of afforestation.

CONCLUSIONS

The reclamation of anthropogenically degraded sites is a very important task, both from an economic and an ecological point of view. The results of the foregoing researches show that piles composed of deep sands and heavy clays, although poor in nutrients and biologically inactive, can be successfully cultivated without special investments by establishing plantations of larch, Weymouth pine, Douglas fir and (Austrian) black pine. All four species developed much better on sands, while Weymouth pine proves the most successful species on clay.

Seventeen years after the establishment of the plantations, the initial stage of soil has formed and native species of ground flora, shrubs and trees are present (Institute of Forestry, Belgrade, 1998: 32–60). Cost: benefit analysis shows that the established plantations fulfil their economic function through the profitable production of timber.

Table 4: Costs and benefits of timber production (in DEM.ha^{-1}, discount rate 12%)

Year	Costs	Benefits (standing timber value)							
		Sand				Clay			
		Douglas fir	Larch	W. pine	B. pine	Douglas fir	Larch	W. pine	B. pine
1	806.10	0.00	0.00	0.00	0.00	0.00	0.00	0.00	0.00
2	294.10	0.00	0.00	0.00	0.00	0.00	0.00	0.00	0.00
3	288.40	0.00	0.00	0.00	0.00	0.00	0.00	0.00	0.00
4	282.70	0.00	0.00	0.00	0.00	0.00	0.00	0.00	0.00
5–16	0.00	0.00	0.00	0.00	0.00	0.00	0.00	0.00	0.00
17	0.00	2190.40	4078.30	3780.20	2641.98	1332.30	1671.10	1874.30	1354.86
Σ	1671.30	2190.40	4078.30	3780.20	2641.98	1332.30	1671.10	1874.30	1354.86

Table 5: Cost: benefit ratio and net present value (per ha)

			B/C				NPV (in DEM)			
State before contamination	Maize production	Intensive	1.10				1020.40			
		Extensive	0.90				-1093.30			
State after recultivation	Wood production	Tree species ⇨	Larch	Dougl. fir	B. pine	W. pine	Larch	Dougl. fir	B. pine	W. pine
		Sand	2.40	1.30	1.58	2.30	2407.00	518.70	970.68	2108.90
		Clay	1.00	0.80	0.81	1.10	-0.20	-339.00	-316.44	203.00

References

Institute of Forestry, Belgrade. 1997. *Recultivation by Afforestation of Minespoil Banks of Opencast Lignite Mine 'Kolubara'*. Ministry of Environmental Protection, Belgrade: Republic of Serbia, 147 pp.

Vuckovic, M. 1989. *Development and Production Characteristics of Austrian Pine in Artificial Stands at Juzni, Kucaj and Goc*. Forestry Faculty, Univ. Belgrade, Belgrade (in Serbian).

Index